# INTRODUCTORY
# TOPOLOGY

## Exercises and Solutions

Second Edition

# INTRODUCTORY
# TOPOLOGY

## Exercises and Solutions

### Second Edition

## Mohammed Hichem Mortad

University of Oran 1, Ahmed Ben Bella, Algeria

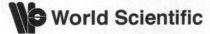
**World Scientific**

NEW JERSEY · LONDON · SINGAPORE · BEIJING · SHANGHAI · HONG KONG · TAIPEI · CHENNAI

*Published by*

World Scientific Publishing Co. Pte. Ltd.

5 Toh Tuck Link, Singapore 596224

*USA office:* 27 Warren Street, Suite 401-402, Hackensack, NJ 07601

*UK office:* 57 Shelton Street, Covent Garden, London WC2H 9HE

**Library of Congress Cataloging-in-Publication Data**
Names: Mortad, Mohammed Hichem, 1978–
Title: Introductory topology : exercises and solutions / by Mohammed Hichem Mortad
    (University of Oran, Algeria).
Description: 2nd edition. | New Jersey : World Scientific, 2016. |
    Includes bibliographical references and index.
Identifiers: LCCN 2016030117 | ISBN 9789813146938 (hardcover : alk. paper) |
    ISBN 9789813148024 (pbk : alk. paper)
Subjects: LCSH: Topology--Problems, exercises, etc.
Classification: LCC QA611 .M677 2016 | DDC 514.076--dc23
LC record available at https://lccn.loc.gov/2016030117

**British Library Cataloguing-in-Publication Data**
A catalogue record for this book is available from the British Library.

Printed in Singapore

# Contents

# Preface

Topology is a major area in mathematics. At an undergraduate level, at a research level, and in many areas of mathematics (and even outside of mathematics in some cases), a good understanding of the basics of the theory of general topology is required. Many students find the course "Topology" (at least in the beginning) a bit confusing and not too easy (even hard for some of them) to assimilate. It is like moving to a different place where the habits are not as they used to be, but in the end we know that we have to live there and get used to it. They are usually quite familiar with the real line $\mathbb{R}$ and its properties. So when they study topology they start to realize that not everything true in $\mathbb{R}$ needs to remain true in an arbitrary topological space. For instance, there are convergent sequences which have more than one limit, the identity mapping is not always continuous, a normally convergent series need not converge (although the latter is not within the scope of this book). So in many references, they use the word "usual topology of $\mathbb{R}$", a topology in which things are as usual! while there are many other "unusual topologies" where things are not so "usual"!

The present book offers a good introduction to basic general topology throughout solved exercises and one of the main aims is to make the understanding of topology an easy task to students by proposing many different and interesting exercises with very detailed solutions, something that it is not easy to find in another manuscript on the same subject in the existing literature. Nevertheless, and in order that this books gives its fruits, we do advise the reader (mainly the students) to use the book in a clever way inasmuch as while the best way to learn mathematics is by doing exercises, the worst way of doing exercises is to read the solution without thinking about how to solve the exercises (at least for some time). Accordingly, we strongly recommend the student to attempt the exercises before consulting the solutions. As a Chinese proverb says: If you give someone a fish, then you have given him to eat for one day, but if you teach him how to fish, then you have given him food for everyday. So we hope the students are going to learn to "fish" using this book.

ix

The present manuscript is mainly intended for an undergraduate course in general topology. It does not include algebraic and geometric topologies. Other topics such as: nets, topologies of infinite products, quotient topology, first countability, second countability and the $T_i$ separation axioms with $i = 0, 3, 4, 5$ are not considered neither or are not given much attention. It can also be used by instructors and anyone who needs the basic tools of topology. Teaching this course several times with many different exercises each year has allowed me to collect all the exercises given in this book. I relied on many references (I cannot remember all of them but most of them can be found in the bibliography) in lectures and tutorials. If there is some source which I have forgotten to mention, then I sincerely apologize for that. Let us now say a few words about the contents of this book. The exercises on the subjects covered in this book can be used for a one semester course of 14-15 weeks.

The book is divided into two parts. In Part 1, each chapter (except for the first one) contains five sections. They are:

(1) **What You Need to Know**: (which has become "Essential Background" in the new editions) In this part, we recall the essential of notions and results which are needed for the exercises. No proofs are given directly. The reader may also wish to consult the following references for further reading: [1], [4], [6], [12], [16], [18], [19], [20], [25], [26], [28] and [29].

(2) **True or False**: In this part some interesting questions are proposed to the reader. They also contain common errors which appear with different students almost every year. Thanks to this section, students should hopefully avoid making many silly mistakes. This part is an important back-up for the "What You Need to Know" section. Readers may even find some redundancy, but this is mainly because it is meant to test their understanding.

(3) **Exercises with Solutions**: The major part and the core of each chapter where many exercises are given with detailed solutions.

(4) **Tests**: This section contains short questions given with just answers or simply hints.

(5) **More Exercises**: In this part some unsolved exercises are proposed to the interested reader.

In Part 2, the reader finds answers to the questions appearing in the section "True or False" as well as solutions to Exercises and Tests.

The prerequisites to use this book are basics of: functions of one variable (some of the several variables calculus is also welcome), sequences and series.

Since the terminology in topology is rich and may be different from a book to another, we do encourage the readers to have a look at the "Notations and Terminology" chapter to avoid an eventual confusion or ambiguity with symbols and notations.

Before finishing, I welcome and I will be pleased to receive any suggestions, questions (as well as pointing out eventual errors and typos) from readers at my email: **mhmortad@gmail.com.**

Last but not least, thanks are due in particular to Dr Lim Swee Cheng and Ms Tan Rok Ting, and all the staff of World Scientific Publishing Company for their patience and help.

*Oran on September the 24<sup>th</sup>, 2013,*
*Mohammed Hichem Mortad*
*The University of Oran (Algeria).*

# Preface to the Second Edition

In this second edition, many changes have been made, some quite significant. For instance, this new edition contains proofs (as exercises) of many results in the old section "What You Need To Know", which has been renamed here as "Essential Background". This new edition has also many additional exercises.

There is another improvement still about the section "Essential Background". Indeed, it has been considerably beefed up as it now includes more remarks and results for readers' convenience. As an illustrative point, we have enlarged the subsection "Fixed Point Theorem" and even added one on applications to integral equations.

I have also added other notions e.g. on the topology of an arbitrary product of topological spaces as well as separation axioms. Despite all these changes, the book still does not cover concepts of Algebraic Topology.

The sections "True or False" and "Tests" have remained as they were apart from a very few changes.

I have also corrected typos and mistakes which appeared in the earlier edition of this manuscrit.

It is worth noticing that some of these points were suggested in the MAA review by Dr Mark Hunacek. I take this opportunity to thank him for his review.

Readers wishing to contact me may do that via my email: **mhmortad@gmail.com.**

Finally, warm thanks are due to all the staff of World Scientific Publishing Company for their patience and help (in particular, Dr Lim Swee Cheng and Ms Tan Rok Ting).

*Oran, Algeria*
*May 08, 2016*
*Prof. Dr M. H. Mortad*

# Notation and Terminology

## 0.1. Notation

- $\mathbb{N}$ is the set of natural numbers, i.e. $\mathbb{N}$ is the set $\{1, 2, \cdots\}$ (note that the French $\mathbb{N}$ contains $0$ too).
- $\mathbb{Z}$ is the set of all integers while $\mathbb{Z}^+$ is the set of positive integers.
- $\mathbb{Q}$ is the set of rational numbers.
- $\mathbb{R}$ is the set of real numbers.
- $\mathbb{C}$ is the set of complex numbers.
- $[a, b]$ is the closed interval with endpoints $a$ and $b$.
- $(a, b)$ is the open interval with endpoints $a$ and $b$.
- The power set of set $X$ is denoted by $\mathcal{P}(X)$ or $2^X$.
- $(a, b)$ also denotes an ordered pair.
- $i$ the complex square root of $-1$.
- $x \mapsto e^x$ the usual exponential function.
- The complement of a set $A$ in $X$ is often denoted by $A^c$ and occasionally it is denoted by $X \setminus A$ or $X - \{a\}$.
- The empty set is denoted by $\varnothing$.
- $d(x, A)$ is the distance between a point $x$ in $X$ to a set $A \subset X$ where $(X, d)$ is a metric space.
- $d(A)$ is the diameter of a set $A$ in a metric space $(X, d)$.
- In a metric space $(X, d)$, the open ball of radius $r > 0$ and center $x \in X$ is denoted by $B(x, r)$.
- In a metric space $(X, d)$, the closed ball of radius $r > 0$ and center $x \in X$ is denoted by $B_c(x, r)$.
- In a metric space $(X, d)$, the sphere of radius $r > 0$ and center $x \in X$ is denoted by $S(x, r)$.

- The interior of a set $A$ in a topological space is denoted by $\overset{\circ}{A}$.
- The closure of a set $A$ in a topological space is denoted by $\overline{A}$.
- $A'$ denotes the derived set of $A$, that is the set of limit points of $A$ (where $A$ is a subset of a topological space $X$).
- The frontier of a set is denoted by Fr or $\partial$.
- card denotes the cardinal of a set.

- $\mathbb{R}_\ell$ is the lower limit topology.
- $\mathbb{R}_K$ is the $K$-topology.
- The set of neighborhoods of a point $x \in X$, where $X$ is a topological space, is denoted by $\mathcal{N}(x)$.
- $C_x$ denotes the (connected) component of $x \in X$ where $X$ is a topological space.
- The real part of a complex number is denoted by Re.
- The imaginary part of a complex number is denoted by Im.
- The conjugate of a complex number $z$ is denoted by $\bar{z}$.
- $C([0,1])$ (or $C([0,1], \mathbb{R})$) is the space of real-valued continuous functions on $[0,1]$ taking values in $\mathbb{R}$. If the field of values is $\mathbb{C}$, then this will be clearly mentioned.
- $C^1([0,1])$ (or $C^1([0,1], \mathbb{R})$) is the space of real-valued continuous functions defined on $[0,1]$, differentiable and having a continuous derivative.
- $(\mathbb{R}, |\cdot|)$ is $\mathbb{R}$ equipped with the standard or the usual metric, i.e. the absolute value function.

## 0.2. Terminology

- There is a general comment which should be remembered by the reader. *Except* in the sections "True or False", if we say $\mathbb{R}$ (or any subset of it) without specifying the topology or the metric, then this means $\mathbb{R}$ endowed with its usual topology or metric. This applies to $\mathbb{C}$, $\mathbb{R}^n$ and $\mathbb{C}^n$ too.
- It will be comprehensible from the context whether "$\subset$" is for comparing two sets or two topologies.
- It will be clear from the context whether $(a, b)$ is the ordered pair or the open interval.
- When it is not too important to specify the metric or the topology, we simply say the topological (or metric) space $X$.
- From time to time the reader will see "(why?)". In such case, this means that this is a question whose answer should be known by the reader. This is used by other authors such as J. B. Conway (see [7]) and also M. Stoll (see [28]). Other expressions such as "(is it not?)" are also used.
- The letters "w.r.t." stand for "with respect to".
- As it is used almost everywhere, "iff" means "if and only if". For the fun, the French use "ssi" for "si et seulement si". Even in Arabic, it has been sorted out by doubling a letter in the end!

- As "iff", there are other words which cannot be found in an English language dictionary but we still use them. For example, "Cauchyness", "Hausdorffness", "clopen", etc...
- WLOG, as it pleases many, stands for "without loss of generality".
- If readers see in the end of an answer "$\cdots$", then this means that details are left to them.

# Part 1

# Exercises

# CHAPTER 1

# General Notions: Sets, Functions et al.

## 1.1. Essential Background

In this section we recall some of the basic material on sets, functions, countability as well as consequences of the least upper bounded property. Other basic concepts will be assumed to be known by the reader.

**1.1.1. Sets.** We start with the definition of a set. We just adopt the naive definition of it as many textbooks do. We shall not go into much detail of Set Theory. It can be a quite complicated theory and a word from J. M. Møller suffices as a warning. He said in [18]: "We don't really know what a set is but neither do the biologists know what life is and that doesn't stop them from investigating it". So, we do advise the student to be careful when dealing with set theory in order to avoid contradictions and paradoxes. Before commencing, we give just one illustrative paradox which is the **Russel's set** (from **Russel's paradox**: *if someone says "I'm lying", is he lying?*): Let

$$R = \{A : A \notin A\}$$

be the set of all sets which are not an element of itself. So is $R \in R$? Or is it not?

**DEFINITION 1.1.1.** *A **set** is a well-defined collection of objects or elements. We denote sets in general by capital letters: A, B, C; X, Y, Z, etc...Elements or objects belonging to the set are usually designated by lowercase letters: a, b, c; x, y, z, etc...*

*If a is an element of A, then we write: $a \in A$. Otherwise, we write that $a \notin A$.*

*If a set does not contain any element, then it is called the **empty** (or **void**) set, and it is denoted by: $\varnothing$ or $\{\}$ (J.B. Conway in [7] likes to denote it by $\square$).*

**EXAMPLES 1.1.2.**
  (1) $\{-4, 1, 2\}$ *is a set constituted of the elements 1, 2 and −4.*
  (2) $\{x : x \geq 0\}$ *is also a set but we cannot write down all its elements.*

One element sets (e.g. $\{2\}$) have a particular name.

**DEFINITION** 1.1.3. *A **singleton** is a set with exactly one element.*

The notion of a "smaller" or "larger" set cannot be defined rigourously even though we will be saying it from time to time. We have a substitute.

**DEFINITION** 1.1.4. *Let $A$ and $B$ be two sets.*

(1) *We say that $A$ is a **subset** of $B$ (or that $B$ is a **superset** of $A$) if every element of $A$ is an element of $B$, and we write*

$$A \subset B \text{ or } B \supset A.$$

*We may also say that $A$ is **contained** (or **included**) in $B$ (or that $B$ **contains** (or **includes**) $A$). The relation "$\subset$" is called **inclusion**.*

(2) *The two sets $A$ and $B$ are equal, and we write $A = B$, if*

$$A \subset B \text{ and } B \subset A.$$

*Otherwise, we write $A \neq B$.*

(3) *We say that $A$ is a **proper subset** of $B$ if $A \subset B$ with $A \neq B$. We denote this by $A \subsetneq B$, and we call "$\subsetneq$" a **proper inclusion**.*

**EXAMPLE** 1.1.5. *We know that:*

$$\mathbb{N} \subset \mathbb{Z} \subset \mathbb{Q} \subset \mathbb{R}.$$

There are other special and important subsets of the real line $\mathbb{R}$.

**DEFINITION** 1.1.6. *A subset $I$ of $\mathbb{R}$ is called an **interval** if:*

$$\forall x, y \in I, \forall z \in \mathbb{R} : (x < z < y \implies z \in I).$$

*Then we write $I = (a, b)$, and we call it an **open interval**. We may also write*

$$(a, b) = \{x \in \mathbb{R} : a < x < b\}.$$

*We can define other intervals in a similar way. For example,*

$$[a, b] = \{x \in \mathbb{R} : a \leq x \leq b\}$$

*(this is called a **closed interval**).*

*The intervals $(a, b]$ and $[a, b)$ are defined analogously. We call them **half-open intervals** (or half-closed, why not?!).*

*Finally, **infinite intervals** $(-\infty, \infty)$, $(a, \infty)$, $[a, \infty)$, $(-\infty, b)$ and $(-\infty, b]$ are defined similarly. For instance,*

$$[a, \infty) = \{x \in \mathbb{R} : x \geq a\} \text{ or } (-\infty, b) = \{x \in \mathbb{R} : x < b\}.$$

**REMARK.** In later chapters, the appellation open and closed intervals will be justified. The reader will also see why e.g. $[a, b)$ cannot be considered as open nor it can be considered as closed. The reader will also see that $(-\infty, \infty)$, $(a, \infty)$ and $(-\infty, b)$ are considered as open intervals, whilst $[a, \infty)$ and $(-\infty, b]$ are considered as closed ones.

**REMARK.** Notice that the empty set may be considered as an interval.

**EXAMPLES** 1.1.7. *(details are to be found in Exercise 1.2.8)*

(1) $\mathbb{Q}$ *or* $\mathbb{R} \setminus \mathbb{Q}$ *are not intervals.*

(2) $\mathbb{R}^*$ *is not an interval.*

(3) $[0, 1] \cup \{2\}$ *is not an interval.*

We can produce new sets from old ones using known operations in Mathematics.

**DEFINITION** 1.1.8. *Let $A$ and $B$ be two sets.*

(1) *The **union** of $A$ and $B$, denoted by $A \cup B$, is defined by*
$$A \cup B = \{x : \ x \in A \ or \ x \in B\}.$$

(2) *The **intersection** of $A$ and $B$, denoted by $A \cap B$, is defined by*
$$A \cap B = \{x : \ x \in A \ and \ x \in B\}.$$

(3) *If $A \cap B = \varnothing$, then we say that the two sets are **disjoint**.*

(4) *If $X$ is some set such that $A \cup B = X$, and $A \cap B = \varnothing$, then we say that $\{A, B\}$ is a **partition** of $X$.*

(5) *The **difference** $A - B$ (or $A \setminus B$) is defined as*
$$A - B = \{x : \ x \in A \ and \ x \notin B\}.$$

We list some straightforward properties.

**PROPOSITION** 1.1.9. *Let $A$ and $B$ be two sets. Then:*

(1) $A \cap \varnothing = \varnothing$.

(2) $A \cup \varnothing = A$.

(3) $A \cap A = A$.

(4) $A \cup A = A$.

(5) $A \cap B = B \cap A$.

(6) $A \cup B = B \cup A$.

**DEFINITION** 1.1.10. *Let $A$ be a subset of a set $X$. Then the set $X - A$ is called the **complement** of $A$, and it is denoted by $A^c$.*

**REMARK.** It is useful to keep in mind that:
$$x \in A^c \iff x \notin A.$$

**REMARK.** We can re-write $A \setminus B$ using complement. We have indeed:
$$A \setminus B = A \cap B^c.$$

**PROPOSITION** 1.1.11. *Let $A$ be a subset of $X$. Then:*
(1) $(A^c)^c = A$.
(2) $A \cup A^c = X$.
(3) $A \cap A^c = \varnothing$.

The next easy two results are fundamental.

**PROPOSITION** 1.1.12. *Let $A$ and $B$ be two sets. Then*
$$A \subset B \iff B^c \subset A^c.$$

**PROPOSITION** 1.1.13. *Let $A$ and $B$ be two sets. Then*
$$A \cap B = \varnothing \iff A \subset B^c \iff B \subset A^c.$$

Going back to intervals, natural questions are: Is the intersection of two intervals an interval? What about their union?

**THEOREM** 1.1.14.
(1) *The intersection of two intervals is an interval.*
(2) *The union of two non-disjoint intervals is an interval.*

**REMARK.** In the second statement of the previous theorem, the hypothesis on the intervals not being disjoint is indispensable as is seen by considering the intervals $[0,1]$ and $[2,3]$.

Before passing to rules of set theory, we have yet two more ways of producing new sets from old ones. Here is the first one:

**DEFINITION** 1.1.15. *Let $A$ and $B$ be two sets. The **cartesian product** of $A$ and $B$, denoted by $A \times B$, is defined to be:*
$$A \times B = \{(a,b) : a \in A, \ b \in B\},$$
*where $(a,b)$ is called an **ordered pair**.*
*If $A = B$, then we write $A \times A = A^2$.*

**REMARK.** It will be clear from the context whether $(a,b)$ denotes an ordered pair or an open interval. For this reason, I prefer the French notation (for instance) for an open interval which is $]a,b[$.

**REMARK.** We can extend the concept of a cartesian product to a finite number of sets $A_i$, $i = 1, \cdots, n$. It is given by
$$A_1 \times \cdots \times A_n = \{(a_1, \cdots, a_n) : a_i \in A_i \text{ for all } i\}.$$
This may also be denoted by $\prod_{i=1}^n A_i$.

**EXAMPLE** 1.1.16. *Let $A = \{0, 1, 2\}$ and $B = \{\alpha, \beta\}$. Then*

$$A \times B = \{(0, \alpha), (1, \alpha), (2, \alpha), (0, \beta), (1, \beta), (2, \beta)\}.$$

**DEFINITION** 1.1.17. *Let $X$ be a set. The set of all possible subsets of $X$ (including $\varnothing$ and $X$ itself) is called the **power set** of $X$. We denote it by $\mathcal{P}(X)$.*

**REMARK**. In some references, they denote the power set by $2^X$ (the reason of this notation will be known in Theorem 1.1.19).

**EXAMPLE** 1.1.18. *Let $X = \{a, b\}$. Then*

$$\mathcal{P}(X) = \{\varnothing, \{a\}, \{b\}, X\}.$$

**REMARKS**.

(1) The readers should not forget that the elements of $\mathcal{P}(X)$ are sets! So if $X = \{a, b\}$, then e.g. we write $\{a\} \in \mathcal{P}(X)$ and not $a \in \mathcal{P}(X)$. We must not write $\{a\} \subset \mathcal{P}(X)$ either! However, we are allowed to write $\{\{a\}\} \subset \mathcal{P}(X)$.

(2) We have seen that if $X$ has two elements, then $\mathcal{P}(X)$ has four. Can we generalize this? We can't yet! Indeed, a set of 2 elements gives a power set of 4 elements. So if a set is of $p$ elements, then is its power set of $2^p$ or $p^2$ elements? Let us therefore pass to three elements. Let $Y = \{a, b, c\}$. Then

$$\mathcal{P}(Y) = \{\varnothing, \{a\}, \{b\}, \{c\}, \{a, b\}, \{a, c\}, \{b, c\}, X\}$$

so that $\mathcal{P}(Y)$ has now 8 elements, i.e. it has $2^3$ elements. In fact, this is true for any $n$ as seen next.

**THEOREM** 1.1.19. *(see Exercise 1.2.10) Let $X$ be a set having "$n$" elements, where $n \in \mathbb{N}$ (we may even allow $n = 0$). Then $\mathcal{P}(X)$ has $2^n$ elements.*

In many situations in mathematics we will have to take unions mixed with intersections (and sometimes products as well). How to remedy this?

**THEOREM** 1.1.20. *Let $A, B, C$ and $D$ be sets. Then:*

(1) $A \cap (B \cup C) = (A \cap B) \cup (A \cap C)$.

(2) $A \cup (B \cap C) = (A \cup B) \cap (A \cup C)$.

(3) $C \setminus (A \cup B) = (C \setminus A) \cap (C \setminus B)$. *In particular,*

$$(A \cup B)^c = A^c \cap B^c.$$

(4) $C \setminus (A \cap B) = (C \setminus A) \cup (C \setminus B)$. *In particular,*

$$(A \cap B)^c = A^c \cup B^c.$$

(5) $A \times (B \cup C) = (A \times B) \cup (A \times C)$.
(6) $(A \times B) \cap (C \times D) = (A \cap C) \times (B \cap D)$.

**REMARK.** The first two properties are called the **distributive laws** while the second two are referred to as the **De Morgan's laws**.

We finish this subsection with a generalization of the union and the intersection of two sets to an arbitrary number of sets.

**DEFINITION** 1.1.21. *Let* $(A_i)_{i \in I}$ *be a family of subsets of* $X$, *where* $I$ *is an arbitrary set. Then the* **union** *of the sets* $(A_i)_{i \in I}$ *is defined by*

$$\bigcup_{i \in I} A_i = \{x \in X : x \in A_i \text{ at least for one } i \in I\}.$$

*The* **intersection** *of the sets* $(A_i)_{i \in I}$ *is defined by*

$$\bigcap_{i \in I} A_i = \{x \in X : x \in A_i \text{ for all } i \in I\}.$$

**EXAMPLES** 1.1.22.

(1) $\bigcap_{n \in \mathbb{N}}(-n, n] = (-1, 1]$
(2) $[a, b] = \bigcup_{x \in [a,b]}\{x\}$.

As in the case of finite unions and intersections, **distributive** and **De Morgan's laws** remain valid in the setting of an arbitrary class of sets. More details are given next.

**THEOREM** 1.1.23. *Let* $(A_i)_{i \in I}$ *be a family of subsets of* $X$, *where* $I$ *is an arbitrary set. Let* $B \subset X$. *Then*

(1)
$$B \cup \left(\bigcap_{i \in I} A_i\right) = \bigcap_{i \in I}(B \cup A_i).$$

(2)
$$B \cap \left(\bigcup_{i \in I} A_i\right) = \bigcup_{i \in I}(B \cap A_i).$$

(3)
$$\left(\bigcup_{i \in I} A_i\right)^c = \bigcap_{i \in I} A_i^c.$$

(4)
$$\left(\bigcap_{i \in I} A_i\right)^c = \bigcup_{i \in I} A_i^c.$$

Lastly, we give a result which is easy to show (yet a proof is given in Exercise 1.2.9). We will be calling on it on many occasions in the present book.

**PROPOSITION** 1.1.24. *Let $A$ be a set. Assume that for each $a \in A$, there is a subset $U_a$ of $A$ containing $a$, i.e. $a \in U_a \subset A$. Then*

$$A = \bigcup_{a \in A} U_a.$$

**1.1.2. Functions.** We give the most basic definition of a function without going into deep details.

**DEFINITION** 1.1.25. *Let $X$ and $Y$ be two sets. A **function** $f$ (or **mapping** or even **map**) from $X$ to $Y$ is a rule which assigns to each element $x \in X$ a unique element $y \in Y$. We generally express this by writing $y = f(x)$.*

*The set $X$ is called the **domain** of $f$ and $Y$ is called the **codomain** of $f$.*

**REMARK.** We usually say that $f$ is a function from $X$ to (or into) $Y$.

**DEFINITION** 1.1.26. *Two functions $f : X \to Y$ and $g : X \to Z$ are said to be **equal** if:*

$$f(x) = g(x), \ \forall x \in X.$$

**DEFINITION** 1.1.27. *The **identity** function on a set $X$ is the function $f : X \to X$ defined by $f(x) = x$. It is usually denoted by $id_X$.*

A related (and *different*) example is:

**DEFINITION** 1.1.28. *Let $A \subset X$. The **inclusion** map on a $A$ is the function $\iota : A \to X$ defined by $\iota(x) = x$.*

An inclusion map is a restriction of the identity in the sense given next.

**DEFINITION** 1.1.29. *Let $f : X \to Y$ be a function and let $A \subset X$. The function $g : A \to Y$ is called a **restriction** of $f$ to $A$ if*

$$f(x) = g(x), \ \forall x \in A.$$

*We may denote $g$ by $f_A$ and we may say: $f$ **restricted** to $A$.*

We now give two important subsets of the domain and the codomain of a given function.

**DEFINITION** 1.1.30. *Let $f : X \to Y$ be a function. Let $A \subset X$ and $B \subset Y$. Then:*

(1) The **image** (or **range**) of $A$, denoted by $f(A)$, is defined to be:
$$f(A) = \{f(x) : \ x \in A\}.$$

(2) The **inverse image** (or the **preimage**) of $B$, denoted by $f^{-1}(B)$, is defined by:
$$f^{-1}(B) = \{x \in X : \ f(x) \in B\}.$$

**REMARKS.**

(1) It is good to remember that the image $f(A)$ is a subset of $Y$ whereas $f^{-1}(B)$ is a subset of $X$.

(2) Also
$$x \in f^{-1}(B) \Longleftrightarrow f(x) \in B$$
and
$$y \in f(A) \Longleftrightarrow \exists x \in A : \ f(x) = y.$$

(3) It may well happen that $f^{-1}(B) = \varnothing$ even if $B \neq \varnothing$ (cf. Proposition 1.1.32), whereas $f(A)$ is never empty unless $A = \varnothing$.

(4) The reader should bear in mind that in general $f(X) \subsetneq Y$.

(5) It is evident that for any $f$: $f^{-1}(Y) = X$.

(6) The inverse image is defined for any function $f$, i.e. $f$ is not assumed to be invertible (see Definition 1.1.37 below), even though in case of invertibility we still denote the inverse by $f^{-1}$ (in such case the two notations coincide).

**EXAMPLES** 1.1.31. Let $f : \mathbb{R} \to \mathbb{R}$ be defined by $f(x) = x^2$. Then:

(1) $f(\{0, \sqrt{2}\}) = \{0, 2\}$. Also $f([-1, 2]) = [0, 4]$.

(2) $f^{-1}(\{4\}) = \{-2, 2\}$. Also, $f^{-1}(\{[-2, 0)\}) = \varnothing$.

We have just observed that $f^{-1}(B)$ may be empty. However, we have:

**PROPOSITION** 1.1.32. Let $f : X \to Y$ be a function and let $B \subset f(X)$. Then
$$f^{-1}(B) \neq \varnothing \Longleftrightarrow B \neq \varnothing.$$

The inverse image is "well-behaved" when it comes to unions and intersections. But, the direct image does not behave that well when it comes to intersections. More details and more properties are given next.

**THEOREM** 1.1.33. (see Exercise 1.2.11) Let $I$ be an arbitrary set. Let $f : X \to Y$ be a function. Also assume that $A$, $A_i \subset X$ and $B$, $B_i \subset Y$. Then:

(1) $f(\bigcup_{i \in I} A_i) = \bigcup_{i \in I} f(A_i)$;

(2) $f(\bigcap_{i \in I} A_i) \subseteq \bigcap_{i \in I} f(A_i)$;

(3) $f^{-1}(\bigcup_{i \in I} B_i) = \bigcup_{i \in I} f^{-1}(B_i)$;

(4) $f^{-1}(\bigcap_{i \in I} B_i) = \bigcap_{i \in I} f^{-1}(B_i)$;

(5) $f^{-1}(B^c) = [f^{-1}(B)]^c$;

(6) $f(f^{-1}(B)) \subset B$;

(7) $A \subset f^{-1}(f(A))$.

**DEFINITION 1.1.34.** *Let $X$, $Y$ and $Z$ be three non-empty sets. Let $f : X \to Y$ and $g : Y \to Z$ be two functions. The **composition** (or **composite**) of $g$ with $f$ is the function denoted by $g \circ f$, and defined from $X$ into $Z$ by*

$$(g \circ f)(x) = g(f(x)).$$

**DEFINITION 1.1.35.** *Let $f : X \to X$ be a function. The **iterated map** (or **iteration**), denoted by $f^{(n)}$ defined from $X$ into $X$, is produced by composing $f$ with itself "$n$" times. In other words, we define $f^{(n)}$ as follows:*

$$f^{(0)} = id_X, \quad f^{(n+1)} = f^{(n)} \circ f.$$

**PROPOSITION 1.1.36.** *(see Exercise 1.2.11) Let $f : X \to Y$ and $g : Y \to Z$ be two functions and let $A \subset Z$. Then*

$$(g \circ f)^{-1}(A) = f^{-1}(g^{-1}(A)).$$

**REMARK.** It is worth noticing that "$\circ$" is not commutative but it is associative. It is also easy to see that for all $f$

$$f \circ id = id \circ f = f$$

where *id* is the identity function on $X$. The next natural question is: When is a function invertible? We state this definition separately.

**DEFINITION 1.1.37.** *A function $f : X \to Y$ is **invertible**, if there is a function $g : Y \to X$ such that:*

$$f \circ g = i_Y \text{ and } g \circ f = i_X,$$

*where $id_X$ and $id_Y$ are the identity functions on $X$ and $Y$ respectively. In other words, $f$ is invertible if*

$$(f \circ g)(y) = y \text{ and } (f \circ g)(x) = x$$

*for all $x \in X$ and all $y \in Y$. The function $g$ is called the **inverse** (or **reciprocal**) **function** of $f$, and it is denoted by $f^{-1}$.*

**REMARK.** In case of invertibility, we may write ($x \in X$ and $y \in Y$)

$$f(x) = y \Longleftrightarrow x = f^{-1}(y).$$

Next, we introduce concepts which help us to check invertibility of functions.

**DEFINITION** 1.1.38. *Let* $f : X \to Y$ *be a function. We say that* $f$ *is* **injective** *(or* **one-to-one***) if:*

$$\forall x, x' \in X : \ (x \neq x' \Longrightarrow f(x) \neq f(x'))$$

*or equivalently:*

$$\forall x, x' \in X : \ (f(x) = f(x') \Longrightarrow x = x').$$

**REMARK.** The inclusion map defined above is also called the **canonical injection**.

We have just noticed that $f(X)$ does not always equal $Y$. But when it does, this has the following appellation.

**DEFINITION** 1.1.39. *Let* $f : X \to Y$ *be a function. We say that* $f$ *is* **surjective** *(or* **onto***) if:*

$$\forall y \in Y, \ \exists x \in X : \ f(x) = y,$$

*or equivalently if* $f(X) = Y$.

**REMARK.** The word "onto" may be a little bit misleading for people who have not done their first degree in English. They might think that a function from $X$ onto $Y$ is just like a function from $X$ to $Y$! This is of course not true as we have just said that: onto=surjective!

The "same expression" of $f(x)$ may produce each time a different result as regards surjectivity and injectivity if we change the domain and the codomain of $f$:

**EXAMPLES** 1.1.40. *Let* $f(x) = x^2$. *Then*

(1) $f : \mathbb{R} \to \mathbb{R}$ *is neither injective nor surjective;*
(2) $f : \mathbb{R}^+ \to \mathbb{R}$ *is injective but not surjective;*
(3) $f : \mathbb{R} \to \mathbb{R}^+$ *is surjective but not injective;*
(4) $f : \mathbb{R}^+ \to \mathbb{R}^+$ *is injective and surjective.*

It can happen that a function is injective and surjective at the same time (as in the last example). This leads to the following fundamental notion.

**DEFINITION** 1.1.41. *Let* $f : X \to Y$ *be a function. We say that* $f$ *is **bijective** (or **one-to-one correspondence**) if $f$ is injective and surjective simultaneously, that is, iff*

$$\forall y \in Y, \ \exists! x \in X : \ f(x) = y$$

*(recall that* $\exists!$ *means "there is a unique").*

One primary use of bijective functions is invertibility.

**THEOREM** 1.1.42. *A function is invertible if and only if it is bijective.*

We now give properties of composite functions as regards injectivity and surjectivity.

**THEOREM** 1.1.43. *(see Exercise 1.2.16) Let* $X$, $Y$ *and* $Z$ *be three non-empty sets. Let* $f : X \to Y$ *and* $g : Y \to Z$ *be two functions. Then:*

(1) *If* $f$ *and* $g$ *are injective, then so is* $g \circ f$.
(2) *If* $f$ *and* $g$ *are surjective, then so is* $g \circ f$.

**THEOREM** 1.1.44. *(see Exercise 1.2.15) Let* $X$, $Y$ *and* $Z$ *be three non empty sets. Let* $f : X \to Y$ *and* $g : Y \to Z$ *be two functions. Then*

(1) *If* $g \circ f$ *is injective, then* $f$ *is injective.*
(2) *If* $g \circ f$ *is surjective, then* $g$ *is surjective.*

**COROLLARY** 1.1.45. *Let* $X$, $Y$ *and* $Z$ *be three non empty sets. Let* $f : X \to Y$ *and* $g : Y \to Z$ *be two functions. Then:*

$$g \circ f \text{ bijective} \implies f \text{ injective and } g \text{ surjective.}$$

The function defined next plays a paramount role in many areas of mathematics.

**DEFINITION** 1.1.46. *Let* $X$ *be a set and let* $A \subset X$. *The **indicator** function (or **characteristic** function) of set* $A \subset X$ *is the function* $\mathbb{1}_A : X \to \{0, 1\}$ *defined by:*

$$\mathbb{1}_A(x) = \begin{cases} 1, & x \in A, \\ 0, & x \notin A. \end{cases}$$

We have gathered basic properties of the indicator function in the next proposition.

**PROPOSITION** 1.1.47. *(see Exercise 1.2.13) Let* $A$ *and* $B$ *be two subsets of* $X$. *Then*

(1) $A \subset B \Leftrightarrow \mathbb{1}_A \leq \mathbb{1}_B$.
(2) $A = B \Leftrightarrow \mathbb{1}_A = \mathbb{1}_B$.
(3) $\mathbb{1}_{A \cap B} = \mathbb{1}_A \mathbb{1}_B$ *(the pointwise product).*

(4) $\mathbb{1}_A^2 = \mathbb{1}_A$.

(5) $\mathbb{1}_{A^c} = 1 - \mathbb{1}_A$.

(6) $\mathbb{1}_{A \cup B} = \mathbb{1}_A + \mathbb{1}_A - \mathbb{1}_{A \cap B}$.

**REMARK.** The second property of the previous result tells us that two sets are equal iff their corresponding indicator functions are equal.

**REMARK.** It is clear that $\mathbb{1}_X = 1$ and that $\mathbb{1}_\varnothing = 0$.

We finish with another useful class of functions.

**DEFINITION** 1.1.48. *Let $I$ be an interval. We say that the function $f : I \to \mathbb{R}$ is **convex** si*

$$\forall x_1, x_2 \in I, \forall \lambda \in [0,1] : f(\lambda x_1 + (1-\lambda)x_2) \leq \lambda f(x_1) + (1-\lambda)f(x_2).$$

*If $(-f)$ is convex, then $f$ is said to be **concave**.*

The convexity is easily characterized using differentiability.

**THEOREM** 1.1.49. *If $f : I \to \mathbb{R}$ is a twice differentiable function, then*

$$f \text{ is convex} \iff f'' \geq 0.$$

**EXAMPLE** 1.1.50. *The real-valued function given by $f(x) = -\ln x$ is convex on $(0, \infty)$.*

Another (important) example is stated separately:

**PROPOSITION** 1.1.51. *Let $a \in \mathbb{R}$. Then $x \mapsto e^{ax}$ is convex on $\mathbb{R}$.*

Finally, we recall two of the fundamental theorems in basic real analysis.

**THEOREM** 1.1.52. *(**Intermediate Value Theorem**) Let $f : [a,b] \to \mathbb{R}$ be a continuous function. Assume that $f(a) < f(b)$. Then*

$$\forall \gamma \in [f(a), f(b)], \ \exists c \in (a,b) : \ f(c) = \gamma.$$

**REMARK.** For a generalization of the previous result see Corollary 6.1.9.

**THEOREM** 1.1.53. *(**Mean Value Theorem**) Let $f : [a,b] \to \mathbb{R}$ be a continuous function. Assume that $f$ is differentiable on $(a,b)$. Then*

$$\exists c \in (a,b) : f(b) - f(a) = (b-a)f'(c).$$

### 1.1.3. Equivalence Relation.

**DEFINITION** 1.1.54. *A **relation** $\mathcal{R}$ on a set $X$ is a subset $A_{\mathcal{R}}$ of $X \times X$. Nevertheless, instead of writing $(x, y) \in A_{\mathcal{R}}$, we will write $x\mathcal{R}y$.*

**DEFINITION** 1.1.55. *Let $\mathcal{R}$ be a relation on $X$. We say that $\mathcal{R}$ is:*
  (1) ***reflexive** if:* $\forall x \in X: x\mathcal{R}x$.
  (2) ***symmetric** if:* $\forall x, y \in X: x\mathcal{R}y \Rightarrow y\mathcal{R}x$.
  (3) ***transitive** if:* $\forall x, y, z \in X: x\mathcal{R}y$ *and* $y\mathcal{R}z \Rightarrow x\mathcal{R}z$.

*If $\mathcal{R}$ is reflexive, symmetric and transitive, then we say that $\mathcal{R}$ is an **equivalence relation** on $X$.*

**EXAMPLE** 1.1.56. *The relation $\mathcal{R}$ defined by:*
$$x\mathcal{R}y \iff \cos^2 y + \sin^2 x = 1$$
*is an equivalent relation on $\mathbb{R}$.*

**DEFINITION** 1.1.57. *Let $\mathcal{R}$ be an equivalence relation on $X$. The **equivalent class** of $x \in X$, denoted by $\dot{x}$ (or $[x]$), is defined by:*
$$\dot{x} = \{y \in X : y\mathcal{R}x\}.$$

*The collection of all equivalent classes of $X$, denoted by $X/\mathcal{R}$ is called the **quotient** of $X$ by $\mathcal{R}$, that is,*
$$X/\mathcal{R} = \{\dot{x} : x \in X\}.$$

Now, we give basic properties of the quotient.

**PROPOSITION** 1.1.58. *Let $\mathcal{R}$ be an equivalence relation on $X$ and let $\dot{x}$ be the equivalent class of $x \in X$. Then:*
  (1) $\forall x \in X: x \in \dot{x}$.
  (2) $\dot{x} = \dot{y}$ *iff $x\mathcal{R}y$.*
  (3) *If $\dot{x} \neq \dot{y}$, then $\dot{x}$ and $\dot{y}$ must be disjoint.*

### 1.1.4. Consequences of The Least Upper Bound Property.
We now turn to important consequences of the Least Upper Bound Property (basic results on the least or greatest bound property are assumed).

**THEOREM** 1.1.59. *(**Archimedian Property**) Let $x, y \in \mathbb{R}$ with $x > 0$. Then there exists $n \in \mathbb{N}$ such that*
$$nx > y.$$

**REMARK**. Another formulation of the previous theorem is to say that $\mathbb{N}$ is not bounded from above.

As a first application of the previous theorem, we introduce a useful class of function.

**DEFINITION** 1.1.60. *The **greatest integer function**, denoted by* $[x]$ *and defined from* $\mathbb{R}$ *to* $\mathbb{Z}$, *is the function which assigns to each* $x$ *the largest integer "$n$" less or equal to* $x$.

**EXAMPLES** 1.1.61.
  (1) $[2.7] = 2$.
  (2) $[1.08] = 1$.
  (3) $[3] = 3$.
  (4) $[-1.6] = -2$.

**PROPOSITION** 1.1.62. *For all* $x \in \mathbb{R}$, *we have*

$$[x] \leq x < [x] + 1.$$

Another (very important) consequence of the Archimedian Property is:

**THEOREM** 1.1.63. *(see Exercise 1.2.32) The set of rational numbers* $\mathbb{Q}$ *is **dense** in* $\mathbb{R}$, *that is, between any two real numbers, there is a rational one.*

Similarly, we can prove the following:

**THEOREM** 1.1.64. *(see Exercise 1.2.32) The set of irrational numbers* $\mathbb{R} \setminus \mathbb{Q}$ *is dense in* $\mathbb{R}$, *that is, between any two real numbers, there is an irrational one.*

**REMARKS**.
  (1) The precise meaning of density will be made clear in its general context in Chapter 3.
  (2) The two previous results are usually expressible respectively as follows (the reason will also be known in Chapter 3):
    (a) Any real number is the limit of a sequence of rational numbers.
    (b) Any real number is the limit of a sequence of irrational numbers.

Another important consequence of the Least Upper Bound Property is the following (a proof is prescribed in Exercise 1.2.25):

**THEOREM** 1.1.65. *(**Nested Interval Property**) Let* $(I_n)_n$ *be a sequence of **nested** closed and bounded intervals of* $\mathbb{R}$, *i.e.* $I_1 = [a_1, b_1]$, $I_2 = [a_2, b_2]$,...*such that* $I_1 \supset I_2 \supset \cdots$. *Then*

$$\bigcap_{n \in \mathbb{N}} I_n \neq \varnothing,$$

*that is, there is at least one point common to each interval $I_n$.*

**REMARK**. The fact that the intervals are closed and bounded is crucial for this result to hold. See Exercise 1.2.27.

This result will be generalized in Corollary 5.1.6 (cf. Theorem 7.1.27).

We finish with a consequence of Theorem 1.1.65 about sequences.

**COROLLARY** 1.1.66. *(**Bolzano-Weierstrass Property**, cf. Definition 5.1.12) Every bounded sequence in $\mathbb{R}$ has a convergent subsequence.*

### 1.1.5. Countability.

**DEFINITION** 1.1.67. *Two sets $A$ and $B$ are said to be **equivalent**, and we write $A \sim B$, if there a bijection between $A$ and $B$.*

**REMARK**. The concept of equivalence of sets is an equivalence relation.

**DEFINITION** 1.1.68. *We say that a set $A$ is **finite** if it is empty or if there is a bijection $f : A \to \{1, 2, \cdots, n\}$ (i.e. $A \sim \{1, 2, \cdots, n\}$). If $A$ is not finite, then we say that $A$ is **infinite**.*

*If a set $A$ is finite, then the **cardinal** of $A$, denoted by $\operatorname{card} A$, is the number of elements of $A$.*

**EXAMPLES** 1.1.69.
  (1) *If $A = \{0, -3, 2, \pi\}$, then $\operatorname{card} A = 4$.*
  (2) *If $A = \varnothing$, then $\operatorname{card} A = 0$.*
  (3) *If $A = \{\{\varnothing\}\}$, then $\operatorname{card} A = 1$.*

**THEOREM** 1.1.70. *If a set $A$ is finite, then there is no bijection of $A$ with any proper subset of $A$.*

**EXAMPLE** 1.1.71. *The set $\mathbb{N}$ is infinite.*

**DEFINITION** 1.1.72. *We say that a set $A$ is **countable** (or **denumerable**) if it is finite or if there is a bijection $f : A \to \mathbb{N}$. A set which is not countable is called **uncountable**.*

*In case $A$ is equivalent to $\mathbb{N}$, then $\operatorname{card} A$ is more commonly known as $\aleph_0$.*

**REMARK**. In most cases, when we say a countable set, then we mean an infinite countable set.

**EXAMPLES** 1.1.73.
  (1) *The set of even integers is countable.*

(2) $\mathbb{Z}$ *is countable.*

When looking for a bijection between $\mathbb{N}$ and a given set $A$, it may happen that we find a function $f$ which is injective but not surjective and vice versa. The next result is therefore quite practical.

**THEOREM** 1.1.74. *Let $A$ be a non-empty set. Then the following statements are equivalent*
  (1) *$A$ is countable;*
  (2) *There exists an injective map $f : A \to \mathbb{N}$;*
  (3) *There exists a surjective map $g : \mathbb{N} \to A$.*

**PROPOSITION** 1.1.75. *(see Exercise 1.2.19) If $X$ is countable, then so is any $A \subset X$.*

**EXAMPLE** 1.1.76. *The set of prime numbers is countable.*

**THEOREM** 1.1.77.
  (1) *A countable union of countable sets remains countable.*
  (2) *A finite (cartesian) product of countable sets is also countable.*

**REMARK.** It is plain that any intersection of countable sets is at most countable!

**COROLLARY** 1.1.78. *The set of rationals $\mathbb{Q}$ is countable.*

It is natural to ask whether $\mathbb{R}$ is countable?

**THEOREM** 1.1.79. *(see Exercise 1.2.22 or Exercise 1.2.29) The interval $[0, 1]$ is uncountable.*

**REMARK.** The cardinality of $[0, 1]$ is usually denoted by $c$. We also say that a set $A$ has the **power of the continuum** if it is equivalent to the interval $[0, 1]$.

It can be shown that the intervals $[a, b]$, $[a, b)$, $(a, b]$ and $(a, b)$ have all the power of the continuum.

It can also be shown that $\mathbb{R}$ is equivalent to $(-1, 1)$, via the function $f : \mathbb{R} \to (-1, 1)$ defined by

$$f(x) = \frac{x}{1 + |x|}.$$

Hence $\mathbb{R}$ too has the power of the continuum.

**COROLLARY** 1.1.80. *$\mathbb{R}$ is uncountable.*

**REMARK.** Another yet simpler proof may be found in Exercise 7.3.35.

**COROLLARY** 1.1.81. *$\mathbb{R} \setminus \mathbb{Q}$ is uncountable.*

The previous result may be generalized to:

**PROPOSITION** 1.1.82. *Let $X$ be an uncountable set and let $A \subset X$ be countable. Then $X \setminus A$ is uncountable.*

**THEOREM** 1.1.83. *([19]) Let $A$ be a non-empty set, and let $\mathcal{P}(A)$ be the set of all subsets of $A$. Then there is no injective map $f : \mathcal{P}(A) \to A$. Also, there is no surjective map $g : A \to \mathcal{P}(A)$.*

**COROLLARY** 1.1.84. *$\mathcal{P}(\mathbb{N})$ is uncountable.*

We finish this section with an important set in analysis. To define it, consider first the set $A_0 = [0, 1]$. The set $A_1$ is obtained from $A_0$ by removing the middle third interval $(\frac{1}{3}, \frac{2}{3})$. To obtain $A_2$, remove from $A_1$ its middle thirds, namely $(\frac{1}{9}, \frac{2}{9})$ and $(\frac{7}{9}, \frac{8}{9})$. Then set

$$A_n = A_{n-1} \setminus \bigcup_{p=0}^{\infty} \left( \frac{1 + 3p}{3^n}, \frac{2 + 3p}{3^n} \right).$$

**DEFINITION** 1.1.85. *The **Cantor set**, denoted by $C$, is the intersection (over $n \in \mathbb{N}$) of the $A_n$ introduced just above, that is:*

$$C = \bigcap_{n \in \mathbb{N}} A_n.$$

**REMARK**. The Cantor set is uncountable as will be seen in later chapters. Its topological properties will be considered throughout the present manuscript.

In Measure Theory, the Cantor set constitutes an example of an uncountable set with zero Lebesgue measure.

## 1.2. Exercises With Solutions

**Exercise** 1.2.1. Show that $\sqrt{2}$ is irrational.

**Exercise** 1.2.2. Show that

$$\frac{\ln 2}{\ln 3} \notin \mathbb{Q}.$$

**Exercise** 1.2.3. Show that $e \notin \mathbb{Q}$.
Hint: you may use the sequences defined by

$$x_n = \sum_{k=0}^{k=n} \frac{1}{k!} \text{ and } y_n = x_n + \frac{1}{n!n}.$$

**Exercise** 1.2.4. Let $A$ and $B$ be two sets. Show that the following statements are equivalent:

    (1) $A \subset B$;

(2) $A \cup B = B$;

(3) $A \cap B = A$.

**Exercise 1.2.5.** Let $A, B, C$ be three sets. Assume that $A \subset B$. Does it follow that

$$A \cap C \subset B \cap C? \ A \cup C \subset B \cup C? \ A \setminus C \subset B \setminus C?$$

**Exercise 1.2.6.** Let $A, B$ and $X, Y$ be sets. Show that:

(1) $A \subset B \Longleftrightarrow B^c \subset A^c$.

(2) $(A \cap B)^c = A^c \cup B^c$.

(3) $(A \cup B)^c = A^c \cap B^c$.

(4) Let $f : X \to Y$ be a given function. Assume that $A, B \subset X$. Show that

$$A \subset B \Longrightarrow f(A) \subset f(B).$$

(5) Now let $f : X \to Y$ be a function but consider $A, B \subset Y$. Show that

$$A \subset B \Longrightarrow f^{-1}(A) \subset f^{-1}(B).$$

**Exercise 1.2.7.** Let $A$ and $B$ be two subsets of $X$. Show that $A \cap B$ and $A \setminus B$ constitute a partition of $A$.

**Exercise 1.2.8.** Show that none of the following sets is an interval:

$$\mathbb{Q}, \ \mathbb{R} \setminus \mathbb{Q}, \ \mathbb{R}^* \text{ and } [0, 1] \cup \{2\}.$$

**Exercise 1.2.9.** Let $A$ be a set. Assume that for each $a \in A$, there is a subset $U_a$ of $A$ containing $a$, i.e. $a \in U_a \subset A$. Show that

$$A = \bigcup_{a \in A} U_a.$$

**Exercise 1.2.10.** Let $X$ be a set such that card$X = n$, where $n \in \mathbb{N}$ (we may even allow $n = 0$). Show that

$$\text{card}\mathcal{P}(X) = 2^n.$$

**Exercise 1.2.11.** Let $I$ be an arbitrary set. Let $f : X \to Y$ and $g : Y \to Z$ be two functions. Also assume that $A, A_i \subset X$; $B, B_i \subset Y$ and $C \subset Z$. Show that the following statements hold

(1) $f(\bigcup_{i \in I} A_i) = \bigcup_{i \in I} f(A_i)$.

(2) $f(\bigcap_{i \in I} A_i) \subseteq \bigcap_{i \in I} f(A_i)$ (cf. Exercise 1.3.3).

(3) $f^{-1}(\bigcup_{i \in I} B_i) = \bigcup_{i \in I} f^{-1}(B_i)$.

(4) $f^{-1}(\bigcap_{i\in I} B_i) = \bigcap_{i\in I} f^{-1}(B_i)$.

(5) $f^{-1}(B^c) = [f^{-1}(B)]^c$.

(6) $(g \circ f)^{-1}(C) = f^{-1}(g^{-1}(C))$.

(7) $f(f^{-1}(B)) \subset B$. Do we have $f(f^{-1}(B)) \supset B$? (cf. Exercise 1.3.3).

(8) $A \subset f^{-1}(f(A))$. Do we have $A \supset f^{-1}(f(A))$? (cf. Exercise 1.3.3).

**Exercise 1.2.12.** Let $f : X \to Y$ be a function. If $f_A$ is the restriction of $f$ to $A \subset X$, then show that for any subset $U$ of $X$ one has

$$f_A^{-1}(U) = A \cap f^{-1}(U).$$

**Exercise 1.2.13.** Let $A, B \subset X$. Show that

(1) $A \subset B \Leftrightarrow \mathbb{1}_A \leq \mathbb{1}_B$.

(2) $A = B \Leftrightarrow \mathbb{1}_A = \mathbb{1}_B$.

(3) $\mathbb{1}_{A\cap B} = \mathbb{1}_A \mathbb{1}_B$ (a pointwise product).

(4) $\mathbb{1}_A^2 = \mathbb{1}_A$.

(5) $\mathbb{1}_{A^c} = 1 - \mathbb{1}_A$.

(6) $\mathbb{1}_{A\cup B} = \mathbb{1}_A + \mathbb{1}_A - \mathbb{1}_{A\cap B}$.

**Exercise 1.2.14.** Let $(A_n)_n$ be a sequence of pairwise disjoint subsets of a set $X$. Show that

$$\mathbb{1}_{(\bigcup_{n=1}^{\infty} A_n)} = \sum_{n=1}^{\infty} \mathbb{1}_{A_n}.$$

**Exercise 1.2.15.** Let $X$, $Y$ and $Z$ be three non empty sets. Let $f : X \to Y$ and $g : Y \to Z$ be two functions.

(1) Show that if $g \circ f$ is injective, then $f$ is injective.

(2) Show that if $g \circ f$ is surjective, then $g$ is surjective.

(3) Infer that

$$g \circ f \text{ bijective} \implies f \text{ injective and } g \text{ surjective.}$$

**Exercise 1.2.16.** Let $X$, $Y$ and $Z$ be three non-empty sets. Let $f : X \to Y$ and $g : Y \to Z$ be two functions. Show that:

(1) If $f$ and $g$ are injective, then so is $g \circ f$.

(2) If $f$ and $g$ are surjective, then so is $g \circ f$.

**Exercise 1.2.17.** Define a relation $\mathcal{R}$ on $\mathbb{R}$ by

$$x\mathcal{R}y \Longleftrightarrow x - y \in \mathbb{Q}$$

for all $x, y \in \mathbb{R}$. Show that $\mathcal{R}$ is an equivalence relation on $\mathbb{R}$.

**Exercise 1.2.18.**

(1) Show that $f : (0, \infty) \to \mathbb{R}$ defined by $f(x) = -\ln x$ is convex.
(2) For which values of $p$ is the function $x \mapsto f(x) = x^p$ (defined on $\mathbb{R}_+^*$) convex?
(3) For which values of $a$, is the function $x \mapsto e^{ax}$ convex on $\mathbb{R}$?

**Exercise 1.2.19.** Show that a subset of a countable set is countable.

**Exercise 1.2.20.**

(1) Show, using the map (or otherwise)
$$f : \mathbb{N} \times \mathbb{N} \to \mathbb{N}$$
$$(n, m) \mapsto f(n, m) = 2^n 3^m,$$
that $\mathbb{N} \times \mathbb{N}$ is countable.
(2) Deduce that if $A$ and $B$ are two countable sets, then so is their cartesian product $A \times B$.

**Exercise 1.2.21.** Show that $\mathbb{Q}$ is countable.

**Exercise 1.2.22.**

(1) Show that $[0, 1]$ is not countable.
(2) Can $\mathbb{R}$ be countable?
(3) What about $\mathbb{R} \setminus \mathbb{Q}$?

**Exercise 1.2.23.** Let $\mathbb{Q}[X]$ be the set of polynomials with *rational* coefficients. Show that $\mathbb{Q}[X]$ is countable.

**Exercise 1.2.24.** Let $X = \{0, 1\}^{\mathbb{N}}$ be the set of all sequences whose elements are 0 or 1. Show that $X$ is uncountable.

**Exercise 1.2.25.** Let $(I_n)_n$ be a sequence of nested closed and bounded intervals of $\mathbb{R}$, i.e. $I_1 = [a_1, b_1]$, $I_2 = [a_2, b_2]$,... such that $I_1 \supset I_2 \supset \cdots$. Show that:
$$\bigcap_{n \in \mathbb{N}} I_n \neq \varnothing.$$

**Exercise 1.2.26.** Show that

(1) $\displaystyle\bigcap_{n \in \mathbb{N}} \left( \frac{-1}{n}, \frac{1}{n} \right) = \{0\};$

(2) $\displaystyle\bigcup_{n \in \mathbb{N}} [-n, n] = \mathbb{R};$

(3) $\displaystyle\bigcap_{n \in \mathbb{N}} [n, \infty) = \varnothing.$

(4) $\displaystyle\bigcup_{n \in \mathbb{N}} [1/n, 1] = (0, 1].$

**Exercise** 1.2.27. Show that the assumption of closed and bounded intervals may not just be dropped in the Nested Interval Property.

**Exercise** 1.2.28. Let $n \in \mathbb{N}$. Find $\bigcap_n A_n$ and $\bigcup_n A_n$ in the following cases:

(1) $A_n = \{1, 2, \cdots, n\}$;
(2) $A_n = (-n, n)$;
(3) $A_n = \left(-\frac{1}{n}, 1 + \frac{1}{n}\right)$;
(4) $A_n = \left[0, 1 - \frac{1}{n}\right]$;
(5) $A_n = \left(-\frac{1}{n}, 1\right)$.

**Exercise** 1.2.29. Using the Nested Interval Property, show that $[0, 1]$ is uncountable.

**Exercise** 1.2.30. (**Young Inequality**) Let $a$ and $b$ be two positive real numbers. Let $p > 1$ and $q > 1$ be such that $\frac{1}{p} + \frac{1}{q} = 1$ ($q$ is called the **conjugate** of $p$). Show that:

$$ab \le \frac{a^p}{p} + \frac{b^q}{q}.$$

**Exercise** 1.2.31. Using the convexity of an appropriate function, re-prove the Young Inequality.

**Exercise** 1.2.32.

(1) Show that the set of rational numbers $\mathbb{Q}$ is dense in $\mathbb{R}$, that is, between any two real numbers, there is a rational one.
(2) Deduce that the set of irrational numbers $\mathbb{R} \setminus \mathbb{Q}$ is dense in $\mathbb{R}$, that is, between any two real numbers, there is an irrational one.

**Exercise** 1.2.33. Let $X = C([0, 1])$ and let $f, g \in X$. Show that

$$\int_0^1 |f(x)g(x)|dx \le \sup_{0 \le x \le 1} |f(x)| \int_0^1 |g(x)|dx.$$

**REMARK.** The main purpose of the foregoing exercise is to familiarize the reader (especially the students) with the result therein. Usually, when it is given like that they think that we have just taken $|f(x)|$ outside the integral which is of course not true. In the solution below we show how to deal with this situation. This is something that must be remembered. A similar idea will be applied to infinite series in due time.

**Exercise** 1.2.34. Let $f : \mathbb{R} \to \mathbb{R}$ be defined by:

$$f(x) = \begin{cases} 0, & x \in \mathbb{Q}, \\ x, & x \notin \mathbb{Q}. \end{cases}$$

Let $a \in \mathbb{R}^*$. Does the limit $\lim_{x \to a} f(x)$ exist?

**Exercise** 1.2.35. Let $f : \mathbb{R} \to \mathbb{R}$ be defined by:

$$f(x) = \begin{cases} 0, & x \in \mathbb{Q}, \\ x, & x \notin \mathbb{Q}. \end{cases}$$

At which point (s) is $f$ continuous?

The same question with the $g$ defined by:

$$g(x) = \begin{cases} 1, & x \in \mathbb{Q}, \\ 0, & x \notin \mathbb{Q}. \end{cases}$$

**Exercise** 1.2.36. Let $I$ be an interval. We say that the function $f : I \to I$ has a **fixed point** if there exists an $x \in I$ such that $f(x) = x$.

Let $f : [a, b] \to [a, b]$ be a continuous function. Show that $f$ has a fixed point.

**REMARK**. The particular choice of $[a, b]$ is not random. Indeed, this interval is compact or complete (as we shall see). We will get back to the concept of the Fixed Point (with conditions on its uniqueness) in Exercise 5.3.36 (for a "compact" setting) and Theorem 7.1.37 (for a "complete" setting).

## 1.3. More Exercises

**Exercise** 1.3.1. Let $X$ and $Y$ be two sets. Show that

$$\mathcal{P}(X) \cap \mathcal{P}(Y) = \mathcal{P}(X \cap Y) \text{ and } \mathcal{P}(X) \cup \mathcal{P}(Y) \subset \mathcal{P}(X \cup Y).$$

Give an example showing that $\mathcal{P}(X) \cup \mathcal{P}(Y) = \mathcal{P}(X \cup Y)$ need not hold in general.

**Exercise** 1.3.2. Let $X$ and $Y$ be two sets and let $A \subset X$. Let $f : X \to Y$ be a function. Do we have

$$f(A^c) = (f(A))^c?$$

**Exercise** 1.3.3. Let $X$ and $Y$ be two sets. Let $f : X \to Y$ be a function.

(1) Show that the following statements are equivalent:
    (a) $f$ is injective
    (b) $f(A \cap B) = f(A) \cap f(B)$ for all $A, B \subset X$
    (c) $f^{-1}(f(C)) = C$ for all $C \subset X$.
(2) Show that $f$ is surjective iff $f(f^{-1}(D)) = D$ for all $D \subset Y$.

**Exercise 1.3.4** ($\pi$ irrational). Let $P$ be a polynomial of degree $2n$. Set
$$F(x) = P(x) - P''(x) + P^{(4)}(x) - \cdots + (-1)^n P^{(2n)}(x).$$

(1) By observing that $P(x) \sin x = (F'(x) \sin x - F(x) \cos x)'$, show that
$$\int_0^\pi P(x) \sin x \, dx = F(0) + F(\pi) \text{ (called \textbf{Hermite Formula})}.$$

(2) Assume that $\pi$ is rational, i.e. $\pi = \frac{a}{b}$ with $a \in \mathbb{N}$ and $b \in \mathbb{N}$.
  (a) Apply the Hermite Formula to $P(x) = \frac{1}{n!} x^n (a - bx)^n$.
  (b) Show that $P(0) = P'(0) = \cdots P^{(n-1)}(0) = 0$ and deduce that $F(0), F(\pi) \in \mathbb{Z}$.
  (c) Now set $I_n = \int_0^\pi P(x) \sin x \, dx$. Prove that $I_n > 0$ and that $\lim_{n \to +\infty} I_n = 0$. Deduce from Question b) that $I_n \in \mathbb{Z}$ and find a contradiction (leading to the irrationality of $\pi$).

**Exercise 1.3.5.** Let $r$ be rational and let $x$ be irrational. Show that $r + x$ is irrational.

**Exercise 1.3.6.** Show that $\{q\sqrt{3} : q \in \mathbb{Q}\}$ is dense in $\mathbb{R} \setminus \mathbb{Q}$.

**Exercise 1.3.7.** Exhibit an explicit dense subset of $\mathbb{C}$.

**Exercise 1.3.8.** Show that the direct image of a countable set *by any function* remains countable. Is this true for inverse images?

**Exercise 1.3.9.** *We say that a number is* **dyadic** *if it is a rational written as* $\frac{m}{2^n}$, $m \in \mathbb{Z}$ *and* $n \in \mathbb{N}$. Show that the set of dyadic numbers is dense in $\mathbb{R}$.

**Exercise 1.3.10.** An **algebraic number** is a root of a polynomial having rational (or integer) coefficients. The set of algebraic numbers certainly includes $\mathbb{Q}$. It also includes many irrational numbers such as $\sqrt{2}$, $\sqrt{3}$, $\frac{1}{\sqrt{2}}$ and $1 + \sqrt[3]{5}$...etc.

A non-algebraic number is called **transcendental**. In number theory, it is known that $\pi$ or $e$ are transcendental.

(1) Show that the set of algebraic numbers is countable.
(2) Is the set of transcendental numbers countable?
(3) Show that the sets of algebraic and transcendental numbers are both dense in $\mathbb{R}$.

**Exercise 1.3.11.** Let $A \subset \mathbb{R}$ be non void and bounded. Show that
$$\sup_{x,y \in A} |x - y| = \sup A - \inf A.$$

**Exercise 1.3.12.** Let $a_1, a_2, \cdots, a_n$ be positive real numbers such that $a_1 a_2 \cdots a_n = 1$. Using induction, show that

$$a_1 + a_2 + \cdots + a_n \geq n.$$

**Exercise 1.3.13.** Let $a_1, a_2, \cdots, a_n$ be positive real numbers. Let $A_n$, $G_n$ and $H_n$ denote the **arithmetic**, **geometric** and **harmonic means** of the numbers $a_1, a_2, \cdots, a_n$, that is,

$$A_n = \frac{a_1 + a_2 + \cdots + a_n}{n}, \quad G_n = \sqrt[n]{a_1 a_2 \cdots a_n}$$

and

$$H_n = \frac{n}{\frac{1}{a_1} + \frac{1}{a_2} + \cdots + \frac{1}{a_n}},$$

respectively. Show that (for all $n \in \mathbb{N}$)

$$A_n \geq G_n \geq H_n$$

(**hint:** You may use Exercise 1.3.12).

# CHAPTER 2

# Metric Spaces

## 2.1. Essential Background

### 2.1.1. Definitions and Examples.

**DEFINITION 2.1.1.** *Let $X$ be a non-empty set. A **metric** (or a **distance**) on $X$ is a function $d : X \times X \to \mathbb{R}^+$ verifying:*
  (1) $d(x, y) = 0 \Leftrightarrow x = y$.
  (2) $\forall x, y \in X : \ d(x, y) = d(y, x)$.
  (3) $\forall x, y, z \in X : \ d(x, z) \leq d(x, y) + d(y, z)$ (**Triangle Inequality**).
*The couple $(X, d)$ is called a **metric space**.*

An easy consequence (but equally important) of the triangle inequality is (see Exercise 2.3.2):

**PROPOSITION 2.1.2.** *(second triangle inequality) Let $(X, d)$ be a metric space. Then*

$$\forall x, y, z \in X : \ |d(x, z) - d(y, z)| \leq d(x, y).$$

A repeated use of the triangle inequality (using induction!) leads to a **generalized triangle inequality**.

**PROPOSITION 2.1.3.** *Let $(X, d)$ be a metric space. Then for all $x_1, x_2, \cdots, x_n \in X$, we have*

$$d(x_1, x_n) \leq d(x_1, x_2) + d(x_2, x_3) + \cdots + d(x_{n-1}, x_n).$$

We start with an extremely important example.

**EXAMPLE 2.1.4** (the discrete metric). *Let $X$ be a non-empty set. Define a map on $X \times X$ by*

$$d(x, y) = \left\{ \begin{array}{ll} 0, & x = y, \\ 1, & x \neq y. \end{array} \right.$$

*Then $d$ is a metric on $X$ called the **discrete metric** (for a proof and some properties of $d$, see Exercise 2.3.5).*

27

REMARK. The discrete metric is not very useful in practise since it can be defined on any non-empty set. However, it can be very valuable as a source for counterexamples as it will be illustrated in many exercises in the sequel.

EXAMPLES 2.1.5.

(1) *The absolute value is a metric. More precisely, the mapping $d : \mathbb{R} \times \mathbb{R} \to \mathbb{R}^+$ defined by $d(x, y) = |x - y|$ is a metric called the **usual** (or **standard**) metric on $\mathbb{R}$. It may be denoted by $|\cdot|$.*

(2) *The modulus of a complex number is also a metric. Indeed, the mapping $d : \mathbb{C} \times \mathbb{C} \to \mathbb{R}^+$ defined by*

$$d(z, z') = |z - z'| = \sqrt{(x - x')^2 + (y - y')^2}$$

*(where $z = x + iy$ and $z' = x' + iy'$) is a metric called the **usual** (or **standard**) metric on $\mathbb{C}$. It may also be denoted by $|\cdot|$ or $|\cdot|_{\mathbb{C}}$.*

(3) *On $\mathbb{R}^n$, we may define different metrics. For instance,*

$$(x, y) \mapsto d_2(x, y) = \sqrt{|x_1 - y_1|^2 + |x_2 - y_2|^2 + \cdots + |x_n - y_n|^2}$$

*where $x = (x_1, x_2, \cdots, x_n)$ and $y = (y_1, y_2, \cdots, y_n)$. This metric is called the **Eucildean metric** on $\mathbb{R}^n$. Other metrics may be defined on $\mathbb{R}^n$ such as*

$$d_1(x, y) = |x_1 - y_1| + |x_2 - y_2| + \cdots + |x_n - y_n|$$

*and*

$$d_\infty(x, y) = \max(|x_1 - y_1|, |x_2 - y_2|, \cdots, |x_n - y_n|).$$

(4) *The previous example may be generalized to infinite sequences, but this will involve an infinite sum, and so a convergence of a series is required. So we introduce*

$$\ell^p = \left\{ (x_n) : \mathbb{N} \to \mathbb{C} : \sum_{n=1}^{\infty} |x_n|^p < \infty \right\},$$

*where $1 \leq p < \infty$. On $\ell^p \times \ell^p$, we define*

$$d_p(x, y) = d_p((x_n), (y_n)) = \left( \sum_{n=1}^{\infty} |x_n - y_n|^p \right)^{\frac{1}{p}}.$$

*Then $d_p$ is a metric on $\ell^p$ (cf. Exercise 2.3.11).*

(5) *Similarly, let $\ell^\infty$ be the space of all bounded sequences in $\mathbb{C}$. Then the function given by*

$$d_\infty(x, y) = d_\infty((x_n), (y_n)) = \sup_{n \in \mathbb{N}} |x_n - y_n|.$$

*defines a metric on $\ell^\infty$.*

Having introduced $\ell^p$ $(1 \leq p \leq \infty)$, we give some of their immediate properties. More properties will be considered in the coming chapters.

**THEOREM 2.1.6.** *Let $1 \leq p, q \leq \infty$. Then:*

(1) *$\ell^p$ is a complex vector space.*
(2) *$\ell^p \subset \ell^q$ if $p < q$.*

**REMARK.** The inclusion in the second statement of the previous theorem is proper. Let $p < s < q$ and

$$x = (1, 2^{-\frac{1}{s}}, \cdots, n^{-\frac{1}{s}}, \cdots).$$

Then $x \in \ell^q$ but $x \notin \ell^p$.

There are useful subspaces of $\ell^\infty$.

**EXAMPLES 2.1.7.**

(1) *$c$ is the vector space of convergent sequences.*
(2) *$c_0$ is the vector space of convergent sequences having the limit zero (notice that $c_0 \subset c$).*
(3) *$c_{00}$ is the vector space of all finitely non-zero sequences (notice also that $c_{00} \subset \ell^p$ for $1 \leq p \leq \infty$).*

We may ask whether a product of two metric spaces remains a metric space (implicit examples already appeared in Examples 2.1.5)? First, we have to equip it with a metric (don't we?).

**PROPOSITION 2.1.8.** *Assume that $(X_k, d_k)$ are metric spaces for all $1 \leq k \leq n$. Let $p \geq 1$. If $x = (x_1, \ldots, x_n)$ and $y = (y_1, \ldots, y_n)$, then*

$$d_p(x, y) = \left( \sum_{k=1}^n [d_k(x_k, y_k)]^p \right)^{\frac{1}{p}}$$

*defines a metric on $X = X_1 \times \cdots \times X_k$, i.e. $(X, d_p)$ is a metric space. Also,*

$$d_\infty(x, y) = \max(d_1(x_1, y_1), \cdots, d_n(x_n, y_n))$$

*defines a metric on $X = X_1 \times \cdots \times X_n$.*

### 2.1.2. Important Sets in Metric Spaces.

**DEFINITION** 2.1.9. *Let $(X, d)$ be a metric space.*

*An **open ball** of center $a \in X$ and radius $r > 0$, denoted by $B(a, r)$, is defined by:*

$$B(a, r) = \{x \in X : \ d(x, a) < r\}.$$

*A **closed ball** of center $a \in X$ and radius $r > 0$, denoted by $B_c(a, r)$, is defined by:*

$$B_c(a, r) = \{x \in X : \ d(x, a) \leq r\}.$$

*A **sphere** of center $a \in X$ and radius $r > 0$, and denoted by $S(a, r)$, is given by:*

$$S(a, r) = \{x \in X : \ d(x, a) = r\}.$$

**REMARK**. A ball in a general metric space is not necessarily round. It can be anything e.g. a singleton, a square, a diamond etc... It can also be of a much more irregular shape (cf. the second example below and take $X$ as you wish).

**EXAMPLES** 2.1.10.

(1) *In usual $\mathbb{R}$, if $r > 0$ and $a \in \mathbb{R}$, then*

$$B(a, r) = (a - r, a + r).$$

(2) *Let $X$ be endowed with the discrete metric. Then (see Exercise 2.3.5)*

$$B(x, r) = \{x\} \ \text{if } r \leq 1 \text{ and } B(x, r) = X \ \text{if } r > 1.$$

**DEFINITION** 2.1.11. *Let $(X, d)$ be a metric space and let $U \subset X$. We say that $U$ is **open** in $(X, d)$ if:*

$$\forall x \in U, \exists r > 0 : \ B(x, r) \subset U.$$

*A subset $V$ of $X$ is said to be **closed** if its algebraic complement $V^c$ is open.*

**EXAMPLES** 2.1.12.

(1) *An open interval is an open set.*
(2) *$[0, 1]$ or $[0, 1)$ is not open in usual $\mathbb{R}$.*
(3) *In usual $\mathbb{R}$ again, $\{0\}$ is not open.*
(4) *In a discrete metric space, all subsets of $X$ (without any exception!) are open (see Exercise 2.3.5). Hence all subsets of $X$ are closed.*

The first example just above may be generalized to the following

**PROPOSITION** 2.1.13. *(see Exercise 2.3.18) Let* $(X, d)$ *be a metric space. Let* $B(x, r)$ *be an open ball of center* $x \in X$ *and radius* $r > 0$. *For each* $y \in B(x, r)$,

$$\exists s > 0: \ B(y, s) \subset B(x, r).$$

*In other words, an open ball is an open set.*

**REMARK.** The converse of the previous result is not always true, that is, an open set is not necessarily an open ball. See Exercise 2.3.18 for a counterexample.

**REMARK.** Open and closed sets do not form a partition of the metric space. There are sets which are neither open nor closed. The reader will see many examples throughout the exercises in Chapters 2 & 3.

There are also sets which are open and closed simultaneously (for example in a discrete metric space, see Exercise 2.3.5). We call them **"clopen"**.

**REMARK.** It is clear that we can define different metrics on the same set $X$. Hence a given subset $A$ of $X$ may be open with respect to a metric and not open with respect to another one. For instance, $[0, 1)$ is neither open nor closed in usual $\mathbb{R}$, but in $\mathbb{R}$ equipped with the discrete metric, $[0, 1)$ is clopen!

**THEOREM** 2.1.14 (See Exercise 2.3.17). *Let* $(X, d)$ *be a metric space. Then*

(1) $\varnothing$ *and* $X$ *are open.*
(2) *The union of **any** collection of open sets remains open.*
(3) *The intersection of a **finite** collection of open sets remains open.*

**REMARK.** We will see in the next chapter that a space in which the properties of the previous theorem are verified will be called a **topological space**. Thus a metric space is an example (very important though) of a topological space.

**EXAMPLES** 2.1.15.

(1) $[0, 1]$ *is closed in usual* $\mathbb{R}$ *for its complement* $(-\infty, 0) \cup (1, \infty)$, *being a union of open sets, is open.*
(2) *A similar idea gives us the closedness of* $\{0\}$ *in usual* $\mathbb{R}$.

We have observed above that an open set is not necessarily an open ball. Nonetheless, we have an interesting and useful result.

**PROPOSITION** 2.1.16. *(Exercise 2.3.19) In a metric space $X$, a set $U \subset X$ is open if and only if it can be written as a union of open balls.*

**COROLLARY** 2.1.17. *Open sets in usual $\mathbb{R}$ are unions of open intervals.*

**REMARK**. A much more refined result can be found in Proposition 3.1.25 (cf. Theorem 3.1.27).

The definition of closed sets combined with elementary set theory yields:

**COROLLARY** 2.1.18. *Let $(X, d)$ be a metric space. Then*

(1) *$\varnothing$ and $X$ are closed.*
(2) *The intersection of **any** collection of closed sets stays closed.*
(3) *The union of a **finite** collection of closed sets remains closed.*

Now we introduce the notion of a bounded set.

**DEFINITION** 2.1.19. *Let $(X, d)$ be a metric space and let $A \subset X$. Define:*
$$d(A) = \sup_{x,y \in A} d(x, y).$$
*Then $d(A)$ is called the **diameter** of $A$.*

**REMARK**. If $A$ is a non-empty set, then $d(A) \leq \infty$.

**REMARK**. We will get back to this notion in the chapter of complete spaces.

**DEFINITION** 2.1.20. *Let $(X, d)$ be a metric space and let $A \subset X$. We say that $A$ is **bounded** if it is contained in a ball of a **finite** radius. Equivalently, $A$ is bounded if its diameter $d(A)$ is finite.*

**EXAMPLE** 2.1.21. *Usual $\mathbb{R}$ is not bounded while $\mathbb{R}$ equipped with a discrete metric is bounded.*

**DEFINITION** 2.1.22. *Let $X$ be a non-empty set. Let $(Y, d)$ be a metric space. A function $f : X \to (Y, d)$ is said to be **bounded** if $f(X)$ is bounded in $(Y, d)$.*

**2.1.3. Continuity in Metric Spaces.** The definition of continuity for a function from $\mathbb{R}$ into $\mathbb{R}$ is known to the reader. It is clear that the natural generalization to metric spaces is the following:

**DEFINITION** 2.1.23. *Let $(X, d)$ and $(Y, d')$ be two metric spaces and let $f : (X, d) \to (Y, d')$ be a function. We say that $f$ is **continuous** at $a \in X$ if:*
$$\forall \varepsilon > 0, \exists \alpha > 0, \ \forall x \in X : \ (d(x, a) < \alpha \implies d'(f(x), f(a)) < \varepsilon).$$

*We say that $f$ is **continuous** on $X$ if it is continuous at each $a \in X$.*

The next definition is just a reformulation of the previous one (using open balls).

**DEFINITION 2.1.24.** *Let $(X, d)$ and $(Y, d')$ be two metric spaces. Let $f : (X, d) \to (Y, d')$ be a function. Then $f$ is **continuous** at a point $a \in X$ if for every ball $B(f(x), \varepsilon)$, there exists a ball $B(x, \alpha)$ such that $f(B(x, \alpha)) \subset B(f(x), \varepsilon)$.*

The previous definition is equivalent to the following result (which will be the definition of continuity in topological spaces):

**PROPOSITION 2.1.25** (See Exercise 2.3.23). *Let $(X, d)$ and $(Y, d')$ be two metric spaces. Let $f : (X, d) \to (Y, d')$ be a function. Then $f$ is continuous iff for every open set $U$ in $(Y, d')$, $f^{-1}(U)$ is open in $(X, d)$.*

**DEFINITION 2.1.26.** *Let $(X, d)$ be a metric space and let $A \subset X$ be non-empty. The **distance** between a point $x \in X$ and $A$, denoted by $d(x, A)$, is defined by:*

$$d(x, A) = \inf_{t \in A} d(x, t).$$

**REMARK.** The function $x \mapsto d(x, A)$ is continuous on $X$ thanks to the following result:

$$\forall x, y \in X : \ |d(x, A) - d(y, A)| \leq d(x, y).$$

We can also define uniform continuity of a function $f$ between two metric spaces.

**DEFINITION 2.1.27.** *Let $(X, d)$ and $(Y, d')$ be two metric spaces and let $f : (X, d) \to (Y, d')$ be a function. We say that $f$ is **uniformly continuous** if:*

$$\forall \varepsilon > 0, \exists \alpha > 0, \ \forall x, x' \in X : \ (d(x, x') < \alpha \implies d'(f(x), f(x')) < \varepsilon).$$

**PROPOSITION 2.1.28.** *Uniform continuity implies continuity (but not vice versa in general, cf. Theorem 5.1.24).*

**DEFINITION 2.1.29.** *Let $(X, d)$ and $(Y, d')$ be two metric spaces. A function $f : (X, d) \to (Y, d')$ is said to be an **isometry** if $f$ is surjective and:*

$$\forall x, y \in X : \ d(x, y) = d'(f(x), f(y)).$$

*We then say that $X$ and $Y$ are **isometric**.*

**REMARKS**.

(1) In some textbooks, they do not assume the surjectivity hypothesis in the definition of an isometry (they then say **an isometry into**).
(2) It is easy to see that an isometry is injective. Therefore, an isometry is bijective.
(3) The inverse of an isometry is an isometry.
(4) It is also clear that an isometry is continuous.

**EXAMPLE** 2.1.30. *Let $f : \mathbb{R}^2 \to \mathbb{C}$ defined by $f(x,y) = x + iy$. Then $f$ is an isometry from $\mathbb{R}^2$ (endowed with the Euclidean metric) onto $\mathbb{C}$ equipped with the usual metric. In other words, $\mathbb{R}^2$ is isometric to $\mathbb{C}$.*

**2.1.4. Equivalent Metrics.** In many cases, there are metrics easier to handle than others (on the same set). Obviously, we should not expect topological properties (continuity, convergence and completeness, among others) to remain true. If, however, the metrics satisfy a certain condition, then this becomes possible.

**DEFINITION** 2.1.31. *Let $d$ and $d'$ be two metrics on the same set $X$. Then we say that $d$ is **equivalent** (we may also say **strongly equivalent** or **Lipschitz equivalent**) to $d'$ if:*

$$\exists \alpha, \beta > 0, \ \forall x, y \in X : \ \alpha d(x,y) \le d'(x,y) \le \beta d(x,y).$$

Let $X$ be a set and let $d$, $d'$ and $d''$ be metrics on $X$ such that $d$ is equivalent to $d'$ and $d'$ is equivalent to $d''$. Is $d$ equivalent to $d''$? The answer is yes as we have

**PROPOSITION** 2.1.32. *Let $X$ be a set. On the set of all possible metrics on $X$, the relation "$d$ equivalent to $d'$ " is an equivalence relation.*

**REMARK**. The metrics $d_p$ with $1 \le p \le \infty$, defined in Examples 2.1.5 (see Exercise 2.3.9), are equivalent.

The same method of Exercise 2.3.9 may be applied to prove the following proposition.

**PROPOSITION** 2.1.33. *The metrics $d_p$ (including $d_\infty$) defined in Proposition 2.1.8 are equivalent.*

There is another type of equivalence of metrics, namely

**DEFINITION** 2.1.34. *Two metrics $d$ and $d'$ on a set $X$ are said to be **topologically equivalent** if: A subset $U$ is open in $(X, d)$ iff it is open in $(X, d')$.*

PROPOSITION 2.1.35. *(a proof is given in Exercise 2.3.28) The (strong) equivalence of two metrics implies their topological equivalence.*

REMARK. The converse of the previous result is not true in general. For a counterexample, see e.g. Exercise 2.3.29.

EXAMPLES 2.1.36.

(1) *The metrics $d_p$ with $1 \leq p \leq \infty$, defined in Examples 2.1.5 which are already equivalent are hence topologically equivalent.*

(2) *In $\mathbb{R}$, the discrete metric is not topologically equivalent to the usual metric for an open set w.r.t. the discrete metric need not be open w.r.t. the usual one. This also implies that the two metrics are not (strongly) equivalent.*

## 2.2. True or False: Questions

QUESTIONS. Comment on the following questions/statements and indicate those which are false and those which are true when this applies. Justify your answers.

(1) Let $X$ be a set with card$X \geq 2$. We can always define a metric on $X$.
(2) If $d : X \times X \to \mathbb{R}$ is a function which satisfies:
   (a) $d(x, y) = 0 \Leftrightarrow x = y$.
   (b) $\forall x, y \in X : d(x, y) = d(y, x)$.
   (c) $\forall x, y, z \in X : d(x, z) \leq d(x, y) + d(y, z)$,
   then $d$ is a positive and hence it is a metric.
(3) The set $\{0\}$ is not open.
(4) The set $\{0\}$ is not open in $(\mathbb{R}, |\cdot|)$ since it is closed.
(5) In a metric space, a ball can contain another ball of strictly bigger radius.
(6) The set $(0, 1)$ is bounded.
(7) Let $(X, d)$ be a metric space and let $A$, $B$ and $C$ be three subsets of $X$. Define (not to be confused with the diameter of a given set)
$$d(A, B) = \inf_{(a,b) \in A \times B} d(a, b).$$
Then
$$d(A, C) \leq d(A, B) + d(B, C).$$
(8) Every closed ball is a closed set. What about the converse?
(9) The sphere is a closed set.
(10) There is a metric space in which all triangles are isosceles.

## 2.3. Exercises With Solutions

**Exercise** 2.3.1. Assume that a function $d$ on $X \times X$ into $\mathbb{R}^+$ verifies
$$\begin{cases} d(x, y) = 0 \Longleftrightarrow x = y, \\ d(x, y) = d(y, x), \forall x, y \in X, \\ d(x, z) \leq d(x, y) + d(y, z), \quad \forall x, y, z \ distinct \ \text{and in } X. \end{cases}$$
Show that $d$ is a metric on $X$.

**Exercise** 2.3.2. Let $(X, d)$ be a metric space. Show that
$$\forall x, y, z \in X : |d(x, z) - d(y, z)| \leq d(x, y).$$

**Exercise** 2.3.3. Are the following functions metrics on $X$?
(1) $d(x, y) = |x^2 - y^2|$, $X = \mathbb{R}$;
(2) $d(x, y) = |x^3 - y^3|$, $X = \mathbb{R}$;

(3) $d(x,y) = e^{x-y}$, $X = \mathbb{R}$;

(4) $d(x,y) = \left| \frac{1}{x} - \frac{1}{y} \right|$, $X = \mathbb{R}^*$;

(5) $d(x,y) = |x - 3y|$, $X = \mathbb{R}$.

**Exercise 2.3.4.** In the usual metric of $\mathbb{R}$, what is the ball corresponding to the open interval $(0,1)$ (i.e. what is its center and what is its radius?).

**Exercise 2.3.5 (the discrete metric).** Let $X$ be a non-empty set. Define a map on $X \times X$ by

$$d(x,y) = \begin{cases} 0, & x = y, \\ 1, & x \neq y. \end{cases}$$

(1) Show that $d$ is a metric on $X$.
(2) Let $r > 0$ and let $x \in X$. Find the open ball $B(x,r)$ and the closed ball $B_c(x,r)$.
(3) Find the sphere $S(x,r)$.
(4) Show that every subset in a discrete metric space is open.
(5) Deduce that every subset in a discrete metric space is closed.
(6) Set $X = \mathbb{R}$. Show that the discrete metric on $\mathbb{R}$ is not equivalent to the usual metric on $\mathbb{R}$.

**Exercise 2.3.6.** Let $(X,d)$ be a metric space. Show that

$$d'(x,y) = \sqrt{d(x,y)}$$

defines a metric on $X$.

**Exercise 2.3.7.** On $\mathbb{N} \times \mathbb{N}$, we define

$$d(x,y) = \begin{cases} 0, & x = y, \\ 3 + \frac{x+y}{xy}, & x \neq y. \end{cases}$$

Show that $d$ is a metric on $\mathbb{N}$.

**Exercise 2.3.8.** On $\mathbb{R}^n \times \mathbb{R}^n$, define the function $d$ by

$$d(x,y) = \sum_{k=1}^{n} |x_k - y_k|$$

for all $x = (x_k), y = (y_k) \in \mathbb{R}^n$. Show that $d$ is a metric on $\mathbb{R}^n$ (called in many references the **taxicab** metric).

**Exercise 2.3.9.** Let $p > 1$ and $q > 1$ be such that $\frac{1}{p} + \frac{1}{q} = 1$. Let $a_1, \cdots, a_n$ and $b_1, \cdots, b_n$ be positive real numbers.

(1) Prove the following **Hölder Inequality**

$$\sum_{k=1}^{n} a_k b_k \leq \left(\sum_{k=1}^{n} (a_k)^p\right)^{\frac{1}{p}} \left(\sum_{k=1}^{n} (b_k)^q\right)^{\frac{1}{q}}.$$

(2) Prove the following so-called **Minkowski Inequality**

$$\left(\sum_{k=1}^{n} (a_k + b_k)^p\right)^{\frac{1}{p}} \leq \left(\sum_{k=1}^{n} (a_k)^p\right)^{\frac{1}{p}} + \left(\sum_{k=1}^{n} (b_k)^p\right)^{\frac{1}{p}}.$$

(3) Let $p \geq 1$. Define the following functions on $\mathbb{R}^n \times \mathbb{R}^n$

$$d_p(x,y) = \left(\sum_{k=1}^{n} |x_k - y_k|^p\right)^{\frac{1}{p}} \quad \text{and} \quad d_\infty(x,y) = \max_{1 \leq k \leq n} \left(|x_k - y_k|\right).$$

   (a) Show that $d_p$ and $d_\infty$ are metrics on $\mathbb{R}^n$.
   (b) Show that $d_p$ and $d_\infty$ are equivalent metrics and deduce
       that
$$\lim_{p \to \infty} d_p(x,y) = d_\infty(x,y).$$

   (c) Is $d_p$ equivalent to $d_q$ where $p \neq q$ and $1 \leq p, q < \infty$?

**REMARKS**.

(1) If $p = 1$, then we get back the taxicab metric whose proof does
    not require the Minkowski's inequality.
(2) If $p = 1$, then in general, we allow $q$ to be $\infty$ and the same
    applies for $q = 1$. This is, however, not discussed in this
    exercise.
(3) If $p = q = 2$, we get back one of the versions of the **Cauchy-Schwarz Inequality**. Therefore, Hölder's Inequality generalizes Cauchy-Schwarz's.
(4) From Question 3, b), we now know why we use the notation
    $d_\infty$.

**Exercise** 2.3.10. Let $M = (X, d)$ be a metric space.

(1) Show that both
    (a) $\delta(x,y) = \min(1, d(x,y))$ and
    (b) $\rho(x,y) = \frac{d(x,y)}{1+d(x,y)}$
    define metrics on $X$ (hint: for $\rho$ you may start by showing that
    $0 \leq x \leq y \Rightarrow \frac{x}{1+x} \leq \frac{y}{1+y}$ and $\frac{x+y}{1+x+y} \leq \frac{x}{1+x} + \frac{y}{1+y}$, $\forall x, y \geq 0$).
(2) Regardless of what $X$ can be, is $(X, \delta)$ bounded? Is $(X, \rho)$
    bounded?

**Exercise 2.3.11.** Let

$$\ell^2 = \left\{ x = (x_n)_n, x_n : \mathbb{N} \to \mathbb{C} : \sum_{n=1}^{\infty} |x_n|^2 < +\infty \right\}.$$

Define a function $d_2$ on $\ell^2 \times \ell^2$ by

$$d_2(x, y) = \sqrt{\sum_{n=1}^{\infty} |x_n - y_n|^2}.$$

(1) Give some elements in $\ell^2$ and others not in it.
(2) Show that $(\ell^2, d_2)$ is a metric space.

**Exercise 2.3.12.** Let $(X_n, d_n)$, $n = 1, 2, \cdots$ be a countable family of metric spaces. Set $X = \prod_{n=1}^{\infty} X_n$. Show that the function $d : X \times X \to \mathbb{R}$ defined by,

$$d(x, y) = \sum_{n=1}^{\infty} \frac{1}{2^n} \times \frac{d_n(x_n, y_n)}{1 + d_n(x_n, y_n)},$$

for each $x = (x_1, \cdots, x_n, \cdots)$ and $y = (y_1, \cdots, y_n, \cdots)$ in $X$, is a metric on $X$.

**Exercise 2.3.13.** Let $(X_1, d_1)$ and $(X_2, d_2)$ be metric spaces. Let the metric

$$d_\infty[(x_1, x_2), (y_1, y_2)] = \max(d_1(x_1, y_1), d_2(x_2, y_2))$$

in $(X_1 \times X_2, d_\infty)$. Let $B_1(x_1, r)$, $B_2(x_2, r)$ and $B((x_1, x_2), r)$ denote the open balls in $(X_1, d_1)$, $(X_2, d_2)$ and $(X_1 \times X_2, d_\infty)$ of centers $x_1$, $x_2$ and $(x_1, x_2)$ respectively and radius $r > 0$. Show that

$$B((x_1, x_2), r) = B_1(x_1, r) \times B_2(x_2, r).$$

**Exercise 2.3.14.** Let $X$ be the space of real-valued continuous functions on $[0, 1]$.

(1) Show that

$$d(f, g) = \int_0^1 |f(x) - g(x)| dx \text{ and } d'(f, g) = \sup_{x \in [0,1]} |f(x) - g(x)|$$

   are two metrics on $X$ ($d'$ is usually called the **supremum metric**).
(2) Show that these two metrics are not equivalent.
(3) Does $d$ remain a metric if $X$ is replaced by the space of Riemann-integrable functions?

**Exercise** 2.3.15. Let $X = C^1([0,1], \mathbb{R})$. Define a function $d$ from $X \times X$ into $\mathbb{R}^+$ by

$$d(f,g) = \sup_{x \in [0,1]} |f'(x) - g'(x)|$$

where $f'$ stands for the derivative of $f$. Is $d$ a metric on $X$?

**Exercise** 2.3.16. Let $M = (X,d)$ be a metric space. Also, let $x, y \in X$ and $r, s > 0$.
  (1) Show that $B(x,r) = B(y,s)$ (for all $x, y$ and all $r, s$) does not always give $x = y$ or $r = s$.
  (2) Give an example when this is correct.

**Exercise** 2.3.17. Let $(X,d)$ be a metric space. Show that:
  (1) $X$ and $\varnothing$ are open sets in $(X,d)$,
  (2) The arbitrary union of open sets in $(X,d)$ is open in $(X,d)$,
  (3) The finite intersection of open sets in $(X,d)$ is open in $(X,d)$. Does this stays true for an infinite intersection?

**Exercise** 2.3.18. Let $(X,d)$ be a metric space. Show that an open ball is an open set. Is the converse always true?

**Exercise** 2.3.19. Show that in any metric space $X$, a subset $U$ of $X$ is open if and only if it can be written as a union of open balls.

**Exercise** 2.3.20. Are the intervals $[a,b]$, $[a,b)$ or $(a,b]$ open in $\mathbb{R}$?

**Exercise** 2.3.21. Associate with $\mathbb{R}^2$ the euclidian metric and denote it by $d$. Define on $\mathbb{R}^2 \times \mathbb{R}^2$ a function $\delta$ by

$$\delta(x,y) = \begin{cases} 0, & x = y, \\ d(x, \mathbf{0}) + d(y, \mathbf{0}), & x \neq y \end{cases}$$

where $\mathbf{0} = (0,0)$.
  (1) Check that $\delta$ is in effect a metric on $\mathbb{R}^2$.
  (2) Let $a \neq \mathbf{0}$. Show that $\{a\}$ is open.
  (3) Is $\{\mathbf{0}\}$ open in $(\mathbb{R}^2, \delta)$?
  (4) What is $\mathbb{R}^2 \setminus \{\mathbf{0}\}$ restricted to $\delta$?

**REMARK.** The metric defined in the previous exercise has a particular name: In the UK, it is called the **British Rail Metric**. The main reason for this designation is that often when one wants to travel from a town to another, then he/she might have to pass by some train station in London which is represented by $\mathbf{0}$ in our exercise. For the same reason, the French call it the **SNCF Metric**. As for the Americans, they call it the **Post Office Metric**. This latter is probably more meaningful than the other two as if one wants to send a letter from his/her place to somewhere else it will have to pass by the post office which is represented by **0** in the exercise.

**Exercise 2.3.22 (Ultrametric space).** Let $X$ be a non empty set. Let $d$ be a function from $X \times X$ into $\mathbb{R}^+$ such that

- $d(x, y) = 0 \Leftrightarrow x = y$,
- $d(x, y) = d(y, x), \forall x, y \in X$,
- $d(x, z) \leq \max(d(x, y), d(y, z)), \forall x, y, z \in X$.

(1) Show that $d$ is a metric, called **ultrametric**.

(2) Show that at least two of $d(x, y)$, $d(y, z)$ and $d(x, z)$ must be equal. Interpret this result geometrically.

(3) Show that in a ultrametric space, every point in an open ball is its center.

(4) Show that open balls are clopen and that so are closed balls too.

**Exercise 2.3.23.** Let $(X, d)$ and $(X', d')$ be two metric spaces and let $f : (X, d) \to (X', d')$ be a function. Show that $f$ is continuous on $X$ iff whenever $U$ is an open set in $X'$, $f^{-1}(U)$ is an open set in $X$.

**Exercise 2.3.24.** Let $(X, d)$ be a metric space. Let

(1) Show that the function $f : X \to \mathbb{R}$ defined for all $x \in X$ by $f(x) = d(x, a)$ is continuous on $X$ where $a \in X$ and $r > 0$.

(2) Let $B \subset X$ be non-empty. Set

$$g(x) = d(x, B) = \inf_{b \in B} d(x, b).$$

Show that $g$ is uniformly continuous on $X$.

**Exercise 2.3.25.** Let $\mathbb{R}^+$ be equipped with the induced usual metric $|\cdot|$. Let $d$ be the *metric*, defined for all $x, y \in \mathbb{R}^+$, by

$$d(x, y) = |\sqrt{x} - \sqrt{y}|.$$

Let $f : (\mathbb{R}^+, |\cdot|) \to (\mathbb{R}^+, d)$ be the identity map.

(1) Show that $f$ is uniformly continuous.

(2) Interpret the result of the previous question differently.

**Exercise 2.3.26.** In usual $\mathbb{R}$, give an example of a function $f$ from $\mathbb{R}$ into $\mathbb{R}$ satisfying each of the following requirements (separately!):

(1) discontinuous everywhere,

(2) continuous at only one point,

(3) continuous at only two points.

**Exercise 2.3.27.** Show that the open ball $B(a, r)$ is open in the metric space $(X, d)$ by using the function $f : x \mapsto f(x) = d(x, a)$ where $a \in X$.

**Exercise 2.3.28.** Let $(X, d)$ and $(X, d')$ be two metric spaces such that $d$ and $d'$ are equivalent. Show that

$$U \text{ open in } (X, d) \Longleftrightarrow U \text{ open in } (X, d').$$

**Exercise 2.3.29.** Let $x$ and $y$ be two reals. Set for all $x$ and $y$

$$\delta(x, y) = |\arctan x - \arctan y|.$$

(1) Show that $\delta$ is a metric on $\mathbb{R}$. Is $\mathbb{R}$ bounded with respect to this metric?
(2) Compare the open balls $B(0, 2)$, $B(0, 4)$ and $B(1, 4)$. What do you observe?
(3) Show that $\delta$ is not equivalent to the usual metric on $\mathbb{R}$.
(4) Is $\delta$ topologically equivalent to the usual metric on $\mathbb{R}$?

**Exercise 2.3.30.** Let $(X, d)$ be a metric space. Let $\rho$ be as in Exercise 2.3.10.

(1) Is $d$ equivalent to $\rho$?
(2) Are they topologically equivalent?

## 2.4. Tests

**Test 1.** Show that even if a metric $d$ is not defined into $\mathbb{R}^+$, the three properties of a metric will guarantee its positivity, i.e.

$$d(x, y) \geq 0 \text{ for all } x, y \in X.$$

**Test 2.** Let $X = \mathbb{R}$. We define on $\mathbb{R} \times \mathbb{R}$, the function $d$ by

$$d(x, y) = \ln(1 + |x - y|), \ \forall x, y \in \mathbb{R}.$$

Show that $d$ is a metric on $X$.

**Test 3.** Let $X = C([0, 1], \mathbb{R})$. Define a function $d$ on $X \times X$ by

$$d(f, g) = \left( \int_0^1 |f(x) - g(x)|^2 dx \right)^{\frac{1}{2}}$$

where $f, g \in X$. Show that $d$ is a metric on $X$.

**Test 4.** Let $X = \{x, y\}$. Let $a > 0$. Assume that a function $d$ on $X \times X$ verifies

$$\begin{cases} d(x, x) = d(y, y) = 0, \\ d(x, y) = d(y, x) = a. \end{cases}$$

Show that $d$ is a metric on $X$.

**Test** 5. Let $(X, d)$ be a metric space and let $A$ be some non-empty set. Let $f : A \to X$ be a *one-to-one* mapping. Set

$$\delta(x, y) = d(f(x), f(y)), \ \forall x, y \in A.$$

Show that $\delta$ defines a metric on $A$.

**Test** 6. Give an example of a metric which is ultrametric and an example of one which is not.

## 2.5. More Exercises

**Exercise** 2.5.1. Let $(X, d)$ be a metric space. Show that

$$d(x_1, x_n) \leq d(x_1, x_2) + d(x_2, x_3) + \cdots + d(x_{n-1}, x_n), \ \forall x_1, x_2, \cdots, x_n \in X.$$

**Exercise** 2.5.2. Let $(X, d)$ be a metric space.
(1) Show that

$$|d(x, z) - d(y, t)| \leq d(x, y) + d(z, t).$$

(2) Why can this result be considered as a generalization of the inequality appearing in Exercise 2.3.2?

**Exercise** 2.5.3. Let $X$ be a non-void set. Assume that a function $d$ on $X \times X$ verifies

$$\begin{cases} d(x, x) = 0, \\ 1 \leq d(x, y) = d(y, x) \leq 2, x \neq y. \end{cases}$$

Show that $d$ is a metric on $X$.

**Exercise** 2.5.4. Let $d$ be a function defined on $\mathbb{C} \times \mathbb{C}$ by

$$d(z, z') = \begin{cases} 0, & z = z', \\ |z| + |z'|, & z \neq z'. \end{cases}$$

(1) Prove that $d$ is a metric on $\mathbb{C}$.
(2) Is $d$ topologically equivalent to the usual metric on $\mathbb{C}$?

**Exercise** 2.5.5. Describe the open sets in the metric spaces of Exercise 2.3.21.

**Exercise** 2.5.6. Let $(X, d)$ be a metric space.
(1) Show that the finite union of bounded sets is bounded.
(2) What about the arbitrary union?
(3) Show that the arbitrary intersection of bounded sets is bounded.

**Exercise** 2.5.7. Is $d$ a metric on $X$ in the following cases:
(1) $d(f, g) = \sup_{x \in [0,1]} |f'(x) - g'(x)| + |f(0) - g(0)|$ where $f'$ stands for the derivative of $f$, $X = C^1([0, 1], \mathbb{R})$ (cf. Exercise 2.3.15).

(2) $d(f,g) = \sup_{x \in [a,b]} |(f(x) - g(x))\omega(x)|$ where $\omega \in X$ (a **weight**) is not vanishing on $[a,b]$, $X = C([a,b])$.

(3) $d(f,g) = \sqrt[p]{\int_0^1 |f(x) - g(x)|^p dx}$ where $p \geq 1$, $X = C([0,1], \mathbb{R})$ (hint: use Exercise 2.3.9).

**Exercise 2.5.8.** Assume that $(X_k, d_k)$ are metric spaces where $1 \leq k \leq n$. Let $p \geq 1$. Show that

$$d(x,y) = \left( \sum_{k=1}^n [d_k(x_k, y_k)]^p \right)^{\frac{1}{p}}$$

defines a metric on $X = X_1 \times \cdots \times X_k$. Does $d$ remain a metric if $0 \leq p < 1$?

**Exercise 2.5.9.** Let $A = (0,2) \times \{1\}$. Show that $A$ is open in $\mathbb{R} \times \{1\}$. Is $A$ open in $\mathbb{R}^2$?

**Exercise 2.5.10.** Let $X$ be a non-empty set. Define a map on $X \times X$ by

$$\delta(x,y) = \begin{cases} 0, & \text{if } x = y, \\ 2, & \text{if } x \neq y. \end{cases}$$

(1) Show that $\delta$ is a metric on $X$.
(2) Is $\delta$ equivalent to the discrete metric? Are they topologically equivalent?

**Exercise 2.5.11.** Show that the metrics defined in Exercise 2.3.14 are not topologically equivalent.

**Exercise 2.5.12.** For all $x, y > 0$, let

$$d(x,y) = |\ln x - \ln y|.$$

(1) Check that $d$ is a metric on $\mathbb{R}_+^*$.
(2) Prove that $d$ is topologically equivalent to the induced usual metric on $\mathbb{R}_+^*$.

**Exercise 2.5.13.** First, we give a definition: *Let $X$ be a vector space on $\mathbb{K}$ (where $\mathbb{K}$ stands for either $\mathbb{R}$ or $\mathbb{C}$). A **norm** on $X$ is a function $\| \cdot \| : X \to \mathbb{R}^+$ satisfying:*

(1) $\|x\| = 0 \iff x = 0$.
(2) $\forall x \in X, \forall \lambda \in \mathbb{K} : \|\lambda x\| = |\lambda| \|x\|$.
(3) $\forall x, y \in X : \|x + y\| \leq \|x\| + \|y\|$ *(called the **triangle inequality**)*.

*The couple $(X, \| \cdot \|)$ is called a **normed vector space** (or a **normed space**).*

(1) On $\mathbb{R}^n$, we define:
$$x \mapsto \|x\|_2 = \sqrt{|x_1|^2 + |x_2|^2 + \cdots + |x_n|^2}$$
where $x = (x_1, x_2, \cdots, x_n)$. Show that $\|\cdot\|_2$ is a norm on $\mathbb{R}^n$ (called the **Euclidean norm** on $\mathbb{R}^n$).

(2) Show also that the following
$$\|x\|_1 = |x_1| + |x_2| + \cdots + |x_n| \text{ and } \|x\|_\infty = \max(|x_1|, |x_2|, \cdots, |x_n|)$$
define two norms on $\mathbb{R}^n$.

**Exercise** 2.5.14. Let $X$ be normed vector space. Let $d : X \times X \to \mathbb{R}^+$ be defined by
$$d(x, y) = \|x - y\| \text{ for all } x, y \in X.$$

(1) Show that $d$ is a metric on $X$ (in other words, every normed vector space is a metric space).
(2) Is the converse true? i.e. can any metric space $X$ be regarded as a normed vector space even if $X$ is a vector space?
(3) Show that if $a \in X$ and $r > 0$, then
$$B(a, r) = \{a\} + B(0, r) = \{a\} + rB(0, 1).$$

A similar result holds for $B_c(a, r)$ and $S(a, r)$.

# CHAPTER 3

# Topological Spaces

## 3.1. Essential Background

### 3.1.1. General Notions.

**DEFINITION 3.1.1.** *Let $X$ be a non-empty set. A **topology** $T$ on $X$ is a subset of $\mathcal{P}(X)$ verifying the following axioms:*

(1) *$\varnothing, X$ are both in $T$.*

(2) *The intersection of two (hence of a **finite** collection of) sets in $T$ remains in $T$.*

(3) *The **arbitrary** union of sets in $T$ is again in $T$.*

*The couple $(X, T)$ is then called a **topological space**, and any element of $T$ is said to be **open**.*

**EXAMPLES 3.1.2.**

(1) *Let $X$ be a set. Then $T = \{\varnothing, X\}$ is a topology on $X$ called the **indiscrete** (or **trivial**) topology. Notice that this topology is not very interesting as there are practically no open sets.*

(2) *Also, take $T = \mathcal{P}(X)$, i.e. the collection of all subsets of $X$. Then $T$ is a topology on $X$. It is called the **discrete topology**.*

(3) *Every metric space is a topological space (Exercise 2.3.17). Hence, the usual metric on $\mathbb{R}$ gives us a topology which we call the **usual** (or **standard**) **topology** of $\mathbb{R}$.*

(4) *Let $X = \{0, 1, 2\}$ and let*

$$T = \{\varnothing, \{0\}, \{2\}, X\}.$$

*Then $T$ is not a topology on $X$.*

(5) *Let $X$ be an infinite set and let $T$ be the family given by*

$$T = \{\varnothing\} \cup \{U \subset X : U^c \text{ finite}\}.$$

*Then $T$ is a topology on $X$ (see Exercise 3.3.31) called the **co-finite topology**. It is a particular case of a more general topology called the **Zariski Topology**.*

**DEFINITION 3.1.3.** *Let $(X, T)$ be a topological space. A set $V \subset X$ is said to be **closed** if $V^c$ is open, that is, if $V^c \in T$.*

**EXAMPLES** 3.1.4.

(1) *In standard* $\mathbb{R}$, $\mathbb{Z}$ *is closed for its complement* $\mathbb{Z}^c$ *is a (here an infinite) union of open intervals, hence it is an open set in* $\mathbb{R}$.

(2) *In a metric space, we have seen that any closed ball is a closed set.*

**REMARK.** As in the case of metric spaces, there are:

(1) Open sets which are not closed.

(2) Closed sets which are not open.

(3) Sets which are neither open nor closed.

(4) Sets which are open and closed at the same time (we call them **"clopen"** ).

By elementary set theory, we have:

**PROPOSITION** 3.1.5. *Let* $(X, T)$ *be a topological space. Then*

(1) $\varnothing$ *and* $X$ *are closed.*

(2) *The union of a **finite** collection of closed sets is closed.*

(3) *The **arbitrary** intersection of closed sets is closed.*

We saw above that every metric space is a topological space. We may therefore ask whether every topological space can be regarded as a metric space in some sense? Giving this a specific terminology then seems to be appropriate.

**DEFINITION** 3.1.6. *Let* $(X, T)$ *be a topological space. We say that* $T$ *is **metrizable** if there is some metric d (on $X$) which gives the topology of* $T$.

**EXAMPLES** 3.1.7.

(1) *A discrete topological space is metrizable. We can easily verify that the discrete metric induces the discrete topology.*

(2) *An indiscrete topological space $X$ (with* $\text{card} X \geq 2$) *is not metrizable.*

**REMARK.** There are theorems giving conditions under which a given topological space is metrizable, such as the "Urysohn Metrization Theorem". But this is not within the scope of the present book. For more details, see e.g. **[19]**.

**DEFINITION** 3.1.8. *Let* $(X, T)$ *be a topological space and let* $x \in X$. *A **neighborhood** of $x$ is any **open** set* $U \in T$ *which contains $x$.*
*The set of neighborhoods of $x$ is denoted by* $\mathcal{N}(x)$.

**REMARK.** The definition we are using here for a neighborhood is adopted by others such as **[19]** or **[20]**. In some other textbooks, neighborhoods are not necessarily open. For further discussion, see the "True or False" Section.

**DEFINITION** 3.1.9. *Let $T$ and $T'$ be two topologies on a set $X$. We say that:*

(1) *$T$ is finer than $T'$ if $T' \subset T$;*
(2) *$T$ is coarser than $T'$ if $T' \supset T$.*

*If $T$ and $T'$ are such that $T' \subset T$ or $T \subset T'$, then we say that $T$ and $T'$ are* **comparable**.

**REMARK.** It is better to remember this definition as the "finer" being the "fatter". Sometimes, the terms "stronger" and "weaker" (or "larger" and "smaller") are used instead of "finer" and "coarser" respectively.

One question may occur to you: Can we characterize a finer or a coarser topology using closed sets? See the "True or False" Section for an answer.

**EXAMPLES** 3.1.10.

(1) *The indiscrete topology is coarser than any other topology which may be defined on the same set.*
(2) *The discrete topology is finer than any other topology which may be defined on the same set.*

**3.1.2. Separation Axioms.** In this subsection, we introduce four important **separation axioms**.

**DEFINITION** 3.1.11. *Let $(X, T)$ be a topological space. We say that $X$ is a:*

(1) **$T_1$**-*space if for all $x, y \in X$, $x \neq y$:*

$$\exists U \in \mathcal{N}(x),\ y \notin U \text{ and } \exists V \in \mathcal{N}(y),\ x \notin V.$$

(2) **$T_2$**-*space or* **Hausdorff** *(or* **separated***) if:*

$$\forall x, y \in X,\ x \neq y,\ \exists (U, V) \in \mathcal{N}(x) \times \mathcal{N}(y) :\ U \cap V = \varnothing.$$

(3) **$T_3$**-*space or* **regular** *if it is $T_1$ and if $x \in X$ and $A$ is a closed set (with $x \notin A$), then*

$$\exists U \in \mathcal{N}(x),\ \exists V \text{ open with } A \subset V :\ U \cap V = \varnothing.$$

(4) **$T_4$**-*space or* **normal** *if for any two closed sets $A$ and $B$ such that $A \cap B = \varnothing$, there are $U, V$ open sets such that*

$$A \subset U,\ B \subset V \text{ and } U \cap V = \varnothing.$$

REMARK. Much of the attention will be given to Hausdorff spaces. Indeed, and without any doubt, they are the most important class of the spaces introduced just above. It is also worth noting that the reader should check "the definitions and the names" in other references as they might be slightly different.

THEOREM 3.1.12. *Let $(X, T)$ be a topological space. Then*

$$X \text{ normal} \implies X \text{ regular} \implies X \text{ Hausdorff} \implies X \text{ is } T_1.$$

REMARK. None of the reverse implications need to hold. The reader is asked to provide counterexamples in Exercise 3.5.25.

EXAMPLES 3.1.13.

(1) *An indiscrete topological space is not Hausdorff.*
(2) *Usual $\mathbb{R}$ is separated (for any $x \neq y$, $x > y$ say, take the disjoint open intervals $U = \left(\frac{x+y}{2}, +\infty\right)$ and $V = \left(-\infty, \frac{x+y}{2}\right)$ containing $x$ and $y$ respectively). More generally, any metric space is Hausdorff (Exercise 3.3.12). Even more generally, we have:*

PROPOSITION 3.1.14. *(cf. Exercise 4.3.33) Every metrizable space is normal.*

PROPOSITION 3.1.15 (see Exercise 3.3.37). *Let $X$ be a topological space which is Hausdorff. Let $x \in X$. Then:*

(1) *The singleton $\{x\}$ (and hence every finite set) is always closed.*
(2) *The intersection of all open sets containing $x$ is $\{x\}$.*

### 3.1.3. Closures, Interiors, Limits Points, et al.

DEFINITION 3.1.16. *Let $(X, T)$ be a topological space and let $A \subset X$. Let $x \in X$.*

(1) *The **closure** of $A$, denoted by $\overline{A}$, is the smallest (w.r.t. "$\subset$") closed set containing $A$. Equivalently, it equals the intersection of all closed sets containing $A$.*
(2) *The **interior** of $A$, denoted by $\overset{\circ}{A}$, is the largest (w.r.t. "$\subset$") open set contained in $A$. Equivalently, it equals the union of all open sets contained in $A$.*
(3) *The **exterior** of $A$, denoted by $\text{ext}(A)$, is the interior of $A^c$.*
(4) *We say that $x$ is a **limit point** of $A$ if:*

$$\forall U \in \mathcal{N}(x): \ U \cap A - \{x\} \neq \varnothing.$$

*The set of limit points of $A$ is denoted by $A'$ (we call it the **derived** set of $A$).*

(5) *If a point $x$ is not a limit point, that is, if*

$$\exists U \in \mathcal{N}(x): \ U \cap A = \{x\},$$

*then we call it an* **isolated point**.

**REMARK.** It is important to keep in mind the following inclusions (which are immediate consequences of the definitions):

$$\overset{\circ}{A} \subset A \subset \overline{A}.$$

**REMARK.** In some textbooks, they use the term **cluster** or an **accumulation** point instead of a limit point.

We have the following relationship between the interior and the closure of a set. It usually allows us to switch from one to the other.

**PROPOSITION** 3.1.17. *(for a proof, see Exercise 3.3.5) Let $X$ be a topological space and let $A \subset X$. Then:*

$$\overline{A^c} = (\overset{\circ}{A})^c \text{ and } \overset{\circ}{A^c} = (\overline{A})^c.$$

**EXAMPLES** 3.1.18.

(1) *In usual $\mathbb{R}$, $\mathbb{Q}' = \mathbb{R}$ and $\overset{\circ}{\mathbb{Q}} = \varnothing$.*

(2) *In usual $\mathbb{R}$, if $A = (0,3) \cup \{4\}$, then $\overset{\circ}{A} = (0,3)$.*

**PROPOSITION** 3.1.19. *Let $(X, T)$ be a topological space and let $A \subset X$. Then*

(1) *$A$ is closed iff $A = \overline{A}$.*

(2) *$A$ is open iff $A = \overset{\circ}{A}$.*

(3) *$x \in \overline{A} \iff \forall U \in \mathcal{N}(x): \ U \cap A \neq \varnothing$.*

(4) *$\overline{A} = A' \cup A$.*

**REMARK.** Let $(X, d)$ be a metric space and let $A \subset X$. In the definitions of limit points, closures, interiors etc..., we can just replace the neighborhood $U$ by an open ball. This comes from the fact that an open set always contains an open ball in a metric space. So, e.g.

$$x \in \overline{A} \iff \forall \varepsilon > 0, \ B(x, \varepsilon) \cap A \neq \varnothing,$$

$$x \in A' \iff \forall \varepsilon > 0: \ B(x, \varepsilon) \cap A - \{x\} \neq \varnothing,$$

and

$$x \in \overset{\circ}{A} \iff \exists r > 0, B(x, r) \subset A.$$

We remain in metric spaces. Is the closure of an open ball equal to the closed ball of the same center and the same radius? In general, only one inclusion is correct.

**PROPOSITION** 3.1.20. *(a proof may be found in Exercise 3.3.23) Let $(X, d)$ be a metric space. If $B(x, r)$ and $B_c(x, r)$ are the open and closed balls respectively of center $x \in X$ and radius $r > 0$, then*

$$\overline{B(x, r)} \subset B_c(x, r).$$

**REMARK.** For counterexamples to the reverse inclusion, see also Exercise 3.3.23.

**REMARK.** There are cases where $\overline{B(x, r)} = B_c(x, r)$ holds in metric spaces. For instance, this is true in $\mathbb{R}^2$ endowed with the euclidian metric (or just $\mathbb{R}$ with the standard metric).

Also, in the setting of normed vector spaces, we *always* have $\overline{B(x, r)} = B_c(x, r)$.

Now, we give more properties of the closures as regards inclusions, unions and intersections, among others.

**THEOREM** 3.1.21. *(a proof is prescribed in Exercise 3.3.7) Let $(X, T)$ be a topological space and let $A, B \subset X$; $A_i \in X$, $i \in I$, where $I$ is arbitrary. Then*

(1) $\overline{\varnothing} = \varnothing$ *and* $\overline{X} = X$.
(2) $\overline{\overline{A}} = \overline{A}$.
(3) $A \subset B \Rightarrow \overline{A} \subset \overline{B}$.
(4) $\overline{A \cup B} = \overline{A} \cup \overline{B}$.
(5) *If $I$ is finite, then*

$$\bigcup_{i \in I} \overline{A_i} = \overline{\bigcup_{i \in I} A_i}.$$

*Otherwise,*

$$\bigcup_{i \in I} \overline{A_i} \subset \overline{\bigcup_{i \in I} A_i}.$$

(6)

$$\bigcap_{i \in I} \overline{A_i} \supset \overline{\bigcap_{i \in I} A_i}.$$

**REMARK.** Exercise 3.3.10 provides counterexamples to the reverse inclusions in (5) and (6).

The analogue of the previous theorem for interiors is given next.

**THEOREM** 3.1.22. *(see Exercise 3.3.8) Let $(X, T)$ be a topological space and let $A, B \subset X$; $A_i \in X$, $i \in I$, where $I$ is arbitrary. Then*

(1) $\overset{\circ}{\varnothing} = \varnothing$ *and* $\overset{\circ}{X} = X$.

(2) $\overset{\circ}{\overset{\circ}{A}} = \overset{\circ}{A}$.

(3) $A \subset B \Rightarrow \overset{\circ}{A} \subset \overset{\circ}{B}$.

(4) $\overset{\circ}{\overbrace{A \cap B}} = \overset{\circ}{A} \cap \overset{\circ}{B}$.

(5) *If $I$ is finite, then*

$$\overset{\circ}{\overbrace{\bigcap_{i \in I} A_i}} = \bigcap_{i \in I} \overset{\circ}{A_i}.$$

*Otherwise,*

$$\overset{\circ}{\overbrace{\bigcap_{i \in I} A_i}} \subset \bigcap_{i \in I} \overset{\circ}{A_i}.$$

(6)

$$\overset{\circ}{\overbrace{\bigcup_{i \in I} A_i}} \supset \bigcup_{i \in I} \overset{\circ}{A_i}.$$

**REMARK.** Exercise 3.3.10 supplies counterexamples to the reverse inclusions in (5) and (6).

Next, we give closures and interiors of all types of intervals in usual $\mathbb{R}$.

**THEOREM** 3.1.23. *(see Exercise 3.3.9) Let $\mathbb{R}$ be endowed with the usual topology and let $a, b \in \mathbb{R}$ such that $a < b$. Then:*

$$\overline{(a, b)} = \overline{[a, b)} = \overline{(a, b]} = \overline{[a, b]} = [a, b],$$

$$\overline{(a, \infty)} = \overline{[a, \infty)} = [a, \infty), \quad \overline{(-\infty, b)} = \overline{(-\infty, b]} = (-\infty, b], \quad \overline{\mathbb{R}} = \mathbb{R},$$

$$\overset{\circ}{[a, b]} = \overset{\circ}{(a, b]} = \overset{\circ}{[a, b)} = \overset{\circ}{(a, b)} = (a, b),$$

*and*

$$\overset{\circ}{(a, \infty)} = \overset{\circ}{[a, \infty)} = (a, \infty), \quad \overset{\circ}{(-\infty, b]} = \overset{\circ}{(-\infty, b)} = (-\infty, b), \quad \overset{\circ}{\mathbb{R}} = \mathbb{R}.$$

The coming result says that the supremum (resp. the infimum) of a set in never far from the closure of the set (in the usual topology).

**PROPOSITION** 3.1.24. *(see Exercise 3.3.21) Let* $\mathbb{R}$ *be equipped with its usual topology. Let* $A$ *be a non-empty and bounded subset of* $\mathbb{R}$. *Then* $\overline{A}$ *too is bounded, and*

$$\inf A, \ \sup A \in \overline{A}.$$

We are used now to the fact that open sets in usual $\mathbb{R}$ are unions of open intervals (Corollary 2.1.17). Can we have a better result than this? For example, $\mathbb{R}$ is an interval, $\mathbb{R}^* = (-\infty, 0) \cup (0, \infty)$ is a union of two intervals which are also disjoint. Also, $\mathbb{R} \setminus \mathbb{Z}$ is open as it can be written as union of (disjoint) open sets. This union is not finite but it is countable which is almost as good. Can we expect this to hold for any open set or are we too greedy? The answer is yes, that is, every open set in $\mathbb{R}$ is expressible as such.

**PROPOSITION** 3.1.25. *(a proof is given Exercise 3.3.13) Every open subset* $U$ *of usual* $\mathbb{R}$ *can be written as a countable union of disjoint open intervals.*

**REMARK**. Unexpectedly, the naive generalization to $\mathbb{R}^n$ is not always true. In fact, just in $\mathbb{R}^2$, an open disc need not be a disjoint union of rectangles of the type $(a, b) \times (c, d)$. More details can be found in Exercise 3.3.14. Thankfully we have a substitute, but first we give a definition.

**DEFINITION** 3.1.26. *We say that a union of rectangles is* **almost disjoint** *if the interiors of the rectangles are disjoint.*

**THEOREM** 3.1.27. *(see e.g. [27]) Each open subset* $U$ *of usual* $\mathbb{R}^n$ *($n \geq 2$) may be written as a countable union of almost disjoint* **closed** *rectangles.*

**REMARK**. The previous two results are useful in Integration Theory.

Before passing to the next concept, we have an interesting result. In the usual topology, take e.g. $\mathbb{R}^*$. Then a successive use of closure-complement and complement-closure yield

$$\overline{\mathbb{R}^*} = \mathbb{R}, \ \mathbb{R}^c = \varnothing, \ \overline{\varnothing} = \varnothing, \ \varnothing^c = \mathbb{R}$$

and

$$\mathbb{R}^{*c} = \{0\}, \ \overline{\{0\}} = \{0\}, \ \{0\}^c = \mathbb{R}^*, \ \overline{\mathbb{R}^*} = \mathbb{R}, \ \mathbb{R}^c = \varnothing.$$

Consequently, a repeated use of "closure-complement" and "complement-closure" of $\mathbb{R}^*$ gives *four different* sets, namely: $\mathbb{R}^*$, $\{0\}$, $\mathbb{R}$ and $\varnothing$.

The natural question is what happens with other sets and in other topologies? Is the number of possibilities finite? Surprisingly, in any set equipped with any topology, the number of possible sets never exceeds 14. Notice, that this is a famous result by Kuratowski more commonly known as the **Kuratowski Closure-Complement Problem**. Despite the fact that this was originally a topological problem, the problem then turned to the algebraic community.

**THEOREM** 3.1.28. *([15]) Let $X$ be a topological space and let $A \subset X$. Then from $A$, one can form no more than 14 different sets by a successive use of "closure-complement" and "complement-closure" operations.*

**REMARK.** The idea of proof is to define a sequence of sets. Then, as we did before the previous theorem, we show that this sequence is periodic. More details may be found in Exercise 3.3.15, where also a set with the maximum of 14 possibilities is given.

**REMARK.** Other authors explored the Kuratowski Problem in a more general setting. For example, what happens if (closure and complement) are replaced or supplemented with other basic topological operations such as: interior, union and intersection? See [22]. See also [9] for an interesting survey.

It may happen that the closure of a set equals the whole space. This is a fundamental concept in mathematics and it has a particular name:

**DEFINITION** 3.1.29. *Let $(X, T)$ be a topological space, and let $A \subset X$. We say that $A$ is **dense** in $X$ if $\overline{A} = X$.*

**REMARK.** We may also say "**everywhere dense**" in lieu of "dense".

**EXAMPLES** 3.1.30.

(1) *In usual $\mathbb{R}$, $\mathbb{Q}$ is dense in $\mathbb{R}$.*
(2) *In a discrete topological space $X$, the only dense set is $X$ itself.*
(3) *In an indiscrete topological space, all subsets (apart from $\varnothing$) are dense.*

**DEFINITION** 3.1.31. *Let $X$ be a topological space and let $A \subset X$. We say that $A$ is **nowhere dense** in $X$ if $\overset{\circ}{\overline{A}} = \varnothing$.*

**REMARK.** We pronounce "nowhere" as "no-where" but not as "now-here"!

**EXAMPLES** 3.1.32.

(1) *In standard $\mathbb{R}$, $\mathbb{Z}$ is nowhere dense in $\mathbb{R}$ for $\overset{\circ}{\overline{Z}} = \varnothing$.*

(2) *In standard* $\mathbb{R}$ *again, the following set*

$$\left\{ \frac{1}{n} : n \in \mathbb{N} \right\}$$

*is also nowhere dense in* $\mathbb{R}$.

(3) $\mathbb{Q}$ *is not nowhere dense in standard* $\mathbb{R}$.

**DEFINITION** 3.1.33. *Let $X$ be a topological space, and let $A \subset X$. The **frontier** (also known as the **boundary**) of $A$ is (the set!) defined by:*

$$\mathrm{Fr}(A) = \overline{A} - \overset{\circ}{A}$$

*(it may also be denoted by $\partial A$).*

**EXAMPLES** 3.1.34.

(1) *In usual* $\mathbb{R}$,

$$\mathrm{Fr}[0, 2) = [0, 2] - (0, 2) = \{0, 2\}.$$

(2) *In a discrete topological space all frontiers are empty!*

**DEFINITION** 3.1.35. *We say that a topological space $(X, T)$ is **separable** if it contains a countable and everywhere dense subset.*

**EXAMPLE** 3.1.36. *The standard* $\mathbb{R}$ *is separable for it contains* $\mathbb{Q}$ *which is countable and dense in* $\mathbb{R}$.

The next result is a useful tool for non-separability in metric spaces.

**PROPOSITION** 3.1.37. *(for a proof see Exercise 3.3.44) Let $(X, d)$ be a metric space. If there is an uncountable $A \subset X$, and for some $r > 0$, we have that*

$$\forall x, y \in A, \ x \neq y : \ d(x, y) \geq r,$$

*then $X$ is not separable.*

**REMARK.** There is a topology associated with a totally ordered set. It may be seen as a generalization of the usual topology of $\mathbb{R}$. We shall not consider this topology in the present book (for more details, see [**19**]).

Having this in mind, we set $\overline{\mathbb{R}} = \mathbb{R} \cup \{-\infty, +\infty\} = [-\infty, +\infty]$. Then we extend the order of $\mathbb{R}$ to $\overline{\mathbb{R}}$ (by staying careful with some arithmetic operations, e.g. $+\infty - \infty$ or $0 \times (\pm\infty)$, although the latter is acceptable in the context of Measure Theory). The set $\overline{\mathbb{R}}$ equipped with this order and the associated topology is called **the extended real line**.

**3.1.4. Bases and Subbases.** First, we give the definition of a basis for a topology.

**DEFINITION** 3.1.38. *Let $X$ be a topological space. A **basis** (or a **base**) for $X$ is a collection $\mathcal{B}$ constituted of open subsets of $X$ such that every open set $U$ in $X$ is the union of elements of $\mathcal{B}$.*

An equivalent definition is given next:

**PROPOSITION** 3.1.39. *A collection $\mathcal{B}$ of open subsets of a topological space $X$ is a basis for $X$ iff for any point $x$ in an open set $U$, there is a $B \in \mathcal{B}$ such that $x \in B \subset U$.*

The next is a criterion for a collection of sets to be a basis.

**THEOREM** 3.1.40. *In order for a family $\mathcal{B}$ of subsets of $X$ to be a basis for a topology on $X$ it is sufficient and necessary to have:*

(1) *$X$ is a union of elements of $\mathcal{B}$, and*
(2) *if $B_1, B_2 \in \mathcal{B}$, then $B_1 \cap B_2$ is a union of elements of $\mathcal{B}$.*

**EXAMPLES** 3.1.41.

(1) *Let $X = \{1, 2, 3\}$ and let $T$ be any topology on $X$. Set $\mathcal{B} = \{\{1, 3\}, \{2, 3\}\}$. Then $\mathcal{B}$ is not a base for $T$.*
(2) *In a metric space, the set of open balls is a basis for it.*
(3) *Let $X$ be a discrete topological space. Then $\mathcal{B} = \{\{x\}\}_{x \in X}$ is a basis for $X$.*

**REMARK.** We can define a topology $T$ **generated** by $\mathcal{B}$ where its open sets are given by the previous proposition. For example, $\{(a, b) : a, b \in \mathbb{R}\}$ is a basis for usual $\mathbb{R}$ which generates the usual topology of $\mathbb{R}$.

There is another related concept.

**DEFINITION** 3.1.42. *Let $X$ be a topological space. A collection $\mathcal{S}$ of open subsets of $X$ is called a **subbasis** (or **subbase**) for $X$ if finite intersections of elements of $\mathcal{S}$ gives a basis for $X$.*

**EXAMPLE** 3.1.43. *The collection of all infinite open intervals is a subbasis for usual $\mathbb{R}$.*

**REMARK.** We shall get back to these two concepts mainly in the next chapter.

The next result tells us how to construct from an arbitrary class of subsets a subbasis.

**THEOREM** 3.1.44. *Each class $C$ of subsets of a non-void set $X$ is the subbasis for a unique topology on $X$ (finite intersections of elements of $C$ form a basis for this topology).*

**DEFINITION** 3.1.45. *The topology defined in the previous theorem is called the* **topology generated** *by $C$.*

The following is another way of producing a topology generated by a collection of sets.

**PROPOSITION** 3.1.46. *Let $C$ be a collection of subsets of $X$. The topology generated by $C$ is the intersection of all topologies on $X$ containing $C$.*

**REMARK**. It is clear that we can define different bases (or subbases) on the same topological space.

### 3.1.5. The Subspace Topology.

**DEFINITION** 3.1.47. *Let $(X, T)$ be a topological space, and let $A \subset X$. Set*
$$T_A = \{A \cap U : U \in T\}.$$
*Then $(A, T_A)$ is a topology on $A$ (for a proof see Exercise 3.3.25). We call it the* **subspace topology**. *It may also be called* **relative** *or* **induced** *topology.*

**REMARK**. Going back to the definition of an isolated point, combined with the subspace topology we may state with ease that: $x$ is an isolated point of a $A$ ($A \subset X$, $X$ is a topological space) iff $\{x\}$ is open in $A$ as a subspace of $X$.

**REMARK**. The set $A = [0, 1)$ is not open in usual $\mathbb{R}$, but it is open in $T_A$. So one has to be careful when using the word "open" when working with the subspace topology. However, there are cases when the "two" open sets coincide. We have

**PROPOSITION** 3.1.48. *Let $X$ be a topological space, and let $A \subset X$ be equipped with the subspace topology. Then if $U$ is open in $A$ and $A$ is open in $X$, then $U$ is open in $X$.*

The next result shows how to find closures and closed sets in the subspace topology.

**PROPOSITION** 3.1.49. *Let $X$ be a topological space and let $A \subset X$ be endowed with the subspace topology. Then:*

(1) *The closed sets in $A$ are of the form $V \cap A$ where $V$ is a closed set in $X$.*

(2) *The closure of $B$ in $A$ is $\overline{B} \cap A$ where $\overline{B}$ is the closure of $B$ in $X$, that is,*

$$\overline{B}^A = \overline{B}^X \cap A.$$

We finish with connections to the separation axioms.

**PROPOSITION 3.1.50.**

(1) *A subspace of a Hausdorff space is Hausdorff.*
(2) *A subspace of a regular space is regular.*
(3) *A **closed** subspace of a normal space is normal.*

**3.1.6. The Product and Quotient Topologies.** We start with defining a topology on a product of two topological spaces. We will treat arbitrary products in the next chapter.

**DEFINITION 3.1.51.** *Let $X$ and $Y$ be two topological spaces. We call an **elementary open** set of $X \times Y$ every set of the form $U \times V$, where $U$ is open in $X$ and $V$ is open in $Y$.*

**DEFINITION 3.1.52.** *Let $X$ and $Y$ be two topological spaces. Let $\mathcal{B}$ be the collection of elementary open sets in $X \times Y$.*

*The **product topology** on $X \times Y$ is the topology having $\mathcal{B}$ as a basis.*

**REMARK.** An open set in $X \times Y$ is therefore a union of elementary open sets. So, an open set in a product space is not necessarily an elementary open! For a counterexample, see the "True or False" Section.

**THEOREM 3.1.53.** *(for a proof, see Exercise 3.3.47) Let $X$ and $Y$ be two topological spaces. Let $A$ and $B$ two subsets of $X$ and $Y$ respectively. Then*

(1) $\overline{A \times B} = \overline{A} \times \overline{B}.$

(2) $\overset{\circ}{\overbrace{A \times B}} = \overset{\circ}{A} \times \overset{\circ}{B}.$

We can use Exercise 2.3.13 to establish the following result:

**THEOREM 3.1.54.** *Let $(X_1, d_1)$ and $(X_2, d_2)$ be metric spaces. Define a metric in $(X_1 \times X_2, d_\infty)$ by*

$$d_\infty[(x_1, x_2), (y_1, y_2)] = \max(d_1(x_1, y_1), d_2(x_2, y_2)).$$

*Then the topology associated with $(X_1 \times X_2, d_\infty)$ is the product topology.*

**REMARK.** Other metrics on $X_1 \times X_2$ such as

$$d_1[(x_1, x_2), (y_1, y_2)] = d_1(x_1, y_1) + d_2(x_2, y_2)$$

or
$$d_2[(x_1, x_2), (y_1, y_2)] = \sqrt{[d_1(x_1, y_1)]^2 + [d_2(x_2, y_2)]^2}$$
also give the product topology on $X_1 \times X_2$.

Before passing to the quotient topology, We give connections to the separation axioms.

**PROPOSITION** 3.1.55.

(1) *A product of Hausdorff spaces is Hausdorff.*
(2) *A product of regular spaces is regular.*

**REMARK**. A product of two normal spaces need not remain normal (by Exercises 3.5.24 & 3.5.25).

We finish this section with the quotient topology.

**PROPOSITION** 3.1.56. *(for a proof, see Exercise 3.3.50) Let $X$ be a topological space and let $\mathcal{R}$ be an equivalence relation on $X$. Let $\varphi : X \to X/\mathcal{R}$ be the quotient map. Let*
$$T = \{A \in X/\mathcal{R} : \varphi^{-1}(A) \text{ is open in } X\}.$$
*Then $T$ is a topology in $X/\mathcal{R}$.*

**DEFINITION** 3.1.57. *The topology defined on $X/R$ in the previous proposition is called the **quotient topology**.*

**REMARK**. The major applications of the quotient topology may be found in Algebraic Topology (see e.g. [**19**]), however, this is not within the scope of the present book.

We shall get back to the quotient topology in the next chapter.

## 3.2. True or False: Questions

QUESTIONS. Comment on the following questions/statements and indicate those which are false and those which are true when this applies. Justify your answers.

(1) Let $X$ be a non-empty set. We can always define a topology on $X$.

(2) To prove a given part, say $T$, is a topological space we need to show (*among others*) that the finite intersection of elements in $T$ is again in $T$. But it is always sufficient to do that for the intersection of two elements. Why is this?

(3) Let $T$ and $T'$ be two topologies on the same set $X$. Then $T \cup T'$ is a topology on $X$.

(4) Let $T$ and $T'$ be two topologies on the same set $X$. Then $T \cap T'$ is a topology on $X$.

(5) Let $X$ be a topological space and let $A \subset X$. The mappings $A \mapsto \overline{A}$ and $A \mapsto \overset{\circ}{A}$ are "monotonic" with respect to "$\subset$".

(6) Let $X$ be a topological space and let $A \subset X$. Then it may well happen that $\overset{\circ}{A} = X$.

(7) Let $X$ be a topological space. Let $A \subset X$. If $A$ is dense in $X$, then $\overset{\circ}{\overline{A}} = \varnothing$.

(8) Let $X$ be a topological space and let $A \subset X$. Then it may well happen that $\overline{A} = \varnothing$.

(9) There is different metric from the discrete one which yields the discrete topology.

(10) Let $X$ be a topological space. Whether we say $X$ is separable or separated, we are talking about the same thing!

(11) Every subspace of a Hausdorff space is Hausdorff.

(12) Let $T$ and $T'$ be two topologies on $X$ such that $T'$ is finer than $T$. If $T$ is Hausdorff, then so is $T'$.

(13) Every subspace of a separable space is separable.

(14) There exists a topological space $X$ in which all subsets (except $\varnothing$) are dense.

(15) Let $X$ be a topological space and let $Y$ be a subset of $X$ endowed with the induced topology of $X$. If $A \subset Y$, then we have

$$\overset{\circ}{A}{}^{Y} = Y \cap \overset{\circ}{A}{}^{X}.$$

(16) There is a definition of a bounded set in an arbitrary topological space.

(17) Let $X$ and $Y$ be two given sets and let $A$ and $B$ be two subsets of $X$ and $Y$ respectively. Then

$$(A \times B)^c = A^c \times B^c.$$

(18) Let $X$ and $Y$ be two topological spaces. If $\Omega = U \times V$ is open (respectively closed) in $X \times Y$, then $U$ is open (respectively closed) in $X$ and $V$ is open (respectively closed) in $Y$.

(19) Let $X$ and $Y$ be two topological spaces. We endow $X \times Y$ with the product topology. Then there is an open set in $X \times Y$ which is not of the form $U \times V$ where $U$ is open in $X$ and $V$ is open in $Y$.

(20) Let $X$ be a topological space and let $A \subset X$. If we unify the interior of $A$ together with its exterior we get back the whole of $X$.

(21) Let $X$ be a topological space and let $A \subset X$. Then $\overset{\circ}{A} = \overline{\overset{\circ}{A}}$.

(22) Let $X$ be a topological space and let $A \subset X$. Then $\overline{A} = \overset{\circ}{\overline{A}}$.

(23) Let $(X, d)$ be a metric space. Let $A$ be a closed set in $X$. Every point in $A$ is a limit point of $A$.

(24) Let $X$ be a topological space and let $A \subset X$. Denote the derived set of $A$, i.e. the set of limit points of $A$, by $A'$. Then $A'$ is always closed.

(25) Let $X$ be a topological space and let $A \subset X$. Then $A$ and $A'$ can be disjoint, comparable and they can be equal as well.

(26) Let $X$ be a topological space and let $A$ and $B$ be two subsets of $X$. Then

$$A \subset B \Longleftrightarrow A' \subset B'.$$

(27) Let $X$ be a topological space and let $A \subset X$. Then

$$\mathrm{Fr}(\overline{A}) = \mathrm{Fr}(A).$$

(28) Let $X$ be a topological space and let $A \subset X$. Then

$$\mathrm{Fr}(\overset{\circ}{A}) = \mathrm{Fr}(A).$$

(29) Let $X$ be a topological space and let $A, B \subset X$. Then

$$\mathrm{Fr}(A \cup B) = \mathrm{Fr}(A) \cup \mathrm{Fr}(B).$$

(30) Let $X$ be a topological space and let $A \subset X$. Then

$$\mathrm{Fr}\, A \subset A.$$

(31) Let $X$ be a topological space and let $A \subset X$. Then

$$\mathrm{Fr}\, A \subset \overline{A}.$$

(32) Let $X$ be a topological space and let $A, B \subset X$ such that $A \subset B$. Then
$$\mathrm{Fr}(A) \subset \mathrm{Fr}(B).$$

(33) Let $(X, d)$ be a metric space and let $x \in X$. Let: $B(x, r)$ be the open ball, $B_c(x, r)$ be the closed ball and $S(x, r)$ be the sphere, all of radius $r > 0$ and center $x$. Then
$$\mathrm{Fr}(B(x, r)) = S(x, r) \text{ or } \mathrm{Fr}(B_c(x, r)) = S(x, r).$$

(34) In usual $\mathbb{R}$, the sets
$$(-1, 1], \ (-2, 2), \ (0, 2], \ [-1, 0] \text{ and } [0, 1]$$
are neighborhoods of 0.

(35) Let $T$ and $T'$ be two topologies on the same set $X$. If $T \subset T'$, then every closed set in $T$ is so in $T'$.

(36) The quotient of a separated space is separated.

## 3.3. Exercises With Solutions

**Exercise 3.3.1.** List all possible topologies which can be defined on the sets $X = \{1\}$ and $Y = \{1, 2\}$.

**Exercise 3.3.2.** Let $X = \{a, b, c, d, e\}$. Define $T \subset \mathcal{P}(X)$ by
$$T = \{\varnothing, \{a\}, \{c, d\}, \{a, c, d\}, \{b, c, d, e\}, X\}.$$

(1) Show that $T$ is a topology on $X$.
(2) What are the closed sets in $T$?
(3) What are the closures and the interiors of $\{a\}$, $\{b\}$ as well as their boundaries?
(4) Give the closure of $\{a, c\}$. What can deduce from that?
(5) Give the neighborhoods of $c$ and $d$.
(6) Is $T$ Hausdorff?

**Exercise 3.3.3.** Let $n \in \mathbb{N}$. Set $A_n = \{n, n+1, n+2, \cdots\}$ and
$$T = \{\varnothing, A_n\}_{n \in \mathbb{N}}.$$

(1) Verify that $T$ is a topology on $\mathbb{N}$.
(2) Is $\{1, 3, 5, 7, \cdots\}$ open in $T$?
(3) Determine $\mathcal{N}(2)$ and $\mathcal{N}(3)$.
(4) Is $T$ separated?
(5) What are the closed sets in $T$?
(6) What is the interior and the closure of $\{4\}$ and of $\{2, 4, 6, 8, \cdots\}$?
(7) What are the dense sets in $T$?

**Exercise** 3.3.4. Let $T$ be a topology on the set $X = \{a, b, c\}$. Show that if the singletons $\{a\}$, $\{b\}$ and $\{c\}$ are open in $T$, then $T$ is the discrete topology.

**Exercise** 3.3.5. Let $(X, T)$ be a topological space and let $A \subset X$. Show that

$$\overline{A^c} = (\overset{\circ}{A})^c \text{ and } (\overline{A})^c = (\overset{\circ}{A^c}).$$

**Exercise** 3.3.6. Let $X$ be a topological space. Show that for every subset $A$ of $X$ and for every open set $U$ we have

$$A \cap U = \varnothing \iff \overline{A} \cap U = \varnothing.$$

**Exercise** 3.3.7. Let $(X, T)$ be a topological space and let $A, B \subset X$; $A_i \in X$, $i \in I$, where $I$ is arbitrary. Show that

(1) $\overline{\varnothing} = \varnothing$ and $\overline{X} = X$.

(2) $\overline{\overline{A}} = \overline{A}$.

(3) $A \subset B \Rightarrow \overline{A} \subset \overline{B}$.

(4) $\overline{A \cup B} = \overline{A} \cup \overline{B}$.

(5) If $I$ is finite, then

$$\bigcup_{i \in I} \overline{A_i} = \overline{\bigcup_{i \in I} A_i}.$$

Otherwise,

$$\bigcup_{i \in I} \overline{A_i} \subset \overline{\bigcup_{i \in I} A_i}.$$

(6)

$$\bigcap_{i \in I} \overline{A_i} \supset \overline{\bigcap_{i \in I} A_i}.$$

**Exercise** 3.3.8. Let $(X, T)$ be a topological space and let $A, B \subset X$; $A_i \in X$, $i \in I$, where $I$ is arbitrary. Show that

(1) $\overset{\circ}{\varnothing} = \varnothing$ and $\overset{\circ}{X} = X$.

(2) $\overset{\circ}{\overset{\circ}{A}} = \overset{\circ}{A}$.

(3) $A \subset B \Rightarrow \overset{\circ}{A} \subset \overset{\circ}{B}$.

(4) $\overset{\circ}{\overparen{A \cap B}} = \overset{\circ}{A} \cap \overset{\circ}{B}$.

(5) If $I$ is finite, then

$$\overset{\circ}{\overparen{\bigcap_{i \in I} A_i}} = \bigcap_{i \in I} \overset{\circ}{A_i}.$$

Otherwise,

$$\bigcap_{i\in I} A_i \subset \bigcap_{i\in I} \overset{\circ}{A}_i.$$

(6)

$$\overline{\bigcup_{i\in I} A_i} \supset \bigcup_{i\in I} \overset{\circ}{A}_i.$$

**Exercise 3.3.9.** Let $a, b \in \mathbb{R}$ with $a < b$. In the usual topology of $\mathbb{R}$, show that the following properties are true:

$$\overline{(a,b)} = \overline{[a,b)} = \overline{(a,b]} = \overline{[a,b]} = [a,b],$$

$$\overline{(a,\infty)} = \overline{[a,\infty)} = [a,\infty), \ \overline{(-\infty,b)} = \overline{(-\infty,b]} = (-\infty,b], \ \overline{\mathbb{R}} = \mathbb{R},$$

$$\overset{\circ}{[a,b]} = \overset{\circ}{(a,b]} = \overset{\circ}{[a,b)} = \overset{\circ}{(a,b)} = (a,b),$$

and

$$\overset{\circ}{(a,\infty)} = \overset{\circ}{[a,\infty)} = (a,\infty), \ \overset{\circ}{(-\infty,b]} = \overset{\circ}{(-\infty,b)} = (-\infty,b), \ \overset{\circ}{\mathbb{R}} = \mathbb{R}.$$

**Exercise 3.3.10.** Let $\mathbb{R}$ be endowed with its standard topology.

(1) Find the closures of $\mathbb{Q}$, $\mathbb{R} \setminus \mathbb{Q}$ and $\{1\} \cup (2,3]$.
(2) Find the interior of $\mathbb{Q}$ and $\mathbb{R} \setminus \mathbb{Q}$.
(3) Give examples of $A$ and $B$ (two subsets of $\mathbb{R}$) such that:

$$\overline{A} \cap \overline{B} \not\subset \overline{A \cap B}.$$

The same question for

$$(A \cup B)^\circ \not\subset \overset{\circ}{A} \cup \overset{\circ}{B}.$$

(4) Let $A_n \subset \mathbb{R}$, $n \in \mathbb{N}$. Do we always have

$$\left( \bigcap_{n\geq 1} A_n \right)^\circ = \bigcap_{n\geq 1} \overset{\circ}{A}_n? \ \overline{\bigcup_{n\geq 1} A_n} = \bigcup_{n\geq 1} \overline{A_n}?$$

**Exercise 3.3.11.** Let $A \subset \mathbb{R}$. Equip $\mathbb{R}$ with the usual topology.

(1) Show that if $\overset{\circ}{A} \neq \varnothing$, then $A$ must be uncountable.
(2) Is this result always true in other topologies on $\mathbb{R}$?

**Exercise 3.3.12.** Show that a metric space is Hausdorff.

**Exercise 3.3.13.** Show that every open subset $U$ of usual $\mathbb{R}$ can be written as a countable union of disjoint open intervals.

**Exercise 3.3.14.** Show that an open disc $D$ in $\mathbb{R}^2$ need not be written as a countable union of disjoint open rectangles.

**Exercise** 3.3.15. Let $X$ be a topological space and let $A \subset X$.

(1) Show that from $A$, we can form no more than 14 different sets by a successive use of "closure-complement" and "complement-closure" operations.

(2) Give an example where the number of 14 different sets is reached.

**Exercise** 3.3.16. Let $A, B \subset \mathbb{R}^n$ with $n \geq 1$. Set

$$A + B = \{a + b : \ a \in A, b \in B\}.$$

Show that $A + B$ is open if $A$ (or $B$) is open in $\mathbb{R}^n$.

**Exercise** 3.3.17. Let $(X, d)$ be a metric space and let $A \subset X$ be an open set. Show that every point in $A$ is a limit point of $A$.

**Exercise** 3.3.18. Let $\mathbb{R}$ be endowed with its usual topology and let $A = \left\{ \frac{1}{n} : \ n \geq 1 \right\}$.

(1) Find the interior, the isolated and the limit point(s) of $A$.

(2) Find $\overline{A}$ and check that $A$ is not closed. What is then the frontier of $A$.

(3) Show that $A$ is nowhere dense in $\mathbb{R}$.

**Exercise** 3.3.19. Give an example of a set $A$ for which the sets $A$, $\overset{\circ}{A}$, $\overline{A}$, $\overline{\overset{\circ}{A}}$ and $\overset{\circ}{\overline{A}}$ are pairwise different.

**Exercise** 3.3.20. Let $X$ be a non-void set. Equip it with the discrete topology and let $A \subset X$. Find $A'$.

**Exercise** 3.3.21. Let $\mathbb{R}$ be equipped with its usual topology. Let $A$ be a non-empty and bounded subset of $\mathbb{R}$.

(1) Show that $\inf A$, $\sup A \in \overline{A}$.

(2) Does this result remain valid in other topologies on $\mathbb{R}$?

(3) Do we have an analogous result for $\overset{\circ}{A}$?

**Exercise** 3.3.22. (cf. Exercise 3.5.7) Let $A$ be a non-void and bounded above subset of $\mathbb{R}$.

(1) Show that in usual $\mathbb{R}$, $\sup A = \sup \overline{A}$.

(2) Show that the previous result need not hold in another topology of $\mathbb{R}$.

(3) In the usual topology of $\mathbb{R}$ again, do we have $\sup \overset{\circ}{A} = \sup A$?

**Exercise** 3.3.23. Let $(X, d)$ be a metric space. Let $r > 0$ and let $x \in X$. Let $B(x, r)$ be the open ball in $X$ of center $x$ and radius $r$ and let $B_c(x, r)$ the corresponding closed ball.

(1) Show that $\overline{B(x,r)} \subset B_c(x,r)$.
(2) Give two examples (one in $\mathbb{R}$ and one in an arbitrary metric space) which show that the backward inclusion is not always true.

**Exercise 3.3.24.** Let $X$ be a non-empty set and let $(Y,T)$ be a topological space. Let $f : X \to Y$ be some function. Show that

$$T' = \{f^{-1}(U) : U \in T\}$$

is a topology on $X$.

**Exercise 3.3.25.** Consider a non-empty set $X$ with a topology $T$. Let $A \subset X$. Show that

$$T_A = \{A \cap U : U \in T\}$$

is a topology on $A$.

**Exercise 3.3.26.** Let $\mathbb{R}$ be endowed with its standard topology. Let $A$ be a topological subspace of $\mathbb{R}$ (except for the last question).
(1) Is $\{3\}$ open in $A = [0,1) \cup \{3\}$?
(2) Are $[0,1)$ and $(0,1)$ open in $A = [0,1]$?
(3) Let $n \in \mathbb{N}$. Is $\{n\}$ open in $A = \mathbb{N}$?
(4) Show that $[0,1]$ and $(2,3)$ are both open in $A = [0,1] \cup (2,3)$. What can you deduce from that?
(5) What is the closure of $\left(0, \frac{1}{2}\right)$ in $A = (0,1]$?
(6) Go back to Exercise 3.3.2 and set $A = \{b,c,d\}$. Give the subspace topology on $A$. Give the closure of $\{b,d\}$ in $A$ by two methods.

**Exercise 3.3.27.** Endow both $A = \mathbb{Q}$ and $B = \mathbb{R} \setminus \mathbb{Q}$ with the induced usual topology of $\mathbb{R}$.
(1) Is $X = A \cap [\sqrt{2}, \pi]$ clopen in $A$?
(2) Is $Y = B \cap [0,2]$ clopen in $B$?
(3) Is $Z = A \cap [\sqrt{2}, \pi)$ closed or open in $A$?
(4) Is $Z' = B \cap [\sqrt{2}, \pi)$ closed or open in $B$?

**Exercise 3.3.28.** Show that in usual $\mathbb{R}$, $[0,1] \cap \mathbb{Q}$ is dense in $[0,1]$.

**Exercise 3.3.29.** Consider the set

$$X = \left\{ x \in [0,1] : x = \sum_{n=1}^{\infty} \frac{\alpha_n}{10^n} \text{ where } \alpha_n = 3 \text{ or } \alpha_n = 5 \right\}.$$

(1) Describe the elements of $X$.
(2) Show that $X$ is not dense in $[0,1]$.

**Exercise 3.3.30.** Let $X$ be a set with card$X \geq 2$. Endow $X$ with the indiscrete topology. Show that $X$ is not metrizable using different approaches.

**Exercise 3.3.31.** Let $X$ be an infinite set and let $T$ be the family given by
$$T = \{\varnothing\} \cup \{U \subset X : U^c \text{ finite }\}.$$
In this exercise, $X$ is taken to be $\mathbb{R}$ except for the last question.

(1) Show that $T$ is a topological space in $\mathbb{R}$.
(2) Describe the closed sets in this topology.
(3) Compare $T$ with the usual topology.
(4) Is it Hausdorff?
(5) Is it metrizable?

(6) (a) Let $A$ be a finite subset of $\mathbb{R}$. Find $\overline{A}$, $\overset{\circ}{A}$ and the boundary of $A$.

    (b) The same questions if we assume that $A$ is infinite.

(7) Is $\mathbb{R}$ separable with respect to $T$?
(8) What does $T$ become if $X$ is a finite set?

**Exercise 3.3.32.** Is
$$T = \{\varnothing, \mathbb{R}\} \cup \{U \subset \mathbb{R} : U^c \text{ infinite}\}$$
a topology on $\mathbb{R}$?

**Exercise 3.3.33.** Let $X$ be an uncountable set. Put
$$T = \{\varnothing\} \cup \{U \subset X : \ U^c \text{ is countable}\}.$$

(1) Show that $T$ a topological space on $X$ (called the **co-countable topology**).
(2) Is it Hausdorff?
(3) Let $A$ be a *proper closed* set of $X$. Show that all subsets of $A$ are closed.
(4) Prove that the countable intersection of open sets in $T$ remains open. Is this result true in usual $\mathbb{R}$?
(5) Show that the finite intersection of non-empty open sets is non-empty. Is this result true in usual $\mathbb{R}$?
(6) Is $\mathbb{Q}$ dense in $\mathbb{R}$ endowed with the co-countable topology? What about $\mathbb{R} \setminus \mathbb{Q}$ or $[0, 1]$?
(7) Give an example of a set $X$ so that $T$ reduces to the discrete topology.

**Exercise 3.3.34.** (**Particular Point Topology**) Let $X$ be a non-empty set. Let $a \in X$ be fixed and set
$$T = \{\varnothing\} \cup \{U \subset X : \ a \in U\}.$$

(1) Check that $T$ is a topology on $X$.
(2) Is $T$ Hausdorff?
(3) Find $\{a\}'$.
(4) Deduce that open sets (except $\varnothing$) are all dense in $T$.
(5) Show that any subset of $X$ is either open or closed (in $T$).
(6) Show that the only clopen subsets of $X$ in $T$ are $\varnothing$ or $X$.
(7) Let $A$ be a set containing $a$. Prove that $A' = X - \{a\}$.
(8) Is $X$ separable? What about $X - \{a\}$ in case $X$ is uncountable?
(9) Show that every proper subset of $X$ is nowhere dense.
(10) What is $T_{\{a\}^c}$, the induced topology on $\{a\}^c$? Is it Hausdorff?

**Exercise 3.3.35.** Let $a > 0$. Let $X = [-a, a]$. We declare
$$U \text{ "open" in } T \iff \{0\} \not\subset U \text{ or } (-a, a) \subset U.$$

(1) Show that $T$ is actually a topology on $X$.
(2) Give all the closed sets in $T$.
(3) Set $A = \{\frac{a}{3}\}$. What is $\overline{A}$?
(4) Show that $0$ is a limit point of any subset of $B \subset (-a, a)$.

**Exercise 3.3.36.** Let $X = [0, 2)$. Define
$$T = \{[0, a) : \ 0 \le a \le 2\}.$$

(1) Show that $T$ is a topology on $X$.
(2) Give an example which shows that the arbitrary intersection of elements of $T$ need not be in $T$ again.
(3) Is $T$ Hausdorff?
(4) What are the closed sets in $T$?
(5) Show that the closure of $A = \left[1, \frac{3}{2}\right]$ in $T$ is $[1, 2)$ and that its interior is the empty set.
(6) Is $X$ separable with respect to $T$?

**Exercise 3.3.37.** Let $T$ be a topology on $X$. Assume that $T$ is Hausdorff and let $x \in X$.

(1) Show that $\{x\}$ (and hence every finite set) is always closed.
(2) Show, by an example, that the separation hypothesis cannot be dispensed with.
(3) Give an example of a *non-separated* space in which singletons are closed.
(4) Show that the intersection of all open sets containing $x$ is $\{x\}$.
(5) Give an example of a *non-separated* space in which the previous result holds.

**Exercise 3.3.38** (Lower Limit Topology). Set
$$\mathcal{B} = \{[a, b) : \ a, b \in \mathbb{R}\}.$$

The topology generated by $\mathcal{B}$ is called the **lower limit topology** and it is denoted by $\mathbb{R}_\ell$ (it is called in some references the **Sorgenfrey Topology**).

(1) Check that $\mathcal{B}$ is in effect a base for $\mathbb{R}$.
(2) Show that $\mathbb{R}$ is strictly coarser than $\mathbb{R}_\ell$.
(3) Give examples of closed, open and clopen sets in $\mathbb{R}_\ell$. What is the nature of $(a, b]$?
(4) Is $\mathbb{R}_\ell$ Hausdorff?
(5) Is $\mathbb{R}_\ell$ separable?

**REMARK**. Likewise, we may define a topology on $\mathbb{R}$ using the base $\mathcal{B} = \{(a, b] : a, b \in \mathbb{R}\}$. This new topology is called the **upper limit topology**.

**Exercise 3.3.39** ($K$-topology). Set $K = \{\frac{1}{n} : n \in \mathbb{N}\}$. The $K$-**topology** on $\mathbb{R}$ is generated by the basis

$$\mathcal{B} = \{(a, b)\} \cup \{(a, b) - K\}$$

where $a, b \in \mathbb{R}$. It is denoted by $\mathbb{R}_K$.

(1) Show that $\mathbb{R}_K$ is strictly finer than $\mathbb{R}$.
(2) Is $\mathbb{R}_K$ Hausdorff?
(3) Are $\mathbb{R}_\ell$ and $\mathbb{R}_K$ comparable?
(4) Is $K$ closed in $\mathbb{R}_K$?
(5) Find $K'$.

**Exercise 3.3.40.**

(1) Show that $\mathcal{B} = \{(a, b) : a, b \in \mathbb{Q}\}$ is a basis generating the usual $\mathbb{R}$.
(2) Show that $\mathcal{B}' = \{[a, b) : a, b \in \mathbb{Q}\}$ is a basis that generates a topology, denoted by $\mathbb{R}_{\mathcal{B}'}$, different from $\mathbb{R}_\ell$.

**Exercise 3.3.41.** Do the details of Example 3.1.43.

**Exercise 3.3.42.** Find

$$d((0, 1) \cap \mathbb{Q}) \text{ and } d((0, 1) \cap \mathbb{R} \setminus \mathbb{Q}).$$

**Exercise 3.3.43.** Let $(X, d)$ be a metric space and let $A$ be a non-empty subset of $X$. Let $x \in X$. Recall

$$d(x, A) = \inf_{a \in A} d(x, a).$$

(1) Justify the existence of $d(x, A)$.
(2) Show that

$$d(x, A) = 0 \iff x \in \overline{A}.$$

Deduce yet another proof that singletons are closed in metric spaces (cf. Exercise 3.3.37).

(3) Show that $d(x, A) = d(x, \overline{A})$.

**Exercise 3.3.44.** Let $(X, d)$ be a metric space. Assume that there is $A \subset X$ which is uncountable, and that for some $\varepsilon > 0$, we have

$$\forall x, y \in A, \ x \neq y : \ d(x, y) \geq \varepsilon.$$

Show that $X$ is not separable.

**Exercise 3.3.45.**

(1) Show that $\ell^p$ is separable when $p \in [1, \infty)$.
(2) Show that $c_0$ is also separable (with respect to the topology of $d_\infty$).

**Exercise 3.3.46.** Prove that $\ell^\infty$ is not separable.

**Exercise 3.3.47.** Let $X$ and $Y$ be two topological spaces. Let $A$ and $B$ two subsets of $X$ and $Y$ respectively.

(1) Show that

$$\overline{A \times B} = \overline{A} \times \overline{B}.$$

(2) Show that

$$\overset{\circ}{\overbrace{A \times B}} = \overset{\circ}{A} \times \overset{\circ}{B}.$$

**Exercise 3.3.48.** Show that the finite product of separable spaces is separable.

**Exercise 3.3.49.** On usual $\mathbb{R}^2$, consider the sets

$$A = \left\{ (x, y) : \ y = \sin \frac{1}{x}, x > 0 \right\}, \ B = \{(x, x) : \ x \in \mathbb{R}\},$$

$C = \{(x, y) \in \mathbb{R}^2 : \ |x| < 2, \ |y| < 3\}$ and $D = \{(1, 1)\} \times C$

($D$ being defined on usual $\mathbb{R}^4$). Determine the interior and the closure of each of these four sets.

**Exercise 3.3.50.** Let $X$ be a topological space and let $\mathcal{R}$ be an equivalence relation on $X$. Let $\varphi : X \to X/\mathcal{R}$ be the quotient map. Set

$$T = \{A \in X/\mathcal{R} : \ \varphi^{-1}(A) \text{ is open in } X\}.$$

Show that $T$ is a topology in $X/\mathcal{R}$.

**Exercise 3.3.51.** Equip $\mathbb{R}$ with its usual topology. Define the equivalence relation $\mathcal{R}$ (see Exercise 1.2.17) on $\mathbb{R}$ by

$$x\mathcal{R}y \iff x - y \in \mathbb{Q}$$

for all $x, y \in \mathbb{R}$.

(1) Show that the quotient topology $\mathbb{R}/\mathbb{Q}$ is not Hausdorff (cf. Exercise 4.5.15).
(2) Show that $\mathbb{R}/\mathbb{Q}$ is in fact the indiscrete topology.

## 3.4. Tests

**Test** 7. Let $(X, T)$ be a topological space. Let $A \subset X$. Assume that
$$\forall x \in A, \exists U \in T : x \in U \subset A.$$
(1) Show that $A$ is open in $(X, T)$.
(2) Does this result remind you of something you are familiar with?

**Test** 8. Is the set $A = \{1, 2^2, 3^2, \cdots\}$ closed in $\mathbb{R}$?

**Test** 9. Does every closed set in $\mathbb{R}$ intersect $\mathbb{Q}$?

**Test** 10. Is the function assigning to each set its closure one-to-one?

**Test** 11. Let $X = \mathbb{N} - \{1\}$. Define $A_n = \{d \in X : d \mid n\}$ for $n \in \mathbb{N}$ where $d \mid n$ stands for $d$ divides $n$. Is $T = \{A_n\}_{n \in \mathbb{N}}$ a topology on $X$?

**Test** 12. Let $X = \{a, b, c\}$. Find the topologies on $X$ having exactly four open sets.

**Test** 13. Give an example of a bounded set having five limit points.

**Test** 14. Let $X$ be a topological space. Let $A \subset X$. Find $A'$ in the case of the co-finite topology.

**Test** 15. Is the topology of Exercise 3.3.36 metrizable?

**Test** 16. Are the co-countable (see Exercise 3.3.33) and the usual topologies comparable on $\mathbb{R}$?

**Test** 17. Is $\mathbb{R}$ separable when endowed with the discrete topology?

**Test** 18. Let
$$T = \{(-a, a) : a \geq 0\} \cup \{\mathbb{R}\}.$$
(1) Check that $T$ is a topology on $\mathbb{R}$.
(2) Find the interior and the closure of $[-1, 2]$.
(3) The same question for $\{0\}$ and $\{1\}$.

**Test** 19. Let $X$ be a topological space and let $A \subset X$. Can $\mathrm{Fr}(A)$ be equal to some open set in $X$?

**Test** 20. Is $\{1, \frac{1}{2}, \frac{1}{3}, \cdots\} \cup \{0\}$ closed in $\mathbb{R}_K$?

**Test** 21. Show that $\{(x, y) \in \mathbb{R}^2 : xy < 2\}$ is open in usual $\mathbb{R}^2$ but it is not an open ball.

## 3.5. More Exercises

**Exercise 3.5.1.** Let $X$ be a non-empty set and let $A, B$ be two non-void proper subsets of $X$. Let $T = \{\varnothing, A, B, X\}$.

(1) Is $T$ always a topology on $X$?

(2) What are the conditions so that $T$ becomes a topology on $X$?

**Exercise 3.5.2.** Do all the questions (except the fifth one) of Exercise 3.3.2 for the couple $(X, T)$ where $X = \{a, b, c, d\}$ and

$$T = \{\varnothing, \{a\}, \{b\}, \{a, b\}, \{a, b, d\}, \{a, b, c\}, X\}.$$

**Exercise 3.5.3.** Let $n \in \mathbb{Z}^+$.

$$T = \{\varnothing\} \cup \{n\mathbb{Z}\}.$$

Show that $T$ is not a topology on $\mathbb{Z}$.

**Exercise 3.5.4.** Let $X$ be an indiscrete topological space and let $A \subset X$. Find $A'$.

**Exercise 3.5.5.** Let $A$ be some subset of a topological space $X$. Show that $A' = \overline{A}'$.

**Exercise 3.5.6.** What is the topology generated by the intervals of the form $[x, x + 1]$ where $x \in \mathbb{R}$?

**Exercise 3.5.7.** Let $A$ be a non-void and bounded below subset of $\mathbb{R}$.

(1) Show that in usual $\mathbb{R}$, $\inf A = \inf \overline{A}$.

(2) Show that the previous result need not hold in another topology of $\mathbb{R}$.

**Exercise 3.5.8.** Set

$$T = \{\varnothing, \mathbb{R}\} \cup \{(a, \infty)\}_{a \in \mathbb{R}}.$$

(1) Show that $T$ is a topology on $\mathbb{R}$.

(2) Is $T$ Hausdorff?

(3) List the closed sets in $T$.

(4) What are the closures of the sets $\{0\}$, $\{0, 5, 11\}$ and $[2, 8)$.

(5) What is the interior of the interval $[0, \infty)$.

(6) Does $T$ remain a topology on $\mathbb{R}$ with $a \in \mathbb{Q}$?

**Exercise 3.5.9.** Let $A = [-2, 2]$ be considered as a topological subspace of $\mathbb{R}$. Which of the following sets are open in $A$ or in $\mathbb{R}$?

(1) $B = \{x \in \mathbb{R} : 1 < |x| < 2\}$;

(2) $C = \{x \in \mathbb{R} : 1 < |x| \leq 2\}$;

(3) $D = \{x \in \mathbb{R} : 1 \leq |x| < 2\}$.

**Exercise 3.5.10.** We say that a topological space $T$ on some set $X$ is **first countable** if it has a countable neighborhood basis at each of its points. We say that $T$ is **second countable** if it has a countable basis. These two concepts belong to what is more known as the **countability axioms**.

(1) Show that any second countable space is necessarily a first countable one. What about the converse? (hint: consider $\mathbb{R}$ equipped with the lower limit topology).

(2) Show that a second countable space is separable. Is the converse true? (hint: consider again $\mathbb{R}$ equipped with the lower limit topology).

(3) Show that the usual topology and more generally, metric spaces, are first countable.

(4) Show that $\mathbb{R}$ is not first countable in the co-finite topology.

(5) Are the topologies of: Exercise 3.3.34, Exercise 3.3.35 first countable? second countable?

**Exercise 3.5.11.** Let $X$ be a topological space. Let $A$ be an open subset of $X$ and let $B \subset X$. Show that

(1) If $A \cap B = \varnothing$, then $A \cap \overline{B} = \varnothing$.

(2) If $B$ is dense in $X$, then $\overline{A \cap B} = \overline{A}$.

(3) Deduce that if $A$ and $B$ are both dense in $X$, then so are $A \cup B$ and $A \cap B$.

**Exercise 3.5.12.** Give $\overline{A}$ and $\overset{\circ}{A}$ where

$$A = \left\{ \frac{1}{n} : n \in \mathbb{N} \right\}$$

with respect to each of the following topologies:

(1) the co-finite topology,

(2) the co-countable topology,

(3) the lower limit topology,

(4) the topology of Exercise 3.3.35 with $a = 1$,

(5) the topology of Test 18.

**Exercise 3.5.13.** Let $X$ be a topological space. Let $A \subset X$.

(1) Show that

$$A \text{ is clopen} \iff \operatorname{Fr}(A) = \varnothing.$$

(2) Is $\operatorname{Fr}(A)$ always nowhere dense?

(3) Show that $\operatorname{Fr}(A)$ is nowhere dense whenever $A$ is either open or closed.

**Exercise 3.5.14.** Let $X$ be a topological space and let $A \subset X$.

(1) Show that $A$ is nowhere dense iff $\overline{A}^c$ is dense in $X$.
(2) Does the arbitrary union of nowhere dense sets remain dense?
(3) Show that the *finite* union of nowhere dense sets remains dense.

**Exercise 3.5.15.** Let $X$ be a topological space. If $A$ and $B$ are nonvoid elements of $X$ such that $A \cap B = \varnothing$, then show that

$$\overset{\circ}{\overline{A}} \cap \overset{\circ}{\overline{B}} = \varnothing.$$

**Exercise 3.5.16.** Let us endow $\mathbb{R}^2$ with the metric defined, for all $x = (x_1, x_2)$ and $y = (y_1, y_2)$, by

$$d_\infty(x, y) = \max(|x_1 - y_1|, |x_2 - y_2|).$$

Let $B_c((0,0), 1)$ be the closed ball in $\mathbb{R}^2$. Let $A = \{(z, z) : z \in \mathbb{R}\}$.

(1) Show that $\overset{\circ}{\overbrace{B_c((0,0), 1)}} = B((0,0), 1)$.

(2) Show that $\overset{\circ}{A} = \varnothing$.

**Exercise 3.5.17.** On $\mathbb{R}^2$, consider the collection $(B(0, r))_r$ where $r$ is a positive number allowed to be $+\infty$ as well.

(1) Show that the collection $(B(0, r))_r$ defines a topology on $\mathbb{R}^2$.
(2) Is this topology Hausdorff?
(3) Is this topology finer than the standard one on $\mathbb{R}^2$?

**Exercise 3.5.18.** Let $X$ be a topological space and let $A$ be a topological subspace of $X$. Is it true that

$$A \text{ Hausdorff} \implies \overline{A} \text{ Hausdorff}?$$

**Exercise 3.5.19.** Show that the product of two separated spaces is separated.

**Exercise 3.5.20.**

(1) Show that the product of two discrete spaces is discrete.
(2) Show that the product of two indiscrete spaces is indiscrete.

**Exercise 3.5.21.**

(1) Show that usual $\mathbb{R} \setminus \mathbb{Q}$ is separable.
(2) Show that usual $\mathbb{C}$ is separable.

**Exercise 3.5.22.**

(1) Let $X$ and $X'$ be two topological spaces. Let $\mathcal{B}$ be a basis for $X$ and let $\mathcal{B}'$ be a basis for $X'$. Show that the collection

$$\mathcal{B}'' = \{B \times B' : B \in \mathcal{B}, \ B' \in \mathcal{B}'\}$$

constitutes a basis for $X \times X'$.

(2) Show that the (countable) collection
$$\mathcal{C} = \{(a,b) \times (c,d) : a,b,c,d \in \mathbb{Q}\}$$
is a basis for $\mathbb{R}^2$.

**Exercise** 3.5.23. Is $\mathbb{R}$ a $T_1$-space with respect to:

(1) the countable complement topology?
(2) the topology of Exercise 3.3.34?
(3) the topology of Exercise 3.3.35?

**Exercise** 3.5.24. Show that the space $\mathbb{R}_\ell$ is normal.

**Exercise** 3.5.25. (cf. Exercise 5.5.14) Show that none of the reverse implications of Theorem 3.1.12 is true. Here are some hints:

(1) $T_1$ but not $T_2$: What about the co-finite topology?
(2) $T_2$ but not $T_3$: What about $\mathbb{R}_K$?
(3) $T_3$ but not $T_4$: What about $\mathbb{R}_\ell^2$?

# CHAPTER 4

# Continuity and Convergence

## 4.1. Essential Background

### 4.1.1. Continuity.

**DEFINITION** 4.1.1. *Let $(X,T)$ and $(Y,T')$ be two topological spaces. We say that the function $f : (X,T) \to (Y,T')$ is **continuous** if*

$$\forall U \in T', \ f^{-1}(U) \in T.$$

**EXAMPLES** 4.1.2.

(1) *Let $(X,T)$ be a topological space, let $f : (X,T) \to (X,T)$ be a function defined by $f(x) = x$. Then $f$ is continuous.*

(2) *Let $(X,T)$ and $(Y,T')$ be two topological spaces, let $f : (X,T) \to (Y,T')$ be a function defined by $f(x) = a$, where $a \in Y$. Then $f$ is continuous.*

We are more used to defining some property locally (at a point, say), then we define it globally. In the previous definition, we did the other way round. We can, however, define continuity at a point (even though we will not be using it in the general context of topological spaces).

**DEFINITION** 4.1.3. *Let $(X,T)$ and $(Y,T')$ be two topological spaces. We say that the function $f : (X,T) \to (Y,T')$. We say that $f$ is **continuous** at $x \in X$ if*

$$\forall U \in \mathcal{N}(f(x)), \ f^{-1}(U) \in \mathcal{N}(x).$$

**REMARK.** We can *show* that $f : (X,T) \to (Y,T')$ is continuous iff it is continuous at each point of $X$.

Bases are important as regards to continuity. Indeed, they allow us to reduce the number or the form of open sets used in proving a function is continuous. More precisely,

**PROPOSITION** 4.1.4. *(cf. the "True or False" Section) Let $X$ and $Y$ be two topological spaces and let $\mathcal{B}$ be a basis for $Y$. Then $f : X \to Y$ is continuous iff the inverse image of each element of $\mathcal{B}$ is an open set in $X$.*

So is the case with subbases.

**PROPOSITION** 4.1.5. *(see Exercise 4.3.3) Let $X$ and $Y$ be two topological spaces and let $S$ be a subbasis for $Y$. Then $f : X \to Y$ is continuous iff the inverse image of each element of $S$ is an open set in $X$.*

The next gives a very interesting class of continuous functions which plays an important role in many situations.

**DEFINITION** 4.1.6. *Let $X$ and $Y$ be two topological spaces. Then each of the following two functions*

$$p : X \times Y \to X$$
$$(x, y) \mapsto p(x, y) = x$$

*and*

$$q : X \times Y \to Y$$
$$(x, y) \mapsto q(x, y) = y$$

*is called a **projection**.*

**THEOREM** 4.1.7. *(see Exercise 4.3.37) Let $X$ and $Y$ be two topological spaces. The two projections defined just above are continuous.*

**PROPOSITION** 4.1.8. *Let $(X, T)$, $(Y, T')$ and $(Z, T'')$ be three topological spaces. Let $f : (X, T) \to (Y, T')$ and $g : (Y, T') \to (Z, T'')$ be two continuous functions. Then $g \circ f : (X, T) \to (Z, T'')$ is continuous.*

The next result gives a characterization of continuity (a proof may be found in Exercise 4.3.4):

**THEOREM** 4.1.9 (cf. Exercise 4.3.5). *Let $(X, T)$ and $(Y, T')$ be two topological spaces. Let $f : (X, T) \to (Y, T')$ be a function. Then the following are equivalent:*

(1) *$f$ is continuous;*
(2) *for any $A \subset X$, $f(\overline{A}) \subset \overline{f(A)}$;*
(3) *for any closed element $V$ in $T'$, $f^{-1}(V)$ is closed in $T$.*

**DEFINITION** 4.1.10. *Let $(X, T)$ and $(Y, T')$ be two topological spaces. We say that the bijective mapping $f : (X, T) \to (Y, T')$ is a **homeomorphism** (or **bi-continuous**) if $f$ and $f^{-1}$ are both continuous.*

*We say that two topological spaces are **homeomorphic** if there exists a homeomorphism between them.*

**EXAMPLES** 4.1.11.

(1) *Let $f : \mathbb{R} \to (-1, 1)$ be defined by:*

$$f(x) = \frac{x}{1 + |x|}.$$

*Then $f$ is a homeomorphism (see Exercise 4.3.12).*

(2) *Every isometry is a homeomorphism.*

(3) *Any two open intervals in standard $\mathbb{R}$ are homeomorphic (see Exercise 4.3.12).*

**REMARK.** Let $X$ and $Y$ be two topological spaces. Let $f : X \to Y$ be a function. Then it is clear that $f$ is a homeomorphism iff the following three conditions hold:

(1) $f$ is a bijection;

(2) for any open $U$ in $Y$, $f^{-1}(U)$ is open in $X$;

(3) for any open $V$ in $X$, $f(V)$ is open in $Y$.

**DEFINITION 4.1.12.** *Let $X$ and $Y$ be two topological spaces. Let $f : X \to Y$ be a function. Then*

(1) *$f$ is said to be **open** if for any open set $U$ in $X$, $f(U)$ is open in $Y$.*

(2) *$f$ is said to be **closed** if for any closed set $V$ in $X$, $f(V)$ is closed in $Y$.*

**EXAMPLE 4.1.13.** *(see Exercise 4.3.37) Let $p$ and $q$ be the projections defined in Definition 4.1.6. Then both $p$ and $q$ are open and neither $p$ nor $q$ is closed.*

**PROPOSITION 4.1.14.** *(see Exercise 4.3.6) Let $f : X \to Y$ be a continuous function. Then:*

(1) *$f$ is closed iff $\overline{f(A)} \subset f(\overline{A})$ for all $A \subset X$.*

(2) *$f$ is open iff $[f^{-1}(B)]^{\circ} \subset f^{-1}(\overset{\circ}{B})$ for all $B \subset Y$.*

The following is a characterization of homeomorphism using open or closed mappings.

**THEOREM 4.1.15.** *Let $X$ and $Y$ be two topological spaces and let $f : X \to Y$ be a function. Then $f$ is a homeomorphism iff $f$ is open, continuous and bijective iff $f$ is closed, continuous and bijective.*

**DEFINITION 4.1.16.** *A **topological property** is a property shared by a given topological space and any other topological space homeomorphic to it.*

**EXAMPLES 4.1.17.**

(1) *We saw above that $\mathbb{R}$ is homeomorphic to $(-1, 1)$. Thus two consequences arise:*

(a) *Boundedness is not a topological property (see also Exercise 4.3.10);*

(b) *The length is not a topological property either.*

(2) *Closedness is not a topological property either. In the usual topology, the closed $\mathbb{R}$ is homeomorphic to the non-closed $\mathbb{R}_+^*$ via the function "$\ln$".*

We have already defined the quotient map. The next results treat its continuity and related results. The first is an immediate consequence of Proposition 3.1.56.

**PROPOSITION** 4.1.18. *Let $X$ be a topological space and let $\mathcal{R}$ be an equivalence relation on $X$. Then the quotient map $\varphi : X \to X/\mathcal{R}$ is continuous.*

**PROPOSITION** 4.1.19. *Let $X$ be a topological space and let $\mathcal{R}$ be an equivalence relation on $X$. Let $\varphi : X \to X/\mathcal{R}$ be the quotient map. The closed sets in the quotient topology are the parts $A$ of $X/\mathcal{R}$ such that $\varphi^{-1}(A)$ is closed in $X$.*

**PROPOSITION** 4.1.20. *Let $X$ and $Y$ be two topological spaces and let $\mathcal{R}$ be an equivalence relation on $X$. Let $\varphi : X \to X/\mathcal{R}$ be the quotient map. A map $g : X/\mathcal{R} \to Y$ is continuous iff $g \circ \varphi$ is continuous.*

We want to investigate continuity of functions as regards to the product topology.

**PROPOSITION** 4.1.21. *(see Exercise 4.3.39) Let $A$, $X$ and $Y$ be three topological spaces. Let $f : A \to X \times Y$ be a function defined by $f(x) = (g(x), h(x))$. Then $f$ is continuous if and only if $g$ and $h$ are so.*

**REMARK.** Can we have a similar result for functions of the type $f : X \times Y \to A$, i.e. if $f$ is continuous at any $x \in X$ and $f$ is continuous at any $y \in Y$, then is $f$ continuous on $X \times Y$? See Exercise 4.3.39 for an answer.

**COROLLARY** 4.1.22. *Let $X, X', Y, Y'$ be topological spaces and let $f : X \to Y$ and $g : X' \to Y'$ be both continuous. Then $h : X \times X' \to Y \times Y'$ defined by*

$$h(x, x') = (f(x), g(x'))$$

*is continuous.*

Before giving interesting consequences of the preceding proposition, we have:

**PROPOSITION** 4.1.23. *(see Exercise 4.3.8) If* $\mathbb{R}$ *is equipped with the usual topology, then the functions*

$$(x,y) \mapsto x+y \text{ and } (x,y) \mapsto xy,$$

*defined from* $\mathbb{R} \times \mathbb{R}$ *into* $\mathbb{R}$, *are both continuous.*

**REMARK.** The same result holds for functions from $\mathbb{C} \times \mathbb{C}$ into $\mathbb{C}$.

**COROLLARY** 4.1.24. *(see Test 34) Let* $X$ *be a topological space and let* $f, g : X \to \mathbb{R}$ *(or* $\mathbb{C}$*) be two continuous functions. Then*
  (1) $f + g : X \to \mathbb{R}$ *or* $\mathbb{C}$,
  (2) $fg : X \to \mathbb{R}$ *or* $\mathbb{C}$ *(when g does not vanish),*
  (3) $f/g : X \to \mathbb{R}$ *or* $\mathbb{C}$
*are all continuous.*

The previous combined with Corollary 4.1.22 easily yield:

**COROLLARY** 4.1.25. *In the usual topology, the function* $f : \mathbb{R}^n \times \mathbb{R}^n \to \mathbb{R}^n$ *given by* $f(x,y) = x+y$ *is continuous.*

Closely related to the previous results, one may wonder whether $X \times Y$ is homeomorphic to $Y \times X$? The answer and more properties are given next.

**PROPOSITION** 4.1.26. *(see Exercise 4.3.13) Let* $X$ *and* $Y$ *be two topological spaces and let* $x \in X$ *and* $y \in Y$. *Then:*
  (1) $X \times Y$ *is homeomorphic to* $Y \times X$.
  (2) $\{x\} \times Y$ *is homeomorphic to* $Y$.
  (3) $X \times \{y\}$ *is homeomorphic to* $X$.

We finish with the notion of graph.

**DEFINITION** 4.1.27. *Let* $f : X \to Y$ *be a function where* $X$ *and* $Y$ *are two topological spaces. The **graph** of* $f$, *denoted by* $G_f$, *is defined as:*

$$G_f = \{(x, f(x)) : x \in X\}.$$

**PROPOSITION** 4.1.28. *(see Exercise 4.3.40) Let* $f : X \to Y$ *be a continuous function between two topological spaces* $X$ *and* $Y$ *where* $Y$ *is Hausdorff. Then its graph,* $G_f$, *is closed.*

**REMARK.** The converse of the result of the previous proposition holds under extra conditions, namely if $X$ and $Y$ are two **Banach** spaces (these are complete normed vector spaces!), and $f$ is linear, then

$$G_f \text{ closed} \implies f \text{ is continuous.}$$

Notice that this is a major theorem of Functional Analysis and it is more known as the **Closed Graph Theorem**. See Exercise 5.5.8 for another result.

### 4.1.2. Convergence.

**DEFINITION** 4.1.29. *Let $(X, T)$ be a topological space. A sequence $(x_n)$ in $X$ **converges** to $x \in X$ if:*

$$\forall U \in \mathcal{N}(x), \ \exists N \in \mathbb{N}, \forall n \in \mathbb{N} \ (n \geq N \implies x_n \in U).$$

**EXAMPLE** 4.1.30. *In any topological space, constant and eventually constant sequences converge.*

**EXAMPLE** 4.1.31. *Let $X = \{1, 2, 3\}$ be equipped with the topology:*

$$T = \{\varnothing, \{2\}, \{1, 2\}, \{2, 3\}, X\}.$$

*Then the sequence defined by $x_n = 2$ converges to 2. It also converges to 1 and 3. Thus a sequence in an arbitrary topological space can converge to more than one limit!*

**REMARK**. The previous very simple example tells us that one has to be careful with sequences in topological spaces. The topology which equips a set may give us "surprises". For example, sequences like $x_n = n$ or $(-1)^n$ may well converge.

We can also define subsequences in an arbitrary topological space (we shall get back to this notion in the next chapter).

**DEFINITION** 4.1.32. *Let $X$ be a topological space and let $(x_n)_n$ be a sequence in $X$. If $(n_k)_k$ is an increasing sequence, then $(y_k)$, where $y_k = x_{n_k}$ (for all $k$), is called a **subsequence** of $(x_n)_n$.*

In a metric space, we have the following definition.

**DEFINITION** 4.1.33. *Let $(X, d)$ be a metric space. A sequence $(x_n)$ in $X$ converges to $x \in X$ if*

$$\forall \varepsilon > 0, \ \exists N \in \mathbb{N}, \forall n \in \mathbb{N} \ (n \geq N \implies d(x_n, x) < \varepsilon).$$

*We also write $x_n \to x$ as $n \to \infty$.*

**REMARK**. The previous definition tells us why Definition 4.1.29 is natural (because $d(x_n, x) < \varepsilon$ just means that $x_n \in B(x, \varepsilon)$).

**REMARK**. The previous definition is just what we mean for the sequence $(d(x_n, x))_n$ to be convergent to 0 in $(\mathbb{R}, |\cdot|)$.

**EXAMPLE** 4.1.34. *In a discrete metric space, the only convergent sequences are the eventually constant ones.*

**THEOREM** 4.1.35. *In a separated (Hausdorff) topological space, a sequence cannot converge to two different points.*

Since any metric space is Hausdorff (Exercise 3.3.12), the previous result implies

**COROLLARY** 4.1.36. *In a metric space, a convergent sequence has a unique limit.*

We now give a fundamental and very practical result in metric spaces (a proof is given in Exercise 4.3.26). For a discussion on this in the setting of arbitrary topological spaces, see the "True or False" section.

**THEOREM** 4.1.37. *Let $(X, d)$ be a metric space. Let $A \subset X$ be non-empty. Then we have:*

(1) $x \in \overline{A} \Longleftrightarrow \exists x_n \in A : x_n \longrightarrow x$.
(2) *A is closed* $\Longleftrightarrow \forall x_n \in A : (x_n \longrightarrow x \Longrightarrow x \in A)$.

**REMARK.** Since $\mathbb{Q}$ is dense in usual $\mathbb{R}$, we can say that for any real $x$, there is a sequence of rationals $(x_n)$ which converges to $x$. A similar conclusion holds for irrationals.

**EXAMPLES** 4.1.38.
(1) *In standard $\mathbb{R}$, $(0, 1]$ is not closed.*
(2) *In standard $\mathbb{R}$ again, $\{\frac{1}{n} : n \in \mathbb{N}\}$ is not closed.*

**REMARK.** Doubtlessly, the second result in the foregoing theorem can also be used to show that a given set is (or is not) open. See Exercises 4.3.28 & 4.3.29.

**4.1.3. Sequential Continuity.** We now come to a practical definition of continuity. But, in a general setting, not as strong as the definition of continuity seen in the beginning of this chapter.

**DEFINITION** 4.1.39. *Let $X$ and $Y$ be two topological spaces and let $f : X \to Y$ be a function. Then we say that $f$ is **sequentially continuous** at $x \in X$ if for any convergent sequence $(x_n)$ to $x$, $(f(x_n))$ converges to $f(x)$.*

*As usual, if $f$ is sequentially continuous at each $x \in X$, then we say that $f$ is sequentially continuous on $X$.*

**REMARK.** Continuity implies sequential continuity but not vice versa. See Exercise 4.3.22. However, the two notions match in a metric space (or more generally in a metrizable space) as we have

**PROPOSITION** 4.1.40 (For a proof see Exercise 4.3.22). *Let $(X, d)$ and $(Y, d')$ be two metric spaces and let $f : (X, d) \to (Y, d')$ be a function. Then $f$ is continuous on $X$ iff it is sequentially continuous on $X$.*

### 4.1.4. A Word on Infinite Product Topology.

We have already defined the topology of a product of two topological spaces. It is clear that this definition extends to a finite number of topological spaces $X_1 \times \cdots \times X_n$, say. The natural question which arises is how we define a topology on an infinite product $X_1 \times X_2 \times \cdots \times \cdots$ (denoted by $\prod_{i \in I} X_i$)? One quick answer is to extend the definition of the finite case to the infinite one. This is true. However, there is another way of defining a topology on an arbitrary product and this proves to be more natural while the other topology is usually used as a source of counterexamples.

First, we generalize the notion of a projection.

**DEFINITION** 4.1.41. *Let $\prod_{i \in I} X_i$ be an arbitrary product of sets. The **projection** $\pi_j$ $(j \in I)$ is defined from $\prod_{i \in I} X_i$ to $X_j$ by*

$$\pi_j(x_1, x_2, \cdots) = x_j.$$

**REMARK.** The set $I$ is arbitrary. So, we are not sure whether $1, 2 \in I$? and yet we write $X_1$ and $X_2$. If this is not the case, just relabel two sets from $(X_i)_{i \in I}$ as such.

Now, we define the two natural topologies on an arbitrary product.

**DEFINITION** 4.1.42. *Let $\prod_{i \in I} X_i$ be an arbitrary product of topological spaces.*

(1) *The topology generated by the basis constituted of all sets of the form $\prod_{i \in I} U_i$ (with $U_i$ being open in $X_i$ for each $i$) is called the **box topology**.*

(2) *The topology generated by the subbasis constituted of unions of collections of the form*

$$\{\pi_j^{-1}(U_j) : \ U_j \ open \ in X_j\}$$

*is called the **product topology**.*

**REMARK.** The product topology is also called the **Tychonoff product topology**.

**PROPOSITION** 4.1.43. *The product topology is the coarsest topology making all the projection maps $\pi_j : \prod_{i \in I} X_i \to X_j$ $(j \in J)$ continuous.*

**THEOREM** 4.1.44. *Let $(X_i)_{i \in I}$ be a collection of topological spaces and let $A_i$ be a subspace topology of $X_i$ for each $i \in I$. Then $\prod_{i \in I} A_i$ is a topological subspace of $\prod_{i \in I} X_i$ (whether we use the box topology or the product topology).*

**THEOREM** 4.1.45. *Let $(X_i)_{i \in I}$ be a collection of separated topological spaces. Then $\prod_{i \in I} X_i$ is also separated (whether we use the box topology or the product topology).*

**THEOREM** 4.1.46. *Let $(X_i)_{i \in I}$ be a collection of topological spaces and let $A_i \subset X_i$ for each $i \in I$. Then*

$$\prod_{i \in I} \overline{A_i} = \overline{\prod_{i \in I} A_i}$$

*(whether we use the box topology or the product topology).*

We have not yet seen a result true with respect to the product topology and not true with respect to the box topology. This is about to end!

**THEOREM** 4.1.47. *Let $\prod_{i \in I} X_i$ be an arbitrary product of topological spaces $(X_i)_{i \in I}$. Consider a function $f : X \to \prod_{i \in I} X_i$ (where $X$ too is a topological space) defined by*

$$f(x) = (f_1(x), f_2(x), \cdots)$$

*where each $f_i$ is a function from $X$ to $X_i$. If $\prod_{i \in I} X_i$ is equipped with the product topology, then*

$$f \text{ is continuous} \iff \text{each } f_i \text{ is continuous.}$$

**REMARK.** The previous result is not true anymore if the product $\prod_{i \in I} X_i$ is given the box topology. A counterexample as well as a proof of the previous theorem may be found in Exercise 4.3.41.

**REMARK.** The reader will also see that the product of compact (resp. connected) spaces remains compact (resp. connected) if we use the product topology, and that none of these results needs to be true if we use the box topology.

## 4.2. True or False: Questions

QUESTIONS. Comment on the following questions/statements and indicate those which are false and those which are true when this applies. Justify your answers.

(1) The identity function between two topological spaces is always continuous.

(2) Let $X$ and $Y$ be two topological spaces. If $f$ is a continuous function from $X$ into $Y$, then for each open set $U$ in $X$, $f(U)$ is open in $Y$.

(3) Let $X$ and $Y$ be two topological spaces. Assume that $f : X \to Y$ is some function and let $f_A$ be its restriction to $A \subset X$. Then

$$f \text{ is continuous} \iff f_A \text{ is continuous.}$$

(4) Let $X$ be a topological space and let $A \subset X$. Then

$$x \in \overline{A} \iff \exists x_n \in A : \ x_n \longrightarrow x.$$

(5) The sequence $\left(\frac{1}{n}\right)$ converges to zero.

(6) Criticize the following proof: In the co-finite topology on $\mathbb{R}$, the sequence $\frac{1}{n}$ converges to any point in $\mathbb{R}$. To prove it, assume that $\frac{1}{n} \not\to x$ for all $x \in \mathbb{R}$. This means that

$$\exists U \in \mathcal{N}(x), \forall N \in \mathbb{N} \ \exists n \ \left(n \geq N \text{ and } \frac{1}{n} \notin U\right).$$

Hence $\frac{1}{n} \in U^c$ for all but finitely many $n$ and this is a contradiction since $U^c$ is finite!

(7) Find fault with the following reasoning: In usual $\mathbb{R}$, consider the set

$$A = \{x_n = n^2 : \ n \in \mathbb{N}\}.$$

Then $A$ is not closed as $(x_n)$ does not converge and hence it cannot have a limit belonging to $A$.

(8) If $f$ is continuous and $-f$ is well-defined, then $-f$ is continuous.

(9) Let $X$ be a topological space and let $(x_n)$ be a convergent sequence in $X$. Then

$$X \text{ is Hausdorff} \iff (x_n) \text{ has a unique limit.}$$

(10) Let $X$ be a topological space. Endow $\mathbb{R}$ with its usual topology. Let $f : X \to \mathbb{R}$ be some function. If we come to show that $f^{-1}((a, b))$ ($a$ and $b$ real numbers with $a < b$) is open in $X$, then $f$ will be declared continuous.

(11) When are two sets homeomorphic? When are they not homeomorphic?

(12) Let $X$ and $Y$ be two topological spaces and let $f$ be a continuous mapping from $X$ into $Y$. Assume that $A \subset X$ is dense. Then $f(A)$ is dense in $f(X)$.

(13) Separability is a topological property.

(14) A bijective mapping between two topological spaces is continuous.

(15) An open, closed and bijective mapping between two topological spaces is continuous.

(16) Let $X$ be a topological space and let $A \subset X$. Let $f : X \to X$ be a bijective mapping. Then

$$f \text{ is a homeomorphism iff } f(\overline{A}) = \overline{f(A)}.$$

(17) Every continuous bijection is a homeomorphism.

(18) Every continuous and open bijection is a homeomorphism.

(19) Hausdorffness is a topological property.

(20) Let $X$ and $Y$ be two topological spaces and let $f : X \to Y$ be a continuous function. Let $A \subset X$. If $x \in A'$, then $f(x) \in (f(A))'$.

(21) Let $f$ be a map between two topological spaces $X$ and $Y$ such that $Y$ is also assumed to be separated. Let $G_f = \{(x, f(x)) : x \in X\} \subset X \times Y$ be the graph of $f$. Then

$$f \text{ is continuous } \Longleftrightarrow G_f \text{ is closed }.$$

(22) The sum of two everywhere discontinuous functions is surely everywhere discontinuous.

(23) Let $p : X \times Y \to X$ be the projection. Then $p$ is continuous and open but not closed. But haven't we said that a homeomorphism is open and equivalently closed! What is wrong with this?

## 4.3. Exercises With Solutions

**Exercise** 4.3.1. Let $X$ and $Y$ be two topological spaces. Let $f : X \to Y$ be a function. Is $f$ continuous in the following cases?

(1) $f(x) = x$, $X$ is the indiscrete topology and $Y = X$ equipped with the discrete topology;

(2) $f(x) = e^x$, $X = (\mathbb{R}, |\cdot|)$ and $Y = \mathbb{R}$ endowed with the discrete topology.

(3) $f(x) = x^2$, $X = (\mathbb{R}, T)$ and $Y = (\mathbb{R}, |\cdot|)$ where

$$T = \{\varnothing, \mathbb{R}\} \cup \{(a, +\infty), \ a \in \mathbb{R}\};$$

(4) $f$ arbitrary, $X$ discrete and $Y$ arbitrary;

(5) $f$ is the constant function;

(6) $f$ is the canonical injection.

**Exercise** 4.3.2. Let $X = \{a, b, c\}$ and let $T = \{\varnothing, \{a\}, \{b\}, \{a, b\}, X\}$ be a topology on $X$. Let $f : X \to X$ be the function defined by

$$f(a) = a, \ f(b) = c \text{ and } f(c) = b.$$

Is $f$ continuous at $a$? at $b$? at $c$?

**Exercise** 4.3.3. Let $X$ and $Y$ be two topological spaces and let $\mathcal{S}$ be a subbasis for $Y$. Show that $f : X \to Y$ is continuous iff the inverse image of each element of $\mathcal{S}$ is an open set in $X$.

**Exercise** 4.3.4. Let $(X, T)$ and $(Y, T')$ be two topological spaces. Let $f : (X, T) \to (Y, T')$ be a function. Show that the following are equivalent:

(1) $f$ is continuous;
(2) for any $A \subset X$, $f(\overline{A}) \subset \overline{f(A)}$;
(3) for any closed element $V$ in $T'$, $f^{-1}(V)$ is closed in $T$.

**Exercise** 4.3.5. Let $X$ and $Y$ be two topological spaces. Let $f : X \to Y$ be a function. Show that

$$f \text{ is continuous} \iff \forall U \subset Y : \ f^{-1}(\mathring{U}) \subset \widehat{f^{-1}(U)}.$$

**Exercise** 4.3.6. Let $f : X \to Y$ be a continuous function. Then:

(1) $f$ is closed iff $\overline{f(A)} \subset f(\overline{A})$ for all $A \subset X$.
(2) $f$ is open iff $[f^{-1}(B)]^{\circ} \subset f^{-1}(\mathring{B})$ for all $B \subset Y$.

**Exercise** 4.3.7. Let $X$ and $Y$ be two topological spaces and let $f : X \to Y$ be a continuous function. Let $A, B \subset X$.

(1) Show that if $\overline{A} = \overline{B}$, then $\overline{f(A)} = \overline{f(B)}$.
(2) Deduce that if $A$ is dense in $X$ and $f(X)$ is dense in $Y$, then $f(A)$ is dense in $Y$.

**Exercise** 4.3.8. Show that if $\mathbb{R}$ is equipped with the usual topology, then the functions

$$(x, y) \mapsto x + y \text{ and } (x, y) \mapsto xy,$$

defined from $\mathbb{R} \times \mathbb{R}$ into $\mathbb{R}$, are both continuous.

**Exercise** 4.3.9. Let $X$ be an uncountable set. We endow $X$ with the co-finite topology (see Exercise 3.3.31) and denote it by $T$. We also equip $X$ with the "co-countable topology" (see Exercise 3.3.33) and we denote it by $T'$. Let $f : T' \to T$ be defined for any $x \in X$ by $f(x) = x$. Is $f$ a homeomorphism?

**Exercise** 4.3.10. Let $(X, d)$ and $(X, d')$ be two metric spaces, where $d'$ is defined for all $x, y \in X$ by

$$d'(x, y) = \frac{d(x, y)}{1 + d(x, y)}.$$

Let $f : (X, d') \to (X, d)$ be defined by $f(x) = x$ for all $x \in X$.

(1) Show that $f$ is a homeomorphism.

(2) Deduce that boundedness is not a topological property.

**Exercise** 4.3.11. Show that Hausdorffness is a topological property.

**Exercise** 4.3.12.

(1) Show that:
   (a) Any two (finite) open intervals in $\mathbb{R}$ are homeomorphic.
   (b) Show that $\mathbb{R}$ is homeomorphic to $(-1, 1)$. Is $\mathbb{R}$ homeomorphic to any open interval in $\mathbb{R}$?
(2) Is $\overline{\mathbb{R}}$ (the extended real line) homeomorphic to $[-1, 1]$?

**Exercise** 4.3.13. Let $X$ and $Y$ be two topological spaces. Show that:

(1) $X \times Y$ is homeomorphic to $Y \times X$.
(2) $\{x\} \times Y$ is homeomorphic to $Y$, with $x \in X$.
(3) $X \times \{y\}$ is homeomorphic to $X$, with $y \in Y$.

**Exercise** 4.3.14. Let $f$ be a continuous function from a topological space $X$ into $\mathbb{R}$. Let $a$ be a real number and set

$$A = \{x \in X : f(x) = a\}.$$

Check that $A$ is closed in $X$.

**Exercise** 4.3.15. Show that following sets are closed in $X$

(1) $A = \{(x, y) \in \mathbb{R}^2 : xy = 1\}$, $X = \mathbb{R}^2$;
(2) $B = \{(x, y) \in \mathbb{R}^2 : x^2 + y^2 \leq 1\}$, $X = \mathbb{R}^2$;
(3) $C = \{A \in \mathcal{M}_n(\mathbb{R}) : \det A = 0\}$, $X = \mathcal{M}_n(\mathbb{R})$, the (vector) space of square matrices of order $n$ with real entries and we associate with it any metric.

**Exercise** 4.3.16. Let $a \in \mathbb{R}$. Let $f : \mathbb{R} \to \mathbb{R}$ be a continuous where $\mathbb{R}$ is endowed with the usual topology. Set

$$A = \{x \in \mathbb{R} : f(x) \leq a\}.$$

(1) Why is $A$ closed in $\mathbb{R}$?
(2) Is the converse always true?

**Exercise 4.3.17.** Let $T$ be a topology on $X$. Let $\mathbb{R}$ be endowed with its usual topology. Let $f : X \to \mathbb{R}$ be a function and let $a \in \mathbb{R}$. Show that

$f$ is continuous iff $f^{-1}((a, +\infty))$ and $f^{-1}((-\infty, a))$ are open in $T$.

**Exercise 4.3.18.** Let $\left(\frac{1}{n}\right)_{n \geq 1}$ be a sequence in $\mathbb{R}$. To what point (s) does $\left(\frac{1}{n}\right)_{n \geq 1}$ converge to (if there is any) with respect to:

(1) the usual topology;
(2) the co-finite topology;
(3) the discrete topology;
(4) the indiscrete topology.

**Exercise 4.3.19.** Set $x_n = \frac{1}{n}$ for all $n \in \mathbb{N}$, and let

$$K = \left\{ \frac{1}{n} : n \in \mathbb{N} \right\}.$$

(1) Prove that $(x_n)$ does not converge to any point of $\mathbb{R}$ in the $K$-topology $\mathbb{R}_K$ (defined in Exercise 3.3.39). What about $(-x_n)$?
(2) Show that $-K$ is not closed in $\mathbb{R}_K$.
(3) Show that $(x_n) \to 0$ in $\mathbb{R}_\ell$. Can $(x_n)$ have another limit in $\mathbb{R}_\ell$?
(4) Is $f : \mathbb{R}_K \to \mathbb{R}_K$, defined by $f(x) = -x$, continuous?
(5) Is $f : \mathbb{R}_\ell \to \mathbb{R}_\ell$, defined by $f(x) = -x$, continuous?

**Exercise 4.3.20.** Let $X$ be an uncountable set endowed with the co-countable topology (see Exercise 3.3.33).

(1) Show that the only convergent sequences are the eventually constant ones.
(2) Set $X = [0, 3]$ and $A = [2, 3]$.
   (a) Show that $1 \in A'$.
   (b) Let $(x_n)$ be a sequence in $A$. Can $(x_n)$ converge to 1?
   (c) What is then the conclusion?

**Exercise 4.3.21.** Let $(X, d)$ be a metric space. Let $(x_n)$ and $(y_n)$ be two convergent sequences (in $(X, d)$) to $x$ and $y$ respectively.

(1) Show that $d(x_n, y_n) \to d(x, y)$.
(2) How can this result be interpreted?

**Exercise 4.3.22 (Continuity Vs. Sequential Continuity).** Let $X$ and $Y$ be two topological spaces and let $f : X \to Y$ be a function.

(1) Show that if $f$ is continuous, then it is sequentially continuous.
(2) Give an example which shows that the converse is not always true.

(3) Assume now that $X$ and $Y$ are metric spaces. Show that $f$ is continuous iff it is sequentially continuous.

**Exercise** 4.3.23. Prove that every *continuous* and *monotonic* map from $\mathbb{R}$ into $\mathbb{R}$ is open.

**Exercise** 4.3.24. Let $P : \mathbb{R} \to \mathbb{R}$ be a polynomial. Show that $P$ is a closed mapping (hint: if $(x_n)$ is a sequence and $P(x_n)$ is bounded, then is $(x_n)$ bounded?).

**Exercise** 4.3.25. Let $X$ and $Y$ be two topological spaces and let $f : X \to Y$ be a continuous function. Assume that $A \subset X$. Show that $f$ remains continuous on $A$.

**Exercise** 4.3.26. Let $(X, d)$ be a metric space. Let $A \subset X$ be non-empty.

(1) Show that:
$$x \in \overline{A} \iff \exists x_n \in A : x_n \longrightarrow x.$$

(2) Deduce that:
$$A \text{ is closed } \iff \forall x_n \in A : (x_n \longrightarrow x \implies x \in A).$$

**Exercise** 4.3.27. Using sequences, check whether the following sets are closed in $X$

(1) $A = (0, 1]$, $X = \mathbb{R}$;
(2) $B = \{\frac{1}{n} : n \geq 1\}$, $X = \mathbb{R}$;
(3) $C = \{(x, y) \in \mathbb{R}^2 : x^2 + y^2 < 1\}$, $X = \mathbb{R}^2$;
(4) $D = \{(x, y) \in \mathbb{R}^2 : 1 < x^2 + y^2 \leq 3\}$, $X = \mathbb{R}^2$?

**Exercise** 4.3.28. Using sequences, show that the set
$$A = \{(x, y) \in \mathbb{R}^2 : 0 < y < 1, xy = 1\}$$
is neither open nor closed in $\mathbb{R}^2$ (endowed with the euclidian metric).

**Exercise** 4.3.29. Is the set $A = \{\frac{1}{n} : n \geq 1\}$ open in $\mathbb{R}$?

**Exercise** 4.3.30. Let $f$ and $g$ be two continuous functions defined on a metric space $X$ into another metric space $Y$.

(1) Let $A = \{x \in X : f(x) = g(x)\}$. Show that $A$ is closed in $X$.
(2) Let $B$ be a dense subset in $X$. Show that if $f$ and $g$ coincide on $B$, then they coincide on $X$.

**Exercise** 4.3.31. Check that the following function is not continuous at $(0, 0)$
$$f(x, y) = \begin{cases} \frac{xy}{x^2 + y^2}, & (x, y) \neq (0, 0), \\ 0, & (x, y) = (0, 0). \end{cases}$$

**Exercise 4.3.32.** Let $f : X \to Y$ be a continuous and one-to-one mapping where $X$ and $Y$ are two topological spaces.

(1) Show that if $Y$ is Hausdorff, then so is $X$.

(2) Give a counterexample showing that the hypothesis of the continuity cannot merely be dropped.

(3) Give a counterexample showing that the hypothesis of the injectivity cannot be completely eliminated.

**Exercise 4.3.33.** Let $(X, d)$ be a metric space. Let $A \neq \varnothing$ and $B \neq \varnothing$ be two closed sets such that $A \cap B = \varnothing$. Define a real-valued function $f$ on $X$ by

$$f(x) = \frac{d(x, A)}{d(x, A) + d(x, B)}.$$

(1) Show that if $g$ is a real-valued continuous function defined on $X$, then its zero set is closed. *Recall that the **zero set** of $g$, denoted by $Z(g)$ as in* [20]*, is defined as*

$$Z(g) = \{x \in X : \ g(x) = 0\}.$$

(2) Show that $f$ is a continuous function on $X$.

(3) Find $f^{-1}(\{0\})$ and $f^{-1}(\{1\})$.

(4) Deduce a converse of the result of Question 1, that is, every closed set can be regarded as the zero of some continuous real-valued function.

(5) (cf. Exercise 4.5.8) Establish the existence of two *disjoint open* sets $U$ and $V$ such that $A \subset U$ and $B \subset V$.

**Exercise 4.3.34.** Let $(f_n)$ be the sequence of functions defined as

$$f_n(x) = \begin{cases} 0, & 0 \leq x \leq \frac{1}{n+1}, \\ 2n(n+1)(x - \frac{1}{n+1}), & \frac{1}{n+1} < x \leq \frac{2n+1}{2n(n+1)}, \\ -2n(n+1)(x - \frac{1}{n}), & \frac{2n+1}{2n(n+1)} < x \leq \frac{1}{n}, \\ 0, & \frac{1}{n} < x \leq 1. \end{cases}$$

(This sequence is usually called the "**moving bump**"). Then all $f_n$ are obviously continuous on $[0, 1]$. Find $d_\infty(f_n, f_m)$ for all $n, m \in \mathbb{N}$ with $n \neq m$, where $d_\infty$ denotes the supremum metric.

**Exercise 4.3.35.** Let $X = C([0, 1], \mathbb{R})$ and

$$A = \{f \in X : \ f(0) = 0\}.$$

Endow $X$ with the metrics $d$ and $d'$ of Exercise 2.3.14. Show that $\overline{A} = A$ with respect to $d'$ and $\overline{A} = X$ with respect to $d$.

**Exercise 4.3.36.** Let $X = C([0, 1], \mathbb{R})$ be endowed with the *supremum metric* and let $A_a = \{f \in X : f(a) = 0\}$ where $a \in [0, 1]$.

(1) Show that $A_a$ is closed in $X$.
(2) Deduce that the set

$$B = \{f \in X : \; f_I = 0\},$$

where $f_I$ is $f$ restricted to $I \subset [0, 1]$, is closed in $X$.

**Exercise** 4.3.37. Let $X$ and $Y$ be two topological spaces.
(1) Show that the projections

$$p : X \times Y \to X$$
$$(x, y) \; \mapsto p(x, y) = x$$

and

$$q : X \times Y \to Y$$
$$(x, y) \; \mapsto q(x, y) = y$$

are continuous.
(2) Are they closed maps? (hint: you may consider the set $A = \{(x, y) \in \mathbb{R}^2 : xy = 1\}$).
(3) Show that $p$ and $q$ are open mappings.

**Exercise** 4.3.38. Let $X$ be a topological space. The **diagonal** of $X$ is defined to be the set

$$\Delta = \{(x, x) : \; x \in X\}.$$

(1) Show that $X$ is Hausdorff iff its diagonal is closed.
(2) Deduce from the preceding question another proof of the first question of Exercise 4.3.30 in a more general context, that is, $f, g : X \to Y$ are continuous, $X$ is any topological space and $Y$ is Hausdorff.

**Exercise** 4.3.39. Let $A$, $X$ and $Y$ be three topological spaces. Let $f : A \to X \times Y$ be a function defined by $f(x) = (g(x), h(x))$.
(1) Show that $f$ is continuous if and only if $g$ and $h$ are so.
(2) Can we have a similar result for functions of the type $f : X \times Y \to A$, i.e. if $f$ is continuous at any $x \in X$ and $f$ is continuous at any $y \in Y$, then $f$ is continuous on $X \times Y$?

**Exercise** 4.3.40. Let $f : X \to Y$ be a map between two topological spaces $X$ and $Y$ where $Y$ is Hausdorff.
(1) Show that if $f$ is continuous, then its graph, $G_f$, is closed.
(2) Give an example of a non-continuous function whose graph is closed.

**Exercise** 4.3.41.

(1) Let $\prod_{i \in I} X_i$ be an arbitrary product of topological spaces $(X_i)_{i \in I}$. Consider a function $f : X \to \prod_{i \in I} X_i$ (where $X$ too is a topological space) be defined by

$$f(x) = (f_1(x), f_2(x), \cdots)$$

where each $f_i$ is a function from $X$ to $X_i$. Show that if $\prod_{i \in I} X_i$ is equipped with the product topology, then

$f$ is continuous $\iff$ each $f_i$ is continuous.

(2) Give a counterexample showing that the previous result is not true anymore if the product $\prod_{i \in I} X_i$ is given the box topology.

## 4.4. Tests

**Test** 22. Let $\mathbb{R} \to \mathbb{R}_\ell$ be the function defined by $f(x) = x^2$ where $\mathbb{R}$ is equipped with the discrete topology and $\mathbb{R}_\ell$ is the lower limit topology on $\mathbb{R}$. Is $f$ a homeomorphism?

**Test** 23. Let $f : \mathbb{R} \to \mathbb{R}^*$ be a continuous function. Let

$$A = \{x \in \mathbb{R} : f(x) = 0\}.$$

Why is $A$ closed in $\mathbb{R}$?

**Test** 24. Is the identity map continuous from $\mathbb{R}$ endowed with the co-finite topology into $\mathbb{R}$ endowed with the topology of Exercise 3.3.34? What about its reciprocal?

**Test** 25. Equip $\mathbb{R}$ with the topology of Exercise 3.3.34. To what point (if there is any) does the sequence defined by $x_n = \frac{1}{n}$ converge to?

**Test** 26. Does the sequence $(-\frac{1}{n})$ converge to 0 in $\mathbb{R}_\ell$?

**Test** 27. Let $X$ be a topological space and $A \subset X$. Let $Y = \{0, 1\}$ be equipped with the topology $T = \{\varnothing, \{1\}, Y\}$. Define a function $f : X \to Y$ by

$$f(x) = \begin{cases} 1, & x \in A, \\ 0, & x \notin A. \end{cases}$$

Show that $f$ is continuous on $X$ iff $A$ is open in $X$.

**Test** 28. Let $X$ be an indiscrete topological space. When is $f : X \to \mathbb{R}$ ($\mathbb{R}$ equipped with the usual topology) continuous?

**Test** 29. In $\mathbb{R}^2$ endowed with the euclidian metric, let

$$A = \left\{ \left(x, \frac{1}{x}\right) : x > 0 \right\} \text{ and } B = \{(y, 0) : y \in \mathbb{R}\}.$$

(1) Show that $A$ and $B$ are closed in $\mathbb{R}^2$.
(2) Is $A + B$ closed?

**Test** 30. Is the "topological property" an equivalence relation?

**Test** 31. Show that $(0,1)$ is homeomorphic to $(0, +\infty)$.

**Test** 32. Let $f : (X, T) \to (X, T)$ be the identity map, i.e. $f(x) = x$. Show that $f$ has a closed graph if and only if $X$ is Hausdorff.

**Test** 33. Prove Corollary 4.1.22.

**Test** 34. Prove Corollary 4.1.24.

## 4.5. More Exercises

**Exercise** 4.5.1. Let $T$ and $T'$ be two topologies on a set $X$. Show that the following statements are equivalent:
  (1) $T \subset T'$;
  (2) The identity map $id : (X, T') \to (X, T)$ is continuous;
  (3) The identity map $id : (X, T) \to (X, T')$ is open;
  (4) The identity map $id : (X, T) \to (X, T')$ is closed.

**Exercise** 4.5.2. Consider the topology of Exercise 3.3.36 and let $(x_n)_n$ be a sequence in $X = [0, 2)$ which is convergent in $\mathbb{R}$ to 1. Show that $(x_n)_n$ converges with respect to $T$ to any point of $[1, 2)$.

**Exercise** 4.5.3. Using Exercise 4.3.37 and known results on the sum and scalar-multiplication of continuous functions, show that real-valued polynomials on $\mathbb{R}^n$ (and taking their values in $\mathbb{R}$) are continuous.

**Exercise** 4.5.4. Let $a \in \mathbb{R}$. Let $f : \mathbb{R} \to \mathbb{R}$ be a continuous where $\mathbb{R}$ is endowed with the usual topology. Set

$$A = \{x \in \mathbb{R} : f(x) > a\}.$$

(1) Show that $A$ is open in $\mathbb{R}$.
(2) Is the converse always true?

**Exercise** 4.5.5. In the usual topology, define a real-valued function $f$ on $\mathbb{R}$ by

$$f(x) = \begin{cases} 1, & x = 0, \\ 0, & x \in \mathbb{R} \setminus \mathbb{Q}, \\ \frac{1}{q}, & x \in \mathbb{Q} \end{cases}$$

where $x$ in the last line is written as $\frac{p}{q}$, $p \in \mathbb{Z}, q \in \mathbb{N}$ and $p$ and $q$ are coprime. This function is called the **Riemann function**.

Show that $f$ is continuous at *every irrational* and it is not continuous at *every rational* number.

**Exercise** 4.5.6. Let $X$ and $Y$ be two topological spaces. Let $A$ and $B$ be two closed sets in $X$ such that $A \cup B = X$. Let $f : A \to Y$ and $g : B \to Y$ be two continuous functions. Assume that $f$ and $g$ coincide on $A \cap B$.

Define a function $h$ as follows

$$h(x) = \begin{cases} f(x), & x \in A, \\ g(x), & x \in B. \end{cases}$$

Show that $h$ is continuous.

**REMARK.** The result in the previous exercise is usually called the **"pasting lemma"**.

**Exercise** 4.5.7. Show that the Cantor set does not have isolated points.

**Exercise** 4.5.8. (cf. Exercise 4.3.33) Let $(X, d)$ be a metric space. Let $A$ and $B$ be two subsets of $X$ such that **either** they are closed and disjoint **or** they satisfy $\overline{A} \cap B = A \cap \overline{B} = \varnothing$. Set

$$U = \{x \in X : d(x, A) < d(x, B)\} \text{ and } V = \{x \in X : d(x, B) < d(x, A)\}.$$

Show that $U$ and $V$ are two disjoint and open sets in $(X, d)$ which contain $A$ and $B$ respectively.

**Exercise** 4.5.9 (cf. Exercise 4.3.23). Show that if $f : A \to \mathbb{R}$ is increasing and open ($A \subset \mathbb{R}$), then $f$ is continuous.

**Exercise** 4.5.10. Prove that two *discrete* spaces are homeomorphic iff they have the same cardinality.

**Exercise** 4.5.11. Show that a circle deprived of one point is homeomorphic to a straight line.

**Exercise** 4.5.12. Let $X$ and $Y$ be two topological spaces, and let $f : X \to Y$ be a continuous function. Show that the graph of $f$ is homeomorphic to $X$.

**Exercise** 4.5.13. (see [8] and the references therein) Let $f, g : \mathbb{R} \to \mathbb{R}$ be two functions.

    (1) Can $f$ and $g$ have closed graph but $f + g$ does not?
    (2) Show that if $f$ and $g$ are *nonnegative* and have a closed graph, then $f + g$ has a closed graph.

**Exercise** 4.5.14. ([17]) Let $X$ and $Y$ be two topological spaces. Let $f : X \to Y$ be any function. Denote its graph by $G_f$.

    (1) Show that if $f$ is an open surjection such that $G_f$ is closed, then $Y$ is Hausdorff.

(2) Show that if $f$ is one-to-one and continuous with $G_f$ closed, then $X$ is Hausdorff.

(3) Deduce that if $f$ is a homeomorphism of $X$ onto $Y$, then both $X$ and $Y$ are Hausdorff.

**Exercise 4.5.15.** Let $X$ be a topological space and let $\mathcal{R}$ be an equivalence relation on $X$. Define

$$G_\mathcal{R} = \{(x, y) \in X^2 : x\mathcal{R}y\}$$

(called the graph of $\mathcal{R}$).

(1) Show that if $X/\mathcal{R}$ is Hausdorff, then $G_\mathcal{R}$ is closed.

(2) Is the converse always true? Justify your answer.

(3) Assume now that $G_\mathcal{R}$ is closed and that the quotient map is open. Show that $X/\mathcal{R}$ is Hausdorff.

**Exercise 4.5.16.** A **direct set** is a couple $(I, \prec)$ where $I$ is a set and $\prec$ is a relation which is: reflexive, transitive and such that: For any $a, b \in I$, there is $c \in A$: $a \prec c$ and $b \prec c$.

A **net** in a set $X$ is a sequence $(x_i)_i \subset X$ indexed by a direct set $(I, \prec)$.

It is clear that this notion generalizes that of a sequence. We also use the definition of a convergent net as the one of a sequence (with the obvious changes). Other topological notions involving sequences may be defined in terms of a net too.

(1) Give an example of a direct set.

(2) Let $X, Y$ be two topological spaces and let $A \subset X$ and $x \in A$. Prove that:

(a) $x$ lies in the closure of $A$ iff $x$ is a limit of a net $(x_i)_{i \in I}$ from $A$.

(b) $X$ is Hausdorff iff every convergent net has a unique limit.

(c) A function $f : X \to Y$ is continuous at $x$ iff for every net $(x_i)_{i \in I}$ converging to $x$, the net $(f(x_i))_{i \in I}$ converges to $f(x)$.

**Exercise 4.5.17.** (**Urysohn Lemma**) Let $X$ be a normal space. Let $A$ and $B$ be two disjoint closed subsets of $X$. Show that there exists a function $f : X \to [0, 1]$ such that

$$f(x) = 0 \text{ for } x \in A \text{ and } f(x) = 1 \text{ for } x \in B.$$

**Exercise 4.5.18.** (**Tietze Extension Theorem**) Let $X$ be a normal space and let $A$ be a closed subspace of $X$. Show that every continuous function $f : A \to [0, 1]$ or $(0, 1)$ or $[0, 1)$ can be extended to the whole of $X$.

**Exercise** 4.5.19. Let $(X_i)$ be a class of discrete topological spaces. Show that $X = \prod_{i \in I} X_i$ (equipped with the box topology) is discrete.

# CHAPTER 5

# Compact Spaces

## 5.1. Essential Background

### 5.1.1. Compactness: General Notions.

**DEFINITION 5.1.1.** *Let $A$ and $I$ be two sets. A collection $\{U_i\}_{i \in I}$ of sets is said to be a **cover** (or a **covering**) for $A$ if*

$$A \subset \bigcup_{i \in I} U_i.$$

*If all $U_i$ are open, then we call the collection $\{U_i\}_{i \in I}$ an **open cover**.*

**EXAMPLES 5.1.2.**

(1) *Let $X$ be a topological space. Then $\{X\}$ is an open (and a closed) cover for $X$ and for any subset $A$ of $X$.*

(2) *In usual $\mathbb{R}$, $\{[-n, n)\}_{n \in \mathbb{N}}$ is a cover for $\mathbb{R}$. Similarly, $\{(-n, n)\}_{n \in \mathbb{N}}$ is an open cover for $\mathbb{R}$.*

**DEFINITION 5.1.3.** *Let $(X, T)$ be a topological space. Then $X$ is called **compact** if every open cover of $X$ has a finite subcollection (also called **subcover**) which still covers $X$.*

**REMARK.** In this book we do not impose the separation (i.e. the Hausdorffness) in the definition of compact sets. This is adopted by many authors. But, in some references, e.g. the French ones, they do add the separation in the definition of compact sets (what we call here compact, they call it **quasi-compact**). One of the reasons, is to exclude spaces not too interesting as regards to compactness such as the indiscrete topological spaces. But this a dilemma as many interesting spaces will be excluded too such as $\mathbb{R}$ in the co-finite topology.

However, many applications arise in the setting of metric spaces where the two definitions coincide.

**REMARK.** We may say a compact space or a compact set (or a compact subspace or a compact subset), but we should not consider this as a source of polemic.

**REMARK.** Let $X$ be a topological space and let $A \subset X$. Open sets in $A$ are of the form $A \cap U$ where $U$ is open in $X$. So if we want to prove that the subspace $A$ is compact, we may just consider covers constituted of open sets in $X$ (instead of open sets in $A$).

**EXAMPLES** 5.1.4.

(1) *In usual* $\mathbb{R}$, $A = [1, \infty)$ *is not compact since* $\{(0, n)\}_{n \in \mathbb{N}}$ *is an open cover for* $A$ *which is not reducible to a finite subcover for* $A$. *Indeed, any union of elements of any finite subcover is of the form* $(0, N)$ *for some natural* $N$.

(2) *Every finite set (in any topological space) is compact. For a proof see Exercise 5.3.8.*

(3) *In a discrete topological space, an infinite subset is never compact (see Exercise 5.3.2).*

Using properties of Set Theory, we can prove the following result:

**PROPOSITION** 5.1.5. *A topological space* $X$ *is compact iff every collection* $(V_i)_{i \in I}$ *of closed subsets of* $X$ *such that* $\bigcap_{i \in J} V_i \neq \varnothing$ *for each finite subset* $J$ *of* $I$, *then* $\bigcap_{i \in I} V_i \neq \varnothing$.

**COROLLARY** 5.1.6. *Let* $X$ *be a compact space and let* $(V_n)$ *be a decreasing sequence of closed and non-empty subsets of* $X$. *Then*

$$\bigcap_{n \in \mathbb{N}} V_n \neq \varnothing.$$

Another (fundamental) example is the following (for a proof see Exercise 5.3.6):

**THEOREM** 5.1.7 (Heine-Borel). *Every closed and bounded interval* $[a, b]$ *in (usual)* $\mathbb{R}$ *is compact.*

In metric spaces, compact sets are bounded.

**PROPOSITION** 5.1.8. *(see Exercise 5.3.17) Let* $(X, d)$ *be a metric space and let* $A \subset X$ *be compact. Then* $A$ *is bounded.*

The next result is devoted to the interaction between compact and closed sets. Proofs are supplied in Exercises 5.3.19 & 5.3.20.

**THEOREM** 5.1.9. *Every closed subspace of a compact space is compact, and every compact subspace of a* **Hausdorff** *space is closed.*

**DEFINITION** 5.1.10. *Let* $X$ *be a topological space and let* $A$ *be a subspace of* $X$. *We say that* $A$ *is* **relatively compact** *if* $\overline{A}$ *is compact.*

**EXAMPLE** 5.1.11. *In the usual topology,* $(-1, 2)$ *is relatively compact in* $\mathbb{R}$ *since* $[-1, 2]$ *is compact.*

Now, we give a definition (more commonly known in $\mathbb{R}$ as the **Bolzano-Weierstrass Property**) of a different concept of compactness in general topological spaces (see [**19**]):

**DEFINITION** 5.1.12. *A topological space $X$ is called **limit point compact** if every infinite subset of $X$ has a limit point.*

The next result relates the concepts of compactness and limit point compactness.

**PROPOSITION** 5.1.13. *Every compact space is limit point compact.*

**REMARK.** The converse of the previous result is not always true. As a counterexample (borrowed from [**19**]): Let $Y = \{a, b\}$ and equip it with an indiscrete topology. Then $X = \mathbb{N} \times Y$ is limit point compact but not compact.

**DEFINITION** 5.1.14. *A topological space $X$ is said to be **locally compact** at $x \in X$ if there exists a compact subspace $A$ of $X$ such that $U \subset A$ where $U$ is a neighborhood of $x$.*
*We say that $X$ is locally compact if it is so at each of its points.*

**EXAMPLES** 5.1.15.

(1) *Obviously, every compact space is locally compact.*
(2) $\mathbb{R}$, *and $\mathbb{R}^n$ in general, are locally compact with respect to their usual topologies.*
(3) *A discrete topological space is locally compact.*

Here are some properties of locally compact spaces:

**PROPOSITION** 5.1.16. *([**19**]) Let $X$ be a Hausdorff space. Then $X$ is locally compact iff for any $x \in X$ and any neighborhood $U$ of $x$, there is a neighborhood of $x$, denoted by $V$, such that $\overline{V}$ is compact and $\overline{V} \subset U$.*

**PROPOSITION** 5.1.17. *(see Exercise 5.3.35)*

(1) *A closed subspace of a locally compact space is locally compact.*
(2) *An open subspace of a locally compact Hausdorff space is locally compact.*

We finish this subsection with an algebraic result on uncountability.

**THEOREM** 5.1.18 (cf. Exercise 7.3.36). *A non-empty compact Hausdorff space $X$ without any isolated point is uncountable.*

**COROLLARY** 5.1.19. *Every closed interval in $\mathbb{R}$ is uncountable.*

**REMARKS.**

(1) Obviously, the previous corollary is not concerned with the very particular closed interval of the form $[a, a]$, $(a \in \mathbb{R})$.

(2) The fact that the result is concerned with closed sets is not a weakness. Indeed, we already know that e.g. $[-1, 1)$ is equivalent to $[-1, 1]$ which is uncountable by the previous corollary. Therefore, $[-1, 1)$ too is uncountable.

(3) The Baire Theorem furnishes another (very short) proof of the uncountability of $[a, b]$.

**5.1.2. Compactness and Continuity.** As is the case of real functions on $[a, b]$, we still have in a more general context that the continuous image of a compact space is compact (what about the inverse image? See the "True or False" section).

**THEOREM** 5.1.20. *(a proof may be found in Exercise 5.3.16) Let $f : X \to Y$ be a continuous map between two topological spaces. If $X$ is compact, then so is its image $f(X)$.*

**COROLLARY** 5.1.21. *Let $(X, T)$ be a compact topological space and let $(Y, d)$ be a metric space. If $f : X \to Y$ is continuous, then it is bounded.*

A proof of the next corollary may be found in Exercise 5.3.18.

**COROLLARY** 5.1.22. *Let $X$ be a compact topological space. If $f : X \to \mathbb{R}$ is continuous, then $\sup f(X)$ and $\inf f(X)$ exist and belong to $f(X)$ (i.e. $f$ attains its bounds on $X$).*

The following examples show that the hypotheses in the previous results cannot merely be dropped:

**EXAMPLES** 5.1.23. *Let $f : X \to Y$ be a function.*

(1) $X = Y = \mathbb{R}$, $f(x) = x$ *is not bounded on the non-compact (non-bounded) $X$.*

(2) $X = (0, 1]$ *and $Y = [1, \infty)$, $f(x) = \frac{1}{x}$ is not bounded on the non-compact (non-closed) $X$.*

(3) $X = \mathbb{R}$ *and $Y = (-1, 1)$, $f(x) = \frac{x}{1+|x|}$ does not attain its bounds on the non-compact (non-bounded) $X$.*

(4) $X = Y = (0, 1)$, $f(x) = x$ *does not attain its bounds on the non-compact (non-closed) $X$.*

We have already defined uniform continuity in metric spaces. The next generalizes a known result from basic real analysis.

**THEOREM** 5.1.24 (**Heine**). *(see Exercise 5.3.40) Let $(X, d)$ and $(Y, d')$ be two metric spaces such that $(X, d)$ is compact. Let $f : (X, d) \to (Y, d')$ be a continuous function. Then $f$ is uniformly continuous.*

Now, we give two results on closed maps.

**PROPOSITION** 5.1.25. *(see Exercise 5.3.21) Let $X$ and $Y$ be two topological spaces such that $Y$ is Hausdorff and $X$ is compact. If $f : X \to Y$ is continuous, then $f$ is a closed map.*

**COROLLARY** 5.1.26. *Let $X$ and $Y$ be two topological spaces such that $Y$ is Hausdorff and $X$ is compact. If $f : X \to Y$ is bijective and continuous, then $f$ is a homeomorphism.*

We have already seen that projections are not closed maps (Exercise 4.3.37). They can be made closed if an extra condition is added.

**PROPOSITION** 5.1.27. *(a proof is given in Exercise 5.3.23) Let $X$ and $Y$ be two topological spaces. If $X$ is compact, then the projection $q : X \times Y \to Y$ is closed.*

**REMARK**. Obviously, the similar result with $Y$ being compact instead of $X$ (and $p$ instead of $q$) is true.

The proof of the foregoing result is based on the following so-called "**Tube Lemma**".

**PROPOSITION** 5.1.28. *(see Exercise 5.3.22) Let $X$ and $Y$ be two topological spaces and let $A \subset X$ be compact. If $y \in Y$ and $W$ is an open set of $X \times Y$ containing $A \times \{y\}$. Then*

$$\exists U \subset X, \; \exists V \subset Y, \; U \text{ and } V \text{ both open}, \; A \subset U, \; y \in V : U \times V \subset W.$$

The Tube Lemma can also be used to prove the following fundamental result on products of compact spaces:

**THEOREM** 5.1.29. *(see Exercise 5.3.24) If $X$ and $Y$ are two compact spaces, then so is their product $X \times Y$. Conversely, if $X \times Y$ is compact, then so are their factors.*

**REMARK**. Somewhat unusually, the easiest part of the proof is: $X \times Y$ compact implying both $X$ and $Y$ are compact.

By induction, we have:

**COROLLARY** 5.1.30. *Let $\{X_i\}_{1 \leq i \leq n}$ be a finite collection of topological spaces. Then*

$$X_1 \times \cdots \times X_n \text{ is compact} \iff \text{each } X_i \text{ is compact}, \; i = 1, \cdots, n.$$

REMARK. It is worth noticing that the arbitrary product of compact spaces remains compact w.r.t. the product topology. This latter generalization is more known as the **Tychonoff Theorem** (see e.g. [16]). However, the arbitrary product of compact spaces need not remain compact w.r.t. the box topology (see Test 46).

In point of fact, Theorem 5.1.7 extends to $\mathbb{R}^n$ (this is also due to Heine-Borel) and we have:

THEOREM 5.1.31. *In the usual topology, every bounded and closed space of $\mathbb{R}^n$ is compact.*

We finish with an interesting result on separability.

PROPOSITION 5.1.32. *(Exercise 5.3.37) Every compact metric space is separable.*

**5.1.3. Sequential Compactness and Total Boundedness.** We have already introduced two "concepts" of compactness. Here is one more.

DEFINITION 5.1.33. *Let $X$ be a metric space. Then $X$ is said to be* **sequentially compact** *if every sequence $(x_n)$ in $X$ has a convergent subsequence in $X$.*

REMARK. A priori, there seems to be no direct link to the definitions of compactness and limit point compactness met above. The next result tells us that, in fact, compactness, limit point compactness and sequential compactness all coincide in a metric space:

THEOREM 5.1.34. *Let $(X, d)$ be a metric space. Then the following are equivalent:*

(1) $X$ *is compact;*
(2) $X$ *is sequentially compact;*
(3) $X$ *is limit point compact.*

The proof of the previous result necessitates the notion of total boundedness, so we recall it here together with some of its properties. We shall need it again in later chapters.

DEFINITION 5.1.35. *Let $(X, d)$ be a metric space. Let $\varepsilon > 0$. An* $\varepsilon$**-net** *for $X$ is a subset $A$ of $X$ verifying $X \subset \bigcup_{x \in A} B(x, \varepsilon)$.*

DEFINITION 5.1.36. *A metric space is* **totally bounded** *(or* **precompact***) if it has a finite $\varepsilon$-net for all $\varepsilon > 0$.*

REMARK. A bounded set is not necessarily totally bounded. For a counterexample, consider the following infinite set

$$A = \{e_1 = (1, 0, 0, \cdots), e_2 = (0, 1, 0, \cdots), e_3 = (0, 0, 1, \cdots), \cdots\}$$

which is obviously a subset of $\ell^2$. Then $A$ is bounded but not totally bounded.

Now we list some properties of totally bounded sets.

THEOREM 5.1.37.

   (1) *A subset of a totally bounded set is totally bounded.*
   (2) *A totally bounded set is bounded.*
   (3) *The closure of a totally bounded set is totally bounded.*

The connection between compactness and total boundedness is elucidated in the next theorem.

THEOREM 5.1.38. *A sequentially compact set is totally bounded.*

Like compactness, sequential compactness is preserved under continuous functions.

PROPOSITION 5.1.39. *(see Exercise 5.3.38) Let* $f : X \to Y$ *be continuous. If $X$ is sequentially compact, then so is $f(X)$.*

## 5.2. True or False: Questions

QUESTIONS. Comment on the following questions/statements and indicate those which are false and those which are true when this applies. Justify your answers.

(1) Let $X$ be a non-empty set. Let $\{U_i\}_{i\in I}$ be a cover for $X$. Then $X = \bigcup_{i\in I} U_i$.

(2) We saw in the first example of Examples 5.1.2 that in any topological space $X$, $\{X\}$ is an open cover for $X$ and for any subset of $X$. It is also clearly a subcover for $X$. Hence $X$ itself and any subset of $X$ is compact (or is it not?).

(3) The fact that we use open covers to define compact spaces is purely conventional. We could adopt the same definition using closed covers (couldn't we?).

(4) In every topological space $X$, the countable intersection of closed, non-empty and decreasing subsets in $X$ is non-empty.

(5) Let $(C_n)$ be a sequence of non-empty decreasing and closed sets in a metric space $X$. If $f : X \to X$ is continuous, then

$$f\left(\bigcap_{n\in\mathbb{N}} C_n\right) = \bigcap_{n\in\mathbb{N}} f(C_n).$$

(6) $\varnothing$ is compact.

(7) The interval $[a, b]$ is always compact.

(8) The closure of a compact subset is compact.

(9) Let $X$ and $Y$ be two topological spaces. Let $f : X \to Y$ be a continuous function and let $A \subset Y$ be compact. Then $f^{-1}(A)$ is compact.

(10) In a topological space, a subspace is compact if and only if it is closed and bounded.

(11) Let $T$ and $T'$ be two topologies on the same set $X$ such that $T \subset T'$. If $X$ is compact with respect to $T'$, it will be so with respect to $T$. What about the converse?

(12) Criticize the following proof of the compactness of $\mathbb{R}$ with respect to the co-finite topology: Let $\{U_i\}_{i\in I}$ be an open cover of $\mathbb{R}$, i.e. $\mathbb{R} \subset \bigcup_{i\in I} U_i$. Now, assume $\mathbb{R} \not\subset \bigcup_{i=1}^{n} U_i$. Then

$$\mathbb{R} \subset \left(\bigcup_{i=1}^{n} U_i\right)^c = \bigcap_{i=1}^{n} U_i^c \text{ (a finite set)},$$

i.e. $\mathbb{R}$ would have to be finite and we arrived at a contradiction. Thus $\mathbb{R}$ is compact with respect to the co-finite topology.

(13) Let $f : [a,b] \to \mathbb{R}$ be a continuous function. Then $f$ is bounded.

(14) $(0,2)$ is relatively compact.

(15) The closed unit ball is always compact.

(16) Every compact set is closed.

(17) Let $(X,d)$ be a metric space and let $A \subset X$. Denote the closed ball of center $x$ and radius $r > 0$ by $B_c(x,r)$. We know from earlier chapters that the union $\bigcup_{x \in A} B_c(x,r)$ need not be closed. Can we expect this union to be closed if $A$ is also assumed to be compact?

(18) Compactness is a topological property.

(19) Sequential compactness is a topological property.

(20) Local compactness is preserved under continuous maps.

(21) Local compactness is a topological property.

(22) The union of two locally compact spaces remains locally compact.

(23) In a metric space, every finite part is totally bounded.

(24) Total boundedness is a topological property.

## 5.3. Exercises With Solutions

**Exercise** 5.3.1. Using (only) the definition of a compact set show that $\mathbb{R}$, $[0, +\infty)$ and $(0,1)$ are not compact in $\mathbb{R}$ (in the usual topology).

**Exercise** 5.3.2. Indicate which of the following sets are compact in $X$

(1) $A = \mathbb{Q}$, $X = \mathbb{R}$;

(2) $A = \{\frac{1}{n} : n \in \mathbb{N}\}$, $X = \mathbb{R}$;

(3) $A = \mathbb{Q} \cap [0,1]$, $X = \mathbb{R}$;

(4) $A = [a,b]$, $B = \mathbb{R}$, $C$ an infinite set; $X = \mathbb{R}$ endowed with the discrete topology;

(5)

$$A = \{(x,y) \in \mathbb{R}^2 : x^2 + y^2 = 1\}, \ B = \{(x,y) \in \mathbb{R}^2 : x^2 + y^2 < 1\},$$
$$C = \{(x,y) \in \mathbb{R}^2 : x^2 + y^2 > 1\}; \ X = \mathbb{R}^2.$$

(6) $A = \{(x,y) \in \mathbb{R}^2 : x \geq 1, \ 0 \leq y \leq \frac{1}{x}\}$; $X = \mathbb{R}^2$;

(7) $A = \{(x, \frac{1}{x}) : 0 < x \leq 1\}$, $B = \{(x, \sin\frac{1}{x}) : 0 < x \leq 1\}$; $X = \mathbb{R}^2$;

(8) $A = \bigcap_{n \in \mathbb{N}} B(0_{\mathbb{R}^2}, 1 + 1/n)$; $X = \mathbb{R}^2$?

  (The topology of Questions 1-3 and 5-8 is the standard one).

**Exercise** 5.3.3.

(1) Show that the following set is not compact in $\mathbb{R}^2$ with respect to the standard topology
$$A = \{(x, y) \in \mathbb{R}^2 : x + y^3 = 1\}.$$

(2) What about the set
$$B = \{(x, y) \in \mathbb{R}^2 : x^4 + y^2 = 1\}?$$

**Exercise** 5.3.4. In the usual topology of $\mathbb{R}$, show that
$$A = \left\{ \frac{1}{n} : n \in \mathbb{N} \right\}$$
is not compact using open covers.

**Exercise** 5.3.5. Let $X$ be a topological Hausdorff space. Let $(x_n)_n$ be a sequence in $X$ which converges to $a$. Set
$$A = \{x_n : n \in \mathbb{N}\} \cup \{a\}.$$
Show that $A$ is compact.

**Exercise** 5.3.6. Show that the closed bounded interval $[a, b]$ ($a$ and $b$ are reals with $a \leq b$) is compact in usual $\mathbb{R}$.

**Exercise** 5.3.7. Consider $\mathbb{R}_K$, the $K$-topology on $\mathbb{R}$.

(1) Is $[0, 1]$ compact in $\mathbb{R}_K$?
(2) What about any set that contains $K = \{\frac{1}{n} : n \in \mathbb{N}\}$?

**Exercise** 5.3.8. Let $X$ be a topological space.

(1) Show that every finite set in $X$ is compact.
(2) Deduce that if $X$ is given the *discrete* topology, then a subset is compact if and only if it is finite.

**Exercise** 5.3.9.

(1) Show that the union of two (and hence of a finite number) of compact sets is compact.
(2) Is this true for an arbitrary union?
(3) Show that the arbitrary intersection of compact sets in a Hausdorff topological space is always compact.

**Exercise** 5.3.10. Let $(X, d)$ be a metric space and let $A \subset X$ be compact. Denote the closed ball of center $x$ and radius $r > 0$ by $B_c(x, r)$.

(1) Show that
$$\bigcup_{x \in A} B_c(x, r) = \{t \in X : d(t, A) \leq r\}.$$

(2) Infer that $\bigcup_{x \in A} B_c(x, r)$ is closed.

**Exercise** 5.3.11. Is $\mathbb{R}$ compact with respect to $X = \mathbb{R}$ endowed with the co-finite Topology?

**Exercise** 5.3.12. Consider $\mathbb{R}$ with respect to the co-countable topology (see Exercise 3.3.33).
  (1) Is $\mathbb{R}$ compact?
  (2) Is $[0, 1]$ compact?
  (3) Let $A$ be a countable set. Is $A$ compact?

**Exercise** 5.3.13. On $\mathbb{R}$, we define

$$T = \{\varnothing\} \cup \{U \subset \mathbb{R} : U^c \text{ is compact in usual } \mathbb{R}\}.$$

  (1) Show that $T$ is a topological space on $\mathbb{R}$.
  (2) Is $T$ Hausdorff?
  (3) Show that $\mathbb{R}$ is separable with respect to $T$.
  (4) Show that $\mathbb{R}$ is compact with respect to $T$.

**Exercise** 5.3.14. By going back to Exercise 3.3.35 and taking $a = 1$, show that $[-1, 1]$ is compact in $T$.

**Exercise** 5.3.15. In the usual topology, explain why $(0, 1)$ is not compact using limit point compactness.

**Exercise** 5.3.16. Let $f : X \to Y$ be a continuous map between two topological spaces. Show that if $X$ is compact, then so is its image $f(X)$.

**Exercise** 5.3.17. Let $(X, d)$ be a metric space and let $A \subset X$ be compact. Prove that $A$ is bounded.

**Exercise** 5.3.18. Let $X$ be a compact topological space and let $f : X \to \mathbb{R}$ be continuous (where $\mathbb{R}$ is equipped with the usual metric). Show that $\sup f(X)$ and $\inf f(X)$ exist and belong to $f(X)$.

**Exercise** 5.3.19. Show that every closed subspace of a compact space is compact.

**Exercise** 5.3.20. Let $X$ be a Hausdorff space and let $A \subset X$ be compact. Let $x \in A^c$.
  (1) Show that there are two open sets $U$ and $V$ s.t.

$$x \in U, \ A \subset V : \ U \cap V = \varnothing.$$

  (2) Deduce that there is an open set $W$ such that $x \in W \subset A^c$.
  (3) Infer that $A$ is closed.

**Exercise 5.3.21.** Let $X$ and $Y$ be two topological spaces such that $Y$ is Hausdorff and $X$ is compact. Assume that $f : X \to Y$ is continuous. Show that $f$ is a closed map.

**Exercise 5.3.22.** Let $X$ and $Y$ be two topological spaces and let $A \subset X$ be compact. Let $y \in Y$ and let $W$ be an open set of $X \times Y$ containing $A \times \{y\}$. Show that

$$\exists U \subset X, \ \exists V \subset Y, \ U \text{ and } V \text{ both open}, \ A \subset U, \ y \in V : U \times V \subset W.$$

**Exercise 5.3.23.** Let $X$ and $Y$ be two topological spaces. Show that if $X$ is compact, then the projection $q : X \times Y \to Y$ is closed.

**Exercise 5.3.24.** Let $X$ and $Y$ be two topological spaces. Show that:

$$X \times Y \text{ is compact} \iff X \text{ and } Y \text{ are both compact}.$$

**Exercise 5.3.25.** Let $A, B \subset \mathbb{R}^n$ with $n \geq 1$. Set
$$A + B = \{a + b : a \in A, b \in B\}.$$

(1) Show that $A + B$ is compact if $A$ and $B$ are so.
(2) Show that $A + B$ is closed if $A$ is compact and $B$ is closed.

**Exercise 5.3.26.** Let $\mathbb{Q}$ be the metric space associated with the metric $d$ defined by $d(x, y) = |x - y|$.

(1) Show that the set $A = \{x \in \mathbb{Q} : 2 < x^2 < 3\}$ is closed and bounded but not compact.
(2) Is $A$ open in $\mathbb{Q}$?

**Exercise 5.3.27.** Consider the following metric defined on $\mathbb{R}$ by
$$d(x, y) = \inf\{|x - y|, 1\}.$$

(1) Is $(\mathbb{R}, d)$ bounded?
(2) Using the sequence defined by $x_n = n$ or otherwise, show that $(\mathbb{R}, d)$ is not compact with respect to this metric.
(3) What can you deduce from the previous question?

**Exercise 5.3.28.**

(1) Show that $\mathbb{R}$ is not sequentially compact with respect to the metric $\delta$ of Exercise 2.3.29.
(2) What can be deduced from the previous question?

**Exercise 5.3.29.** Let $X = C([0, 1], \mathbb{R})$ be equipped with the supremum metric.

(1) What is $\dim X$?
(2) Show that unit closed ball is not sequentially compact in $X$.

(3) Can the unit closed ball be compact in $X$?

(4) Is $X$ locally compact?

**Exercise** 5.3.30. We endow $\mathbb{R}$ with two topologies, one is the cofinite one and we denote it by $X$, the other is the discrete one and we denote it by $Y$. Let $f : X \to Y$ be a function defined for all $x \in \mathbb{R}$ by $f(x) = x^2$.

(1) Is $f(\mathbb{R})$ compact?

(2) What can you deduce from the previous question?

**Exercise** 5.3.31. In the usual topology, is $[0,1]$ homeomorphic to $[0,\infty)$? to $(0,1]$?

**Exercise** 5.3.32. Let $X$ be a compact *metric* space and let $f : X \to X$ be continuous. Assume that $(A_n)$ is a sequence of decreasing nonvoid and closed sets in $X$. Show that

$$f(\bigcap_{n \in \mathbb{N}} A_n) = \bigcap_{n \in \mathbb{N}} f(A_n).$$

**Exercise** 5.3.33.

(1) Indicate among the following spaces $X$ those which are locally compact and those which are not:

  (a) $X$ is a compact topological space,

  (b) $X = \mathbb{R}$ in the usual topology,

  (c) $X = \mathbb{Q}$ in the usual topology,

  (d) $X = \mathbb{R} \setminus \mathbb{Q}$ in the usual topology,

  (e) $X$ is a discrete topological space,

  (f) $X$ equipped with the topology of Exercise 3.3.34.

(2) Give an example of a Hausdorff space which is not locally compact and one which is locally compact.

**Exercise** 5.3.34. Give an example showing that the continuous image of a locally compact space need not be locally compact.

**Exercise** 5.3.35. Let $X$ be a locally compact space.

(1) Show that every closed subspace in $X$ is locally compact.

(2) Show that every open subspace in $X$ is locally compact provided $X$ is *Hausdorff*.

(3) In usual $\mathbb{R}^2$, say why $\{(0,0)\}$ and $\{(x,y) \in \mathbb{R}^2 : x > 0\}$ are locally compact.

(4) In the usual topology again, is

$$\{(0,0)\} \cup \{(x,y) \in \mathbb{R}^2 : x > 0\}$$

locally compact?

(5) What is then the conclusion?

**Exercise** 5.3.36. Let $(X, d)$ be a *compact* metric space. Let $f : X \to X$ be a function satisfying

$$d(f(x), f(y)) < d(x, y), \ \forall x \neq y.$$

(1) Can we say that there is always a $k \in [0, 1)$ such that

$$d(f(x), f(y)) \leq k d(x, y), \ \forall x, y \in X.$$

(2) Show that $f$ has a unique **fixed point**, that is, there is one and only one point $x \in X$ such that $f(x) = x$.

**Exercise** 5.3.37. Let $(X, d)$ be a *compact* metric space. Prove that $(X, d)$ is separable.

**Exercise** 5.3.38. Let $f : X \to Y$ be continuous. Show that if $X$ is sequentially compact, then so is $f(X)$.

**Exercise** 5.3.39. When is $(X, d)$ totally bounded if $X$ is an arbitrary set and $d$ is the discrete metric?

**Exercise** 5.3.40. Let $(X, d)$ and $(Y, d')$ be two metric spaces such that $(X, d)$ is compact. Let $f : (X, d) \to (Y, d')$ be a continuous function. Show that $f$ is uniformly continuous.

## 5.4. Tests

**Test** 35. Let $(X, d)$ be a metric space and let $A, K \subset X$ be such that $K$ is compact and $A$ is closed. Is $A \cap K$ compact?

**Test** 36. Using open covers, show that $A = (0, 1] \cup \{2\}$ is not compact w.r.t. the usual $\mathbb{R}$.

**Test** 37. In the topology of Exercise 3.3.34, is $\{a\}$ compact? What about $X$? What can you deduce from that?

**Test** 38. Let $(X, T)$ be separated topological space. Let $(X, S)$ be a compact topological space. Show that if $T \subset S$, then $T = S$.

**Test** 39. Show that $\mathbb{N}$ is not compact with respect to the topology of Exercise 3.3.34 (with $X = \mathbb{N}$ and $a = 1$).

**Test** 40. Is $\mathbb{R}$ compact in $\mathbb{R}_\ell$?

**Test** 41. Is $\mathbb{Q}$ relatively compact in $\mathbb{R}$ with respect to the co-finite topology?

**Test** 42. Is the unit sphere in $\mathbb{R}^3$ homeomorphic to $\mathbb{R}^2$?

**Test** 43. Let $X$ be a topological space. Let $A$ be a finite subset of $X$. Show that $A$ is sequentially compact.

**Test** 44. Does the quotient of a compact space remain compact?

**Test** 45. Define a function $f$ on $[0,1]$ by

$$f(x) = \begin{cases} 0, & x = 0, \\ x \ln x, & 0 < x \leq 1. \end{cases}$$

Is $f$ uniformly continuous on $[0,1]$?

**Test** 46. Let $A_i = \{1, -1\}$ be equipped with the discrete topology for $i \in I$ ($I$ is an arbitrary set).

(1) Is $\prod_{i \in I} A_i$ compact with respect to the product topology?
(2) Is $\prod_{i \in I} A_i$ compact with respect to the box topology?

## 5.5. More Exercises

**Exercise 5.5.1.** Consider the topology of Exercise 3.3.36. Is $[0, 2)$ compact in this topology?

**Exercise 5.5.2.** Show that the set

$$A = \{(x, y) \in \mathbb{R}^2 : x^2 - xy + y^2 \leq 1\}$$

is compact in $\mathbb{R}^2$.

**Exercise 5.5.3.** Let $\mathbb{N}$ be endowed with the co-finite topology. Show that the set of even integers is compact in this topology. Is it closed?

**Exercise 5.5.4.** Find an example of two compact subspaces $A$ and $B$ (in a non-Hausdorff space) such that $A \cap B$ is not compact.

**Exercise 5.5.5.** Let $T$ and $T'$ be two topologies on the same set $X$ which is assumed to be compact and Hausdorff w.r.t. both $T$ and $T'$. Prove that $T$ and $T'$ are either equal or not comparable.

**Exercise 5.5.6.** Let $(X, d)$ be a metric space. Assume that $A$ and $B$ are two disjoint non empty subsets of $X$ such that $A$ is closed and $B$ is compact. Show that

$$d(A, B) > 0$$

where $d(A, B)$ has already been defined in the "True or False" Section of Chapter 2. Give an example showing that two (only) closed sets are not sufficient for the result to hold (look for examples around $\mathbb{N}$).

**Exercise 5.5.7.** Show that the Cantor set is compact. Show also that it is uncountable.

**Exercise** 5.5.8. Let $X$ and $Y$ be two *separated* topological and *compact* spaces. Let $G_f$ be the graph of $f$. Show that $f$ is continuous if and only if $G_f$ is closed in $X \times Y$.

**Exercise** 5.5.9. ([8]) Let $f : \mathbb{R} \to \mathbb{R}$ be a function having a closed graph. Show that if $K$ is a compact set in $\mathbb{R}$, then $f^{-1}(K)$ is closed in $\mathbb{R}$.

**Exercise** 5.5.10. Show that $C([0,1], \mathbb{R})$ (with respect to the supremum metric) is not compact using open covers.

**Exercise** 5.5.11. On $\mathbb{R}$, consider the equivalence relation $\mathcal{R}$ (cf. Exercise 1.2.17) defined by:

$$x\mathcal{R}y \iff y - x \in \mathbb{Z}.$$

Denote the quotient of $\mathbb{R}$ by $\mathcal{R}$ this time by $\mathbb{R}/\mathbb{Z}$ and let $\varphi : \mathbb{R} \to \mathbb{R}/\mathbb{Z}$ be the quotient map. Let $C = \{z \in \mathbb{C} : |z| = 1\}$ and let $f : \mathbb{R} \to C$ be defined by $f(t) = e^{2i\pi t}$, where $C$ is endowed with the induced topology of $\mathbb{C}$. Show that there is a homeomorphism $g : \mathbb{R}/\mathbb{Z} \to C$ such that $f = g \circ \varphi$.

**Exercise** 5.5.12. Let $(X, d)$ be a compact metric space. Denote the collection of closed sets in $(X, d)$ by $Cl(X)$. Define a function $d$ on $Cl(X) \times Cl(X)$ by

$$d_H(A, B) = \max(\sup_{a \in A} d(a, B), \sup_{b \in B} d(b, A))$$

where $d(x, C) = \inf_{c \in C} d(x, c)$.

Show that $d_H$ is a metric on $Cl(X)$ ($d_H$ is called the **Hausdorff metric**).

**REMARK**. If the compactness hypothesis is dropped, then $d_H$ need not remain a metric anymore. For instance, the Hausdorff "metric" applied to the sets $\{1\}$ and $(-\infty, 1]$ is infinite, hence it is necessary to have bounded sets. It is also necessary to have closed sets (take $A = [-1, 1]$ and $B = (-1, 1)$).

**Exercise** 5.5.13. We know that "*a non-empty compact Hausdorff space $X$ without any isolated point is uncountable*". Give an example showing that the hypothesis *Hausdorff* cannot be dispensed with.

**Exercise** 5.5.14. Show that every compact Hausdorff space is normal.

**Exercise** 5.5.15. Let $(X, d)$ be a *compact* metric space. Let $f$ be an isometry from $X$ *into* $X$. Prove that $f$ is onto (hint: show that $X \subset f(X)$).

**Exercise** 5.5.16. Let $X$ be a locally compact topological space. Set $\tilde{X} = X \cup \{\infty\}$ where $\infty$ is something which does not belong to $X$. Then $\tilde{X}$ is called the **Alexandroff one-point compactification** of $X$. Then a topology can be defined on $\tilde{X}$ by declaring a set open iff it is either open in $X$ or it is of the form $\tilde{X} \setminus K$ where $K$ is compact in $X$. This will be the topology associated with $\tilde{X}$ by default.

A known result then says that $X$ is homeomorphic to $\tilde{X} \setminus \{\infty\}$.

(1) Prove that $\tilde{X}$ is compact.
(2) Verify that $X$ is closed in $\tilde{X}$ iff it is compact.
(3) Deduce that if $X$ is not compact, then it is dense in $\tilde{X}$.
(4) Show that if $Y$ and $Z$ are two compact spaces, then a homeomorphism $f : Y \setminus \{y\} \to Z \setminus \{z\}$ (where $y \in Y$ and $z \in Z$) can be extended to a homeomorphism $g : Y \to Z$ by setting $g(y) = z$.
(5) Show that in the usual topology, the Alexandroff one-point compactification of $(-1, 0]$ is $[-1, 0]$ and that of $\mathbb{R}$ is homeomorphic to the unit circle in $\mathbb{R}^2$.

**Exercise** 5.5.17. Show that bounded subsets of $\mathbb{R}^n$ are totally bounded.

**Exercise** 5.5.18. First, we give the following definition:

**DEFINITION** 5.5.1. *Let $X$ be a metric space and let $A \subset X$. Let $\mathcal{U} = \{U_i\}_{i \in I}$ be an open cover for $A$. A real number $\varepsilon > 0$ is said to a* **Lebesgue number** *for $\mathcal{U}$ if for any $x \in A$, there is some set $U$ ($U \ni x$) in $\mathcal{U}$ verifying $B(x, \varepsilon) \subset U$.*

Prove that any open cover of a sequentially compact metric space has a Lebesgue number.

# CHAPTER 6

# Connected Spaces

## 6.1. Essential Background

### 6.1.1. Connectedness.

**DEFINITION** 6.1.1. *A topological space $X$ is said to be **connected** if the only closed and open sets (that is, the only clopen sets) in $X$ are $\varnothing$ and $X$ itself.*

**EXAMPLES** 6.1.2.

(1) *Evidently, every indiscrete topological space is connected.*
(2) *A discrete topological space $X$ is never connected unless card $X = 1$ (in which case the discrete and indiscrete spaces coincide!).*
(3) *In any topological space $X$, the empty set and any singleton are always connected.*

There exist equivalent definitions of connectedness. Before giving them, we first have a definition.

**DEFINITION** 6.1.3. *An **open partition** is a couple of non-empty open subspaces $A$ and $B$ of a topological space $X$ such that $A \cap B = \varnothing$ and $A \cup B = X$.*

**REMARK.** The concept of a **closed partition** is defined similarly (just replace any "open word" by a "closed one"). In fact, from easy algebraic considerations, a closed partition is equivalent to an open partition.

**THEOREM** 6.1.4. *Let $X$ be a topological space. Then the following statements are equivalent*

(1) *$X$ is connected.*
(2) *Any continuous function from $X$ to the discrete space $\{0, 1\}$ is constant.*
(3) *$X$ does not admit any open (or closed!) partition.*

**REMARK.** Notice that there is nothing special about the set $\{0, 1\}$, any $\{a, b\}$ will do!

**REMARK.** We can rephrase the second statement of the previous theorem by saying that $X$ is connected iff there is no continuous function from $X$ *onto* $\{0, 1\}$.

The following result characterizes connected subspaces of the usual real line.

**THEOREM** 6.1.5. *(see Exercise 6.3.7 for a proof) Every interval is connected. Conversely, the only connected subspaces of $\mathbb{R}$ are the intervals.*

We know from sometime that the direct image of an interval by a continuous function is an interval. Does this remain valid in a general setting? The answer is yes.

**THEOREM** 6.1.6. *(a proof is supplied in Exercise 6.3.8) Let $X$ and $Y$ be two topological spaces. Let $f : X \to Y$ be continuous. If $X$ is connected, then $f(X)$ is connected too.*

**COROLLARY** 6.1.7. *Connectedness is a topological property.*

**COROLLARY** 6.1.8. *Let $X$ and $Y$ be two topological spaces. Let $f : X \to Y$ be continuous. If $X$ is connected, then $G_f$, i.e. the graph of $f$, is connected.*

**COROLLARY** 6.1.9. *Let $X$ be a topological space which is connected. If $f : X \to \mathbb{R}$ is continuous, then $f(X)$ is an interval.*

**REMARK.** The previous corollary generalizes the Intermediate Value Theorem.

Another consequence of Theorem 6.1.6 is:

**PROPOSITION** 6.1.10. *(a proof can be found in Exercise 6.3.22) Let $C$ be the circle $\{z \in \mathbb{C} : |z| = 1\}$. Then $C$ is not homeomorphic to $\mathbb{R}^2$.*

**COROLLARY** 6.1.11. *(see Exercise 6.3.22 and cf. Exercise 6.3.21) $\mathbb{R}^2$ is not homeomorphic to $\mathbb{R}$.*

We finish this subsection with results on unions and products of connected spaces. Proposition 6.1.12 below is also useful for the ensuing subsection.

**PROPOSITION** 6.1.12. *(a proof is given in Exercise 6.3.9) Let $X$ be a topological space and let $(A_i)_{i \in I}$ be a class of connected subspaces of $X$ such that $\bigcap_{i \in I} A_i \neq \varnothing$. Then $\bigcup_{i \in I} A_i$ is connected.*

**REMARK.** If $\bigcap_{i \in I} A_i = \varnothing$, then nothing can be said about the connectedness of $\bigcup_{i \in I} A_i$, even for a finite number of sets (see Exercise 6.3.1, Questions (4) and (6)).

**PROPOSITION 6.1.13.** *(a proof may be found in Exercise 6.3.11)* *If $X$ and $Y$ are two non-empty connected sets such that $\overline{X} \cap Y \neq \varnothing$, then $X \cup Y$ is connected.*

**REMARK.** The condition $\overline{X} \cap Y \neq \varnothing$ in the preceding proposition cannot be weakened to $\overline{X} \cap \overline{Y} \neq \varnothing$. For a counterexample see Question (4) of Exercise 6.3.1.

A natural question is how stable is connectedness as regards (finite) with products? The answer is given next (and a proof is supplied in Exercise 6.3.12).

**THEOREM 6.1.14.** *Let $X$ and $Y$ be two topological spaces. Then*

$$X \times Y \text{ is connected} \Longleftrightarrow X \text{ and } Y \text{ are connected.}$$

By induction, we have:

**COROLLARY 6.1.15.** *(cf. Exercise 6.5.5) Let $X_1, X_2, \cdots X_n$ be topological spaces. Then $\prod_{i=1}^{n} X_i$ is connected iff each $X_i$ is connected $(i = 1, \cdots, n)$.*

We close this subsection with an interesting result. But first, we give a definition.

**DEFINITION 6.1.16.** *A non-empty compact, connected and Hausdorff space is called **continuum**.*

**THEOREM 6.1.17.** *(**Sierpinski**) No continuum can be written as a countable union of closed non-empty and disjoint sets.*

**REMARK.** The Sierpinski Theorem was proved in 1918. Its original proof may be found in [**23**].

### 6.1.2. Components.

**THEOREM 6.1.18.** *Let $X$ be a topological space and let $x \in X$. Then there exists a maximal or largest (with respect to "$\subset$") connected subspace of $X$, denoted by $C_x$, containing $x$. Moreover, $C_x$ is closed in $X$.*

**DEFINITION 6.1.19.** *Let $X$ be a topological space, and let $x \in X$. The maximal, closed and connected subspace $C_x$ containing $x$ is called a (connected) **component** of $X$.*

**REMARK.** We may easily define an equivalence relation on $X$ as:

$$x \mathcal{R} y \iff y \in C_x.$$

Hence, the components of $X$ constitute a partition of $X$.

**EXAMPLE** 6.1.20. $\mathbb{R} \setminus \{1\}$ *has two components, namely* $(-\infty, 1)$ *and* $(1, \infty)$.

**DEFINITION** 6.1.21. *A topological space in which the components are all singletons is said to be* **totally disconnected**.

**EXAMPLE** 6.1.22. *A discrete topological space is totally disconnected.*

### 6.1.3. Path-connectedness.

**DEFINITION** 6.1.23. *Let $X$ be a topological space. We say that $X$ is* **path-connected** *if for all $x, y \in X$, there exists a continuous function $f : [0, 1] \to X$ such that $f(0) = x$ and $f(1) = y$.*
*This continuous function is usually called a* **path** *from $x$ to $y$.*

**REMARK.** It must be remembered that $[0, 1]$ in the previous definition is always equipped with its usual topology.

Here is a quite practical lemma.

**LEMMA** 6.1.24. *Let $X$ be a topological space in which $f$ is a path joining $x$ and $y$, and $g$ is a path joining $y$ and $z$. Then the function*

$$h(t) = \begin{cases} f(2t), & 0 \leq t \leq \frac{1}{2}, \\ g(2t - 1), & \frac{1}{2} \leq t \leq 1, \end{cases}$$

*defines a path joining $x$ and $z$.*

**EXAMPLES** 6.1.25.
  (1) $\mathbb{R}$, *and in general, $\mathbb{R}^n$ ($n \geq 1$) are path-connected in the standard topology.*
  (2) $\mathbb{R}^*$ *is not path-connected.*

**THEOREM** 6.1.26. *The continuous image of a path-connected set is path-connected.*

The notion of convexity gives a wide range of examples of path-connected sets:

**DEFINITION** 6.1.27. *Let $X$ be a real vector space and let $A \subset X$. We say that $A$ is* **convex** *if:*

$$\forall x, y \in A, \forall t \in [0, 1] : \ (1 - t)x + ty \in A.$$

**EXAMPLES** 6.1.28.

(1) *Intervals are convex.*
(2) $\mathbb{R}$ *is convex.*
(3) *A (real) linear subspace is convex.*
(4) *The closed unit ball is convex.*

Thanks to the next result, the previous four examples are all path-connected:

**THEOREM** 6.1.29. *Any convex set is path-connected.*

Before continuing, we digress a little bit. We have met the notion of convex sets as well as that of convex functions. Is there a link between them? The answer is yes.

**PROPOSITION** 6.1.30. *Let $X$ be a real vector space and let $C$ be a convex part of $X$. Let $f : C \to \mathbb{R}$ be a function. Then $f$ is convex iff $\{(x,y) \in X \times \mathbb{R} : f(x) \leq y\}$ is a convex subset of $X \times \mathbb{R}$.*

The next result shows the relationship between connectedness and path-connectedness.

**THEOREM** 6.1.31. *(see Exercise 6.3.15) Every path-connected space is connected.*

**REMARK.** The converse of the previous theorem is not always true. Then (with respect to the Euclidean metric) the set

$$A = \left\{ (x,y) \in \mathbb{R}^2 : \ 0 < x \leq 1, \ y = \sin\left(\frac{1}{x}\right) \right\}.$$

is connected and hence so is $\overline{A}$. But it can be shown that $\overline{A}$ is not path-connected. Details are to be found in Exercise 6.3.24.

The set $\overline{A}$ is a famous set in Topology, more commonly known as the **Topologist Sine Curve**.

Nonetheless, there are sufficient conditions making a connected space path-connected (in usual $\mathbb{R}^n$), one of them is openness. We have:

**THEOREM** 6.1.32. *Every connected open subset of $\mathbb{R}^n$ is path-connected.*

## 6.2. True or False: Questions

QUESTIONS. Comment on the following questions/statements and indicate those which are false and those which are true when this applies. Justify your answers.

(1) Let $T$ and $T'$ be two topologies on the same set $X$. Assume that $T \subset T'$. Then

$$(X, T) \text{ is connected} \iff (X, T') \text{ is connected.}$$

(2) A closed subspace of a connected space is connected.

(3) The boundary of a connected set is itself connected.

(4) Let $X$ be a topological space and let $A \subset X$. If $\overline{A}$ is connected, then $A$ is connected. What about the converse?

(5) The union of connected sets is always connected.

(6) The intersection of connected sets is always connected.

(7) If a set $A$ is connected, then so is its interior $\overset{\circ}{A}$.

(8) Let $X$ be a real vector space and let $A \subset X$. Then

$$A + A = 2A$$

where $2A = \{2a : a \in A\}$.

(9) Let $f : [-1, 1] \to \mathbb{R}$ be a continuous function. Assume that $f(-1)f(1) < 0$. Then there exists some $\alpha \in (-1, 1)$ such that $f(\alpha) = 0$.

(10) Every polynomial on $\mathbb{R}$ having an odd degree has at least one real root.

(11) Let $X$ be a topological space and let $A \subset X$. If $\overline{A}$ is path-connected, then so is $A$.

(12) In the usual topology, $\mathbb{R}$ is homeomorphic to $\mathbb{R}^2$.

(13) The preimage of a connected set by a continuous function is connected.

(14) If $X$ and $Y$ are homeomorphic, then there is a one-to-one correspondence between their components.

(15) The quotient of a connected (path-connected respectively) topological space is connected (path-connected respectively).

(16) Let $X$ be a topological space. Let $C_x$ be the component of $x \in X$. Then

$$X \text{ is connected} \iff C_x = X.$$

(17) $\mathbb{R}^*$ is not connected as it has two components $(0, \infty)$ and $(-\infty, 0)$. But we know that the components are closed whereas here they are not closed in $\mathbb{R}$. Is there anything wrong with this reasoning?

## 6.3. Exercises With Solutions

**Exercise** 6.3.1. Is $A$ connected in the topological space $T$ in the following cases?

(1) $A = \{1, 2, 3, 4\}$, $T = \{\varnothing, \{1\}, \{2, 3\}, \{1, 2, 3\}, A\}$;
(2) $A$ is a subset of some set (with $\mathrm{card}A \geq 2$) endowed with the discrete topology (denoted by $T$);
(3) $A = \mathbb{R}$ and $T$ the associated co-finite topology. Investigate also the case of an infinite $A$ and the case of a finite $A$;
(4) $A = B((0, 1), 1) \cup B((0, -1), 1)$,
(5) $A = B_c((0, 1), 1) \cup B_c((0, -1), 1)$;
(6) $A = B((0, 1), 1) \cup B_c((0, -1), 1)$;
(the last three sets in $\mathbb{R}^2$ with respect to the euclidian metric).

**Exercise** 6.3.2. Let $X = \mathcal{M}_n(\mathbb{R})$, the (vector) space of square matrices of order $n$ with real entries. Let $A$ be the subset of $X$ of invertible matrices. Is $A$ connected?

**Exercise** 6.3.3. Show that $\mathbb{Q}$ is not connected using different methods.

**Exercise** 6.3.4. In the induced usual topology, are the following sets connected

(1) $\mathbb{R} \setminus \mathbb{Q}$;
(2) $\{\frac{1}{n} : n \geq 1\}$;
(3) $[0, 1) \cup (1, 2)$;
(4) $\mathbb{N}$?

**Exercise** 6.3.5. Show that $[0, 2)$ is connected with respect to the topology of Exercise 3.3.36.

**Exercise** 6.3.6. What are the components of

(1) $A = \mathbb{C} \setminus \mathbb{R}$;
(2) $A = \{(x, y) \in \mathbb{R}^2 : x \neq y\}$;
(3) $A = B((0, 1), 1) \cup B((0, -1), 1)$, $B = B_c((0, 1), 1) \cup B_c((0, -1), 1)$ in usual $\mathbb{R}^2$?

**Exercise** 6.3.7. In the usual topology of $\mathbb{R}$, show that every interval is connected, and that no subset of $\mathbb{R}$ other than an interval is connected.

**Exercise** 6.3.8. Let $X$ and $Y$ be two topological spaces and let $f : X \to Y$ be a continuous function. If $X$ is connected, then so is $f(X)$.

**Exercise 6.3.9.** Let $X$ be a topological space and let $(A_i)_{i \in I}$ be a collection of connected subspaces of $X$ such that $\bigcap_{i \in I} A_i \neq \varnothing$. Show that $\bigcup_{i \in I} A_i$ is connected.

**Exercise 6.3.10.** Let $A$ be a connected set in a topological space $X$.

(1) If $A \subset B \subset \overline{A}$, then show that $B$ is connected.
(2) Deduce that if $A$ is connected, then so is $\overline{A}$.

**Exercise 6.3.11.** Let $X$ and $Y$ be two non-empty connected sets such that $\overline{X} \cap Y \neq \varnothing$. Show that $X \cup Y$ is connected.

**Exercise 6.3.12.** Let $X$ and $Y$ be two topological spaces. Show that $X \times Y$ is connected iff $X$ and $Y$ are connected.

**Exercise 6.3.13.** Let $\mathbb{R}$ be endowed with the $K$-topology.
(1) Show that $(-\infty, 0)$ and $(0, \infty)$ inherit their usual topology as subspaces of $\mathbb{R}_K$.
(2) Deduce that $\mathbb{R}$ is connected in $\mathbb{R}_K$.

**Exercise 6.3.14.** Show that there is no continuous function $f : \mathbb{R} \to \mathbb{R}$ such that

$$f(\mathbb{Q}) \subset \mathbb{R} \setminus \mathbb{Q} \text{ and } f(\mathbb{R} \setminus \mathbb{Q}) \subset \mathbb{Q}.$$

**Exercise 6.3.15.** Prove that every path-connected space is connected.

**Exercise 6.3.16.** Let $\mathbb{N}$ be endowed with the co-finite topology. Show that $\mathbb{N}$ is not path-connected.

**Exercise 6.3.17.** Using *only* the definition of a path-connected set, show that $\mathbb{R}^*$ is not path-connected.

**Exercise 6.3.18.**
(1) Show that every convex part is path-connected.
(2) Deduce that $\mathbb{R}^n$ and the closed and open balls on $\mathbb{R}^n$ are all connected ($n \geq 1$).

**Exercise 6.3.19.** Let $n \geq 1$.
(1) Is $\mathbb{R}^n \setminus \{0\}$ path-connected?
(2) Let
$$S^{n-1} = \{x \in \mathbb{R}^n : x^2 = 1\}.$$
Is $S^{n-1}$ path-connected?

**Exercise 6.3.20.** Let $\mathbb{T} = S(O_{\mathbb{R}^2}, 1)$ be the unit sphere in $\mathbb{R}^2$. Let $A$ be the **annulus** on $\mathbb{R}^2$, i.e. $A = \{(x, y) \in \mathbb{R}^2 : a^2 \leq x^2 + y^2 \leq b^2\}$ where $0 < a < b$.

(1) Verify that the following functions are continuous $f : [a, b] \times \mathbb{T} \to A$ and $g : A \to [a, b] \times \mathbb{T}$ defined as

$$f(z, x, y) = (zx, zy) \text{ and } g(x, y) = \left( \sqrt{x^2 + y^2}, \frac{(x, y)}{\sqrt{x^2 + y^2}} \right).$$

Compute $f \circ g$ and $g \circ f$. What can you deduce from this question?

(2) Deduce from the previous equation that the annulus is a path-connected part of $\mathbb{R}^2$.

**Exercise** 6.3.21. Let $X$ and $Y$ be two topological spaces. Let $f : X \to Y$ be a homeomorphism. Let $a \in X$.

(1) Show that $\overline{f} : X \backslash \{a\} \to Y \backslash \{f(a)\}$ remains a homeomorphism.
(2) Deduce that there cannot be a homeomorphism between $\mathbb{R}$ and $\mathbb{R}^2$.

**Exercise** 6.3.22. Let

$$C = \{(x, y) \in \mathbb{R}^2 : x^2 + y^2 = 1\}$$

(which, in $\mathbb{C}$, is also given by $\{z \in \mathbb{C} : |z| = 1\}$).

(1) Show that $C$ is connected.
(2) Show that there cannot be a continuous injective function from $C$ into $\mathbb{R}$.
(3) Infer that $\mathbb{R}^2$ is not homeomorphic to $\mathbb{R}$ (cf. Exercise 6.3.21).

**Exercise** 6.3.23. Consider the letters X and S as subsets of $\mathbb{R}^2$. Can we say that X and S are homeomorphic? What about E and W?

**Exercise** 6.3.24. In the Euclidean metric, let

$$A = \left\{ (x, y) \in \mathbb{R}^2 : 0 < x \leq 1, \ y = \sin \left( \frac{1}{x} \right) \right\}.$$

(1) Show that $A$ is connected.
(2) Find $\overline{A}$.
(3) Prove that $\overline{A}$ is not path-connected.

## 6.4. Tests

**Test** 47. Show that a topological space with one point is always connected. What about a set with two points?

**Test** 48. Is $\mathbb{R}$ connected with respect to the lower limit topology of $\mathbb{R}$?

**Test** 49. Show that $\mathbb{R}_\ell$ is totally disconnected.

**Test** 50. Is $\mathbb{R}$ connected with respect to the co-countable topology? What about $\mathbb{Q}$?

**Test** 51. Let $X$ be a topological space. Suppose that $X$ contains a connected dense subspace. Show that $X$ is connected.

**Test** 52. What are the components of $\mathbb{R}$ with the respect to the co-finite topology?

**Test** 53. In the usual topology, can $[-1, 1]$ be homeomorphic to the unit circle?

**Test** 54. Let $A, B \subset \mathbb{R}^n$ with $n \geq 1$. Set
$$A + B = \{a + b : a \in A, b \in B\}.$$
Show that $A + B$ is connected if both $A$ and $B$ are connected in $\mathbb{R}^n$.

## 6.5. More Exercises

**Exercise** 6.5.1. Set
$$A = \{\mathbb{R}\} \cup \{U \subset \mathbb{R} : U^c \text{ connected with respect to usual } \mathbb{R}\}.$$
Is $T$ a topology on $\mathbb{R}$?

**Exercise** 6.5.2. Let $X$ be a connected space. Assume that $\mathrm{card} X \geq 2$. Show that if for each $x \in X$, $\{x\}$ is closed, then $\mathrm{card} X = \infty$.

**Exercise** 6.5.3. Using Exercise 5.5.16, give an alternative way of proving that the unit circle is connected.

**Exercise** 6.5.4. Is $\mathbb{R}$ connected with respect to the topology of Exercise 5.3.13?

**Exercise** 6.5.5. Show that an arbitrary product of connected sets is not necessarily connected w.r.t. the box topology, but it is connected with respect to the product topology.

**Exercise** 6.5.6. In the usual topology of $\mathbb{R}^2$, say whether M and N are homeomorphic? What about B and V? or Y and T?

**Exercise** 6.5.7. Let
$$A = (\mathbb{R} \times \mathbb{Q}) \cup (\mathbb{Q} \times \mathbb{R}).$$
Show that $A$ is path-connected.

**Exercise** 6.5.8. Prove that the Cantor set is totally disconnected.

**Exercise** 6.5.9. Let $X$ be a topological space. We say that $X$ is **locally connected** at a point $x \in X$ if for each neighborhood of $U$ of $x$, there exists a *connected* neighborhood $V$ of $x$ such that $V \subset U$.

(1) In the usual topology, are $\mathbb{R}$, $\mathbb{R}^*$ and $\mathbb{Q}$ locally connected?
(2) Show that the every open set in a locally connected set remains locally connected.
(3) Prove that the components of a locally connected space are open.

# CHAPTER 7

# Complete Metric Spaces

## 7.1. Essential Background

**7.1.1. Completeness.** In this chapter, we deal with the extremely important concept of complete spaces. The reader must have certainly come across this notion in basic real analysis. One of the main features of complete spaces is that in such spaces, the Cauchyness of a sequence is sufficient to declare its convergence (without knowing its limit a priori).

**DEFINITION 7.1.1.** *Let $(X, d)$ be a metric space and let $(x_n)$ be a sequence in $X$. We say that $(x_n)$ is **Cauchy** if:*

$$\forall \varepsilon > 0, \exists N \in \mathbb{N}, \forall n, m \in \mathbb{N} : (n, m \geq N \Longrightarrow d(x_n, x_m) < \varepsilon).$$

**REMARK.** The previous means that

$$\lim_{n,m \to \infty} d(x_n, x_m) = 0$$

(in the sense of a **double limit**).

The following is an equivalent definition:

**PROPOSITION 7.1.2.** *Let $(X, d)$ be a metric space and let $(x_n)$ be a sequence in $X$. Then $(x_n)$ is Cauchy iff*

$$\forall \varepsilon > 0, \exists N \in \mathbb{N}, \forall n, p \in \mathbb{N} : (n \geq N \Longrightarrow d(x_{n+p}, x_n) < \varepsilon).$$

**REMARK.** Let $(X, d)$ be a metric space and let $A \subset X$ be equipped with the induced metric, denoted by $d_A$. Let $(x_n)$ be a sequence in $A$. Then $(x_n)$ is Cauchy in $(X, d)$ iff $(x_n)$ is Cauchy in $(A, d_A)$.

**PROPOSITION 7.1.3.** *A convergent sequence in a metric space is Cauchy.*

The converse of the previous proposition is not always true. But when it is, then this gives rise to a fundamental notion.

**DEFINITION 7.1.4.** *A metric space $(X, d)$ is said to be **complete** if every Cauchy sequence in $(X, d)$ converges in $(X, d)$, i.e. its limit w.r.t. $d$ belongs to $X$.*

**EXAMPLES** 7.1.5.

(1) $(\mathbb{Q}, |\cdot|)$ *is not complete (see Exercise 7.3.2).*
(2) $(\mathbb{R}, |\cdot|)$ *is complete (see Exercise 7.3.5).*
(3) $(\mathbb{C}, |\cdot|)$ *is complete (see Exercise 7.3.6).*
(4) $([0, 1), |\cdot|)$ *is not complete.*
(5) *Any discrete metric space is complete (see Exercise 7.3.3).*

**PROPOSITION** 7.1.6. *If $(X, d)$ and $(X, d')$ are two metric spaces such that $d$ and $d'$ are equivalent, then $(X, d)$ is complete iff $(X, d')$ is complete.*

**REMARK.** If $d$ and $d'$ are (only) topologically equivalent, then the completeness of $(X, d)$ need not imply that of $(X, d')$. For instance in $\mathbb{R}$, consider the usual metric $d$ and the metric $d'$ of Exercise 2.3.29 (which are topologically equivalent). Then $(\mathbb{R}, d)$ is complete while $(\mathbb{R}, d')$ is not (see Test 56).

We have already defined what it means for a set to be bounded. For sequences, this is just a matter of rephrasing.

**DEFINITION** 7.1.7. *Let $(X, d)$ be a metric space. A sequence $(x_n)$ in $(X, d)$ is **bounded** if:*

$$\exists M \geq 0, \ \exists a \in X, \ \forall n \in \mathbb{N}: \ d(x_n, a) \leq M.$$

**PROPOSITION** 7.1.8. *(cf. the "True or False" Section) Let $(X, d)$ be a metric space. A Cauchy sequence $(x_n)$ is always bounded.*

Before turning to more examples, we give a few more properties of the diameter of a set (already defined in Chapter Two). Most of these properties could have been stated there anyway.

**PROPOSITION** 7.1.9. *Let $(X, d)$ be a metric space and $A, B \subset X$. Then:*

(1) $d(A) = 0 \iff A$ *reduces to a singleton or $A = \varnothing$.*
(2) $A \subset B \implies d(A) \leq d(B)$.
(3) $d(A) = d(\overline{A})$.

The next is a quite useful characterization of Cauchyness using the notion of diameter.

**PROPOSITION** 7.1.10. *Let $(X, d)$ be a metric space and let*

$$A_n = \{x_k : \ k \geq n\}$$

*be a sequence of subsets of $X$. Then*

$$(x_n) \ \text{is Cauchy} \iff \lim_{n \to \infty} d(A_n) = 0.$$

**REMARK.** It is perhaps better to visualize the sets $(A_n)$ as:

$$A_1 = \{x_1, x_2, \cdots\}, \ A_2 = \{x_2, x_3, \cdots\}, \ A_3 = \{x_3, x_4, \cdots\}, \cdots$$

Now, we give examples on completeness in functional spaces. The next example is important enough to be stated separately:

**EXAMPLE** 7.1.11. *(for a proof and more discussion, see Exercise 7.3.14 and the "True or False" Section) Let $X = C([0,1], \mathbb{R})$ be equipped with the metric defined, for all $f, g \in X$, by*

$$d_1(f, g) = \int_0^1 |f(x) - g(x)| dx.$$

*Then $(X, d_1)$ is not complete.*

If we keep the same space $X$ but we equip it this time with the supremum metric, then we obtain the following fundamental result:

**THEOREM** 7.1.12. *(see Exercise 7.3.15) Let $X = C([0,1], \mathbb{R})$ be equipped with the supremum metric $d_\infty$. Then $(X, d_\infty)$ is complete.*

Now, we keep the same metric but we change the space $X$. The result is again different.

**EXAMPLE** 7.1.13. *(Exercise 7.3.17) Let $X$ be the set of all polynomials (of any degree) defined on $[0,1]$. Endow $X$ with the supremum metric $d_\infty$. Then $(X, d_\infty)$ is not complete.*

The next result will be recalled and generalized in the next chapter. Why we are stating it here is that it allows us to prove that some metric spaces are not complete (Exercise 7.3.17 among others).

**THEOREM** 7.1.14. *(**Weierstrass**) If $f : [a, b] \to \mathbb{R}$ is a continuous function, then there exists a sequence $(P_n)$ of polynomials on $[a, b]$ such that*

$$\lim_{n \to \infty} d_\infty(f, P_n) = \lim_{n \to \infty} \sup_{x \in [a,b]} |f(x) - P_n(x)| = 0.$$

Now, we pass to the completeness of $\ell^p$ spaces.

**THEOREM** 7.1.15. *Let $1 \le p < \infty$. Then*

(1) *$(\ell^p, d_p)$ is complete (cf. Exercise 7.3.10).*
(2) *$(\ell^\infty, d_\infty)$ is complete.*

If a sequence has a convergent subsequence, then this does not always mean that the whole sequence converges (e.g. $(-1)^n$). But, if the sequence is further assumed to be Cauchy, then the outcome is different. We have:

**PROPOSITION** 7.1.16. *(a proof is provided in Exercise 7.3.4) Let* $(X, d)$ *be a metric space. Let* $(x_n)$ *be a Cauchy sequence in* $X$. *If* $(x_n)$ *has a subsequence* $(x_{n_k})$ *converging to* $x \in X$, *then* $(x_n)$ *converges to* $x \in X$ *too.*

Closedness and completeness are intimately related. The two notions are even equivalent in some context.

**THEOREM** 7.1.17. *(a proof is prescribed in Exercise 7.3.20) Let* $(X, d)$ *be a metric space and let* $A \subset X$. *Then:*
 (1) *If* $A$ *is complete, then it is closed in* $X$.
 (2) *If* $A$ *is closed and* $X$ *is complete, then* $A$ *is complete.*

**REMARK.** Combining the two results in the preceding theorem, we may state with ease that: *In a complete metric space, a subset is complete if and only if it is closed.*

A consequence of the last theorem and the last example of Examples 7.1.5 gives:

**COROLLARY** 7.1.18. *In a discrete metric space, all subsets are complete.*

The coming theorem gives a relationship between the concepts of compactness and completeness.

**THEOREM** 7.1.19. *Let* $X$ *be a metric space.*
 (1) *If* $X$ *is compact, then it is complete.*
 (2) *If* $X$ *is totally bounded and complete, then it is compact.*

**REMARK.** We may then easily show that if $X$ is a complete metric space, then $A \subset X$ is compact iff $A$ is closed and totally bounded.

>From the previous chapters, we have been used to ask about unions, intersections and products of certain spaces. We do the same here on completeness.

**PROPOSITION** 7.1.20. *(see Exercise 7.3.21) The finite union and the arbitrary intersection of complete spaces remain complete.*

**THEOREM** 7.1.21. *(see Exercise 7.3.22) Let* $(X, d)$ *and* $(Y, d')$ *be two metric spaces. Endow the product space* $Z = X \times Y$ *with the metric* $\delta$ *defined by*
$$\delta((x, y), (x', y')) = d(x, x') + d'(y, y').$$
*Then* $(Z, \delta)$ *is complete iff both* $(X, d)$ *and* $(Y, d')$ *are complete.*

**REMARK.** The previous result can be generalized to a finite product using induction.

REMARK. There is nothing special about the chosen metric in the previous theorem. We could have used for instance

$$\delta((x,y),(x',y')) = \max(d(x,x'),d'(y,y'))$$

or

$$\delta((x,y),(x',y')) = \sqrt{[d(x,x')]^2 + [d'(y,y')]^2}$$

(which, from Proposition 2.1.33, are known to be equivalent). In fact, by Proposition 7.1.6, the completeness of a space implies that of the other and vice versa (when the metrics are equivalent). So, we have:

COROLLARY 7.1.22. *Using any* $d_p$ *($1 \leq p \leq \infty$) from Proposition 2.1.8,* $(\mathbb{R}^n, d_p)$ *is complete.*

Another result in this spirit is given next.

PROPOSITION 7.1.23. *(a proof is given in Exercise 7.3.29) Let* $(X,d)$ *and* $(X',d')$ *be two isometric metric spaces. Then* $(X,d)$ *is complete if and only if* $(X',d')$ *is complete.*

REMARK. A slightly weaker result holds, that is, if there exists some one-to-one correspondence $f$ between $(X,d)$ and $(X',d')$ such that

$$\exists \alpha, \beta > 0, \ \forall x,y \in X : \ \alpha d(x,y) \leq d'(f(x),f(y)) \leq \beta d(x,y),$$

then $X$ is complete iff $X'$ is so.

Example 2.1.30 tells us that Euclidean $\mathbb{R}^2$ is isometric to $\mathbb{C}$. Hence we have:

COROLLARY 7.1.24. *(cf. Exercise 7.3.6)* $(\mathbb{C}, |\cdot|_{\mathbb{C}})$ *is complete.*

Also (using again any $d_p$ from Proposition 2.1.8):

COROLLARY 7.1.25. $(\mathbb{C}^n, d_p)$ *is complete.*

The next theorem is of great interest. It tells us when to extend a function $f$ defined on a subset $A$ (of $X$) to the whole of $X$.

THEOREM 7.1.26. *Let* $(X,d)$ *and* $(Y,d')$ *be two metric spaces. Assume that* $A \subset X$ *is dense and that* $(Y,d')$ *is complete. Let* $f : (A,d) \to (Y,d')$ *be uniformly continuous on* $A$. *Then there exists a unique function* $g : (X,d) \to (Y,d')$ *which is also uniformly continuous and such that* $g(x) = f(x)$ *for each* $x \in A$.

REMARK. The proof of uniqueness is a simple application of Exercise 4.3.30.

REMARK. In normed vector spaces, the continuity is equivalent to uniform continuity whenever the function or map is linear. Thus a very similar result holds in normed vector spaces by assuming that the function is linear and continuous.

REMARK. The completeness of $(Y, d)$ and the uniform continuity of $f$ may not merely be dropped. See the section "True or False" for counterexamples.

We finish this subsection with two fundamental theorems.

THEOREM 7.1.27. *(due to **Cantor**, a proof is given in Exercise 7.3.31) Let $(X, d)$ be a metric space. Then $(X, d)$ is complete iff for each sequence $(A_n)$ of non-empty and closed subsets of $X$ verifying:*

(1) $A_{n+1} \subset A_n$ for all $n$;
(2) $d(A_n) \to 0$ as $n$ goes to $\infty$;

we have $\displaystyle\bigcap_{n=1}^{\infty} A_n \neq \varnothing$.

REMARK. We can easily see that if $(A_n)$ is a sequence of nonempty subsets of a metric space such that $\lim_{n \to \infty} d(A_n) = 0$, then the intersection $\bigcap_{n \in \mathbb{N}} A_n$ has at most one point. Under the circumstances of Theorem 7.1.27, the intersection $\bigcap_{n \in \mathbb{N}} A_n$ has at least one point. Combining these two observations, we are now sure that this intersection contains *one and only one point*.

The previous result combined with Corollary 5.1.6 yields

COROLLARY 7.1.28. *Every compact space is complete.*

THEOREM 7.1.29. *(due to **Baire**, see Exercise 7.3.33) Let $(X, d)$ be a complete metric space. The countable intersection of open dense subsets of $X$ is dense in $X$.*

COROLLARY 7.1.30. *(see Exercise 7.3.34 for a proof) If $(X, d)$ is a complete metric space, and if $(V_n)_n$ is a sequence of closed subsets of $X$ such that $\bigcup_{n=1}^{\infty} V_n = X$, then*

$$\overline{\bigcup_{n=1}^{\infty} \mathring{V}_n} = X.$$

REMARK. We may also say that a topological space is a **Baire space** if the countable union of closed sets with empty interior is also of empty interior.

REMARK. The Baire Theorem is of paramount importance as it has tremendous applications in Functional Analysis. It is used for instance in the proof of two major results there, namely the Banach-Steinhaus and the Open Mapping theorems.

When first doing continuity and differentiability of functions from $\mathbb{R}$ into $\mathbb{R}$, we met some continuous functions with one or two points at which they were not differentiable. It was almost an event when we had a continuous function with more than three points of non-differentiability. An astonishing example by Weierstrass (dating back to 1874) shows that there is a continuous function which is nowhere differentiable. Here is the example (details may be found in e.g. [28]):

$$f(x) = \sum_{n=0}^{\infty} \frac{\cos(a^n \pi x)}{2^n}$$

where $a$ is an odd positive integer satisfying $a > 3\pi + 2$.

Then one would say, this is a very rare example! But then many examples had followed. The sensational fact is that most continuous functions are nowhere differentiable as we have (the proof also relies on the Baire Theorem):

THEOREM 7.1.31. *(see e.g. [19]) Let $f : [0,1] \to \mathbb{R}$ be a continuous function. Given $\varepsilon > 0$, there exists a continuous and **nowhere differentiable** function $g : [0,1] \to \mathbb{R}$ such that*

$$|f(x) - g(x)| < \varepsilon$$

*for all $x \in [a,b]$.*

### 7.1.2. Fixed Point Theorem.

DEFINITION 7.1.32. *Let $(X, d)$ be a metric space. A function $f : (X, d) \to (X, d)$ is called a **contraction** (or a **contracting mapping**) if:*

$$\exists k \ s.t. \ 0 \leq k < 1 : \ d(f(x), f(y)) \leq k d(x, y), \ \forall x, y \in X.$$

REMARK. If $k$ is positive but not necessarily smaller than one, then we may say that the previous function satisfies the **Lipschitz condition**.

EXAMPLES 7.1.33.
(1) $x \mapsto \frac{x}{2}$ *is clearly a contraction from $\mathbb{R}$ into $\mathbb{R}$.*
(2) $x \mapsto x + \frac{1}{x}$ *is not a contraction from $[1, \infty)$ into $[1, \infty)$. See Exercise 7.3.26.*

PROPOSITION 7.1.34. *A contraction is uniformly continuous, hence continuous.*

The next is a characterization of Lipschitzity on $\mathbb{R}$ using differentiability. It allows us to have a wide choice of Lipschitz functions (on $\mathbb{R}$).

**PROPOSITION 7.1.35.** *(see Exercise 7.3.23). Let $I$ be an interval and let $f : I \to \mathbb{R}$ be a continuous function which is also differentiable on $\overset{\circ}{I}$. If $k \geq 0$, then $f$ satisfies the Lipschitz condition iff for all $x \in \overset{\circ}{I}$, we have $|f'(x)| \leq k$.*

**REMARK.** It is clear that in the previous proposition if $k < 1$, then we have a contracting mapping (don't we?).

The next definition has already been met. We recall it for convenience.

**DEFINITION 7.1.36.** *Let $X$ be a set and let $f : X \to X$ be a function. We say that $x \in X$ is a **fixed point** of $f$ if $f(x) = x$.*

The next is a fundamental result in Mathematical Analysis and it is due to S. Banach (sometimes, we say Picard-Banach Theorem).

**THEOREM 7.1.37.** *(**Fixed Point Theorem**) Let $(X, d)$ be a complete metric space. Let $f : (X, d) \to (X, d)$ be a contracting mapping. Then $f$ has a unique fixed point.*

**REMARK.** The hypothesis "$(X, d)$ being complete" is crucial for the result to hold (see Exercise 7.3.27 for a counterexample). So is the assumption "$f$ being a contraction". Indeed, if $(X, d)$ is a complete metric space such that a function $f : X \to X$ satisfies

$$d(f(x), f(y)) < d(x, y), \; \forall x, y \in X \; (x \neq y),$$

then $f$ may not have any fixed point (see Exercise 7.3.28 for a counterexample).

The previous inequality may give a unique fixed point if "$X$ complete" is replaced by "$X$ compact". This has already been seen in Exercise 5.3.36.

**REMARK.** The proof of the previous may be found in Exercise 7.3.24. The proof is constructive. It gives the way of finding the fixed point by the so-called **method of successive approximations** which roughly speaking consists of starting by a point $x_0 \in X$, then by a recursive relation of the type $x_{n+1} = f(x_n)$ we find $x_1$, $x_2$, etc...until we approach the exact solution. Then, the fixed point is the limit of the sequence $(x_n)$ w.r.t. the metric equipping $X$.

This method is well known in many areas such as differential equations (in particular, **Picard's iteration method**). An illustrative exercise is Exercise 7.3.38. See also the next subsection.

Another application of the Fixed Point Theorem to (non-linear) systems of equations may be found in Exercise 7.3.41. See also [6]. The interested reader may also wish to consult [5], [13], [14] and [20].

**REMARK.** There are other versions of the Fixed Point Theorem such as Brouwer's or the more general version of it by Schauder.

The Fixed Point Theory has become such a vast branch of mathematical analysis that books (e.g. [10]) and research journals have been named after it.

**REMARK.** If $f : \mathbb{R} \to \mathbb{R}$ is a polynomial and we want to find its roots approximatively, i.e. the solutions of $f(x) = 0$, then one of the ways to do this is to set $g(x) = x + af(x)$, where $a \in \mathbb{R}^*$. Next, we try to choose $a$ in a way that $g$ becomes a contraction.

Now, by the Fixed Point Theorem, $g$ will have a unique fixed point $\alpha$, i.e. $g(\alpha) = \alpha$. This leads to $f(\alpha) = 0$, as required. An illustrative example may be found in Exercise 7.3.25.

>From the proof of the Fixed Point Theorem, we may obtain the following estimation of the error:

**PROPOSITION 7.1.38.** *Let $(X, d)$ be a complete metric space. Let $f : (X, d) \to (X, d)$ be a contracting mapping. If $x$ is the (unique) fixed of $f$, then*

$$d(x_n, x) \leq \frac{k^n}{1 - k} d(x_1, x_0)$$

*(where $x_n$ is as in the proof of Theorem 7.1.37).*

It may happen that a given function $f$ is not a contraction but $f^{(n)}$ (its iterated map) is a contraction for some $n \in \mathbb{N}$. Then obviously $f^{(n)}$ will have a fixed point (this already known from Theorem 7.1.37). The good news is that $f$ too has a unique fixed point. We have:

**PROPOSITION 7.1.39.** *(a proof may be found in Exercise 7.3.26) Let $X$ be a complete metric space. Let $f$ be a mapping defined on $X$ into $X$. We assume that for some integer $n \geq 1$, $f^{(n)}$ is a contraction. Then $f$ has a unique fixed point.*

An application of the previous result may be found in Exercise 7.3.26.

### 7.1.3. A Word on Integral Equations.
As alluded to above, another major application of the Fixed Point Theorem is to solve integral equations. Before stating some of these applications, we first recall some essential background here.

**DEFINITION** 7.1.40. *An **integral equation** of a function $u(x)$ is defined by*

$$(E) \cdots f(x) = g(x)u(x) + \lambda \int_a^b K(x,t)u(t)dt,$$

*where $\lambda \neq 0$ and $f(x)$, $g(x)$ and $K(x,t)$ are all given functions.*

(1) *We say that (E) is of the **first kind** if $g(x) = 0$; otherwise we say that (E) is of the **second kind**.*

(2) *We say that (E) is a **Fredholm** integral equation if $a, b$ are both constants; otherwise (that is, if at least one of "a" or "b" is a function of $x$) we say that (E) is a **Volterra** integral equation.*

(3) *The function $K$ of two variables is called the **kernel**.*

There are many methods for solving Fredholm or Volterra integral equations (the interested reader may wish to consult [30], [31] and the highly applied book [11] for further reading). In this subsection, we only focus our attention on the method of the fixed point.

We start with the Fredholm integral equation.

**THEOREM** 7.1.41. *(see Exercise 7.3.39) Let $C([a,b], \mathbb{R})$ be endowed with the supremum metric. Let*

$$u(x) = g(x) + \lambda \int_a^b K(x,t)u(t)dt$$

*where $K : [a,b] \times [a,b] \to \mathbb{R}$ and $g : [a,b] \to \mathbb{R}$ are given continuous functions, and $\lambda$ is real. Then the previous equation admits a unique solution $u \in C([a,b], \mathbb{R})$ provided that:*

$$|\lambda| < \frac{1}{M(b-a)}$$

*where $M = \sup_{a \leq x, t \leq b} |K(x,t)|$.*

**REMARK**. The way of finding the solution using the method of successive approximations is to pick any $u_0$ in $X$, then find the successive values of $u_1, u_2, \cdots, u_n, \cdots$ using

$$u_{n+1}(x) = g(x) + \lambda \int_a^b K(x,t)u_n(t)dt$$

(remembering that $|\lambda| < \frac{1}{M(b-a)}$).

We now turn to Volterra integral equations (notice that this is also a consequence of Proposition 7.1.39).

**THEOREM 7.1.42.** *Let $X := C([a,b], \mathbb{R})$ be endowed with the supremum metric, denoted by $d_\infty$. Consider the Volterra equation of the second type:*

$$u(x) = f(x) + \lambda \int_a^x K(x,t)u(t)dt...(E)$$

*where $x \in [a,b]$. Then (E) has a unique solution $u$ in $X$.*

**REMARK.** Unlike Fredholm's equations, Volterra's have solutions for an arbitrary $\lambda$ not only the "small" ones. In the proof of the previous result (to be found in Exercise 7.3.40), we will define $T : X \to X$ by

$$T(u)(x) = f(x) + \lambda \int_a^x K(x,t)u(t)dt.$$

Then we show that $T^n$ is a contraction for large enough $n$ (and regardless of the value of $\lambda$). By Proposition 7.1.39, $T$ will have a unique fixed point (for any $\lambda$).

Practically, to reach a desired large enough $n$, notice that the iteration $u_{n+1} = T(u_n)$ gives

$$u_{n+1} = T(u_n) = T(T(u_{n-1})) = T^2(u_{n-1}) = \cdots = T^n(u_1).$$

Hence to obtain the good enough $n$, it suffices to do as many iterations as needed.

## 7.2. True or False: Questions

QUESTIONS. Comment on the following questions/statements and indicate those which are false and those which are true if it applies. Justify your answers.

(1) Let $(X, d)$ be a metric space and let $A, B \subset X$. Then
$$A = B \Longleftrightarrow d(A) = d(B).$$

(2) Let $(X, d)$ be a metric space. It is known that $(x_n)$ is a Cauchy sequence if
$$\forall \varepsilon > 0, \exists N \in \mathbb{N}, \forall n, m \in \mathbb{N} \ (n, m \geq N \Longrightarrow d(x_n, x_m) < \varepsilon).$$

We can take $\varepsilon \geq 0$ and $d(x_n, x_m) \leq \varepsilon$ in the previous definition.

(3) How do we show that a given metric space is complete?

(4) How do we show that a given metric space is *not* complete?

(5) If a given space is complete, then what is the best way to exploit this hypothesis (as regards sequences)?

(6) Why is it important to consider complete spaces?

(7) $(0, \infty)$ is not complete.

(8) In a complete metric space, the intersection of two dense sets is never empty.

(9) The union of two non-complete sets can be complete.

(10) Let $(X, d)$ be a metric space and let $(x_n)$ be a sequence in $(X, d)$. Then
$$(x_n) \text{ is Cauchy} \Longleftrightarrow (x_n) \text{ is bounded}.$$

(11) Let $(X, d)$ and $(X, d')$ be two metric spaces. Assume that $d$ and $d'$ are equivalent. Let $(x_n)$ be a sequence in $X$. Then is it true that $(x_n)$ is Cauchy with respect to $d$ if and only if $(x_n)$ is Cauchy with respect to $d'$?

(12) Let $(X, d)$ and $(X, d')$ be two metric spaces. Assume that $d$ and $d'$ are not equivalent. Let $(x_n)$ be a sequence in $X$. If $(x_n)$ is Cauchy with respect to $d$, then can it be so with respect to $d'$ and vice versa?

(13) Let $X = C([0, 1], \mathbb{R})$ endowed with the *metric*
$$d(f, g) = \int_0^1 |f(x) - g(x)| dx.$$

Let
$$f_n(x) = \begin{cases} 0, & 0 \leq x \leq \frac{1}{2} - \frac{1}{n}, \\ nx + (1 - \frac{1}{2}n), & \frac{1}{2} - \frac{1}{n} \leq x \leq \frac{1}{2}, \\ 1, & \frac{1}{2} \leq x \leq 1. \end{cases}$$

This sequence has as its pointwise limit the function

$$f(x) = \begin{cases} 0, & 0 \leq x < \frac{1}{2}, \\ 1, & \frac{1}{2} \leq x \leq 1. \end{cases}$$

Criticize the following reasoning: The sequence $(f_n)$ is a Cauchy sequence of continuous functions. Since $f$ is not continuous, $(X, d)$ is not complete.

(14) Keeping everything as in the previous question, say whether the following reasoning is correct: The sequence $(f_n)$ is a Cauchy sequence of continuous functions. Since $f$ is not continuous and $d(f_n, f) \to 0$, $(X, d)$ is not complete.

(15) Cauchyness is a topological property.

(16) Cauchyness is conserved by uniform continuity.

(17) Completeness is preserved by uniform continuity.

(18) Completeness is a topological property.

(19) Let $X$ and $Y$ be two metric spaces and let $A \subset X$ be dense in $X$. Let $f : A \to Y$ be a continuous function. Then $f$ has a continuous extension $\tilde{f} : X \to Y$ such that $\tilde{f}_{|A} = f$.

(20) Let $(X, d)$ be a metric space. State some properties that illustrate the analogy which exists between compact and complete spaces.

(21) Let $(X, d)$ be a metric space. Assume that $f : (X, d) \to (X, d)$ is a contraction. If $d'$ is another metric on $X$, equivalent to $d$, then $f : (X, d') \to (X, d')$ too is a contraction? What if $(X, d)$ is complete?

## 7.3. Exercises With Solutions

**Exercise 7.3.1.** Let $(X, d)$ be a metric space and $A, B \subset X$. Show that:

(1) $d(A) = 0 \iff A$ reduces to a singleton or $A = \varnothing$.

(2) $A \subset B \implies d(A) \leq d(B)$.

(3) $d(A) = d(\overline{A})$.

**Exercise 7.3.2.** Among the following sets indicate those which are complete (the metric is the usual one)

(1) $\mathbb{Q}$;

(2) $\mathbb{R} \setminus \mathbb{Q}$;

(3) $\mathbb{Q} \cap [3, 4]$;

(4) $(0, +\infty)$;

(5) $\{n : n \geq 1\}$;

(6) $\{(-1)^n : n \geq 1\}$;

(7) $\{\frac{1}{n} : n \in \mathbb{N}\} \cup \{0\}$;

(8) $\{(x, y) \in \mathbb{R}^2 : x > 1, y \geq \frac{1}{x-1}\}$?

**Exercise 7.3.3.** Show that a discrete metric space is always complete.

**Exercise 7.3.4.** Let $(X, d)$ be a metric space. Let $(x_n)$ be a Cauchy sequence in $X$. Show that if $(x_n)$ has a subsequence converging to $x \in X$, then $(x_n)$ converges to $x \in X$ too.

**Exercise 7.3.5.** Show that $(\mathbb{R}, |\cdot|_{\mathbb{R}})$ is complete.

**Exercise 7.3.6.** Show that $(\mathbb{C}, |\cdot|_{\mathbb{C}})$ is complete.

**Exercise 7.3.7.**

(1) Show that every compact metric space is complete.
(2) Is the converse always true?
(3) Deduce from the first question another proof of the completeness of usual $\mathbb{R}$.
(4) Can we say that every complete space is locally compact and vice versa?

**Exercise 7.3.8.** Define the following *metric* on $\mathbb{R}$

$$(x, y) \mapsto d(x, y) = \left| \frac{x}{1 + |x|} - \frac{y}{1 + |y|} \right|.$$

Show that $(\mathbb{R}, d)$ is not complete (hint: you may use the sequence defined by $x_n = n$, $n \in \mathbb{N}$).

**Exercise 7.3.9.** Show that $(\mathbb{N}, d)$, the metric space defined in Exercise 2.3.7, is complete.

**Exercise 7.3.10.** Show that $(\ell^2, d_2)$ is complete (with the "$d_2$" defined in Exercise 2.3.11).

**Exercise 7.3.11.** Let $1 \leq p < \infty$.

(1) Show that $(c_0, d_\infty)$ is a complete space.
(2) Show that $\ell^p \neq c_0$.
(3) Show that $\ell^p \subset c_0$.
(4) Prove that $\ell^p$ is dense in $(c_0, d_\infty)$.
(5) Show that $c_{00}$ is dense in $\ell^p$.
(6) Is $c_{00}$ dense in $\ell^\infty$? If not, what is its closure in $\ell^\infty$?

**Exercise 7.3.12.** In $(\ell^2, d_2)$ as defined in Exercise 7.3.10, show that

$$A = \{(x_n)_n \in \ell^2 : \exists N \in \mathbb{N}, x_n = 0, \forall n \geq N\}$$

is not complete ($A$ is just some $c_{00}$!).

**Exercise 7.3.13.** We endow $\mathbb{R}$ with following *metric*

$$d(x, y) = |e^x - e^y|.$$

Show that $\mathbb{R}$ is not complete with respect to $d$ (hint: you may use the sequence defined by $x_n = -n$, $n \in \mathbb{N}$).

**Exercise 7.3.14.** Let $X = C([0, 1], \mathbb{R})$. We equip it with the *metric* defined, for all $f, g \in X$, by

$$d(f, g) = \int_0^1 |f(x) - g(x)| dx.$$

Consider the sequence of *continuous* functions

$$f_n(x) = \begin{cases} 0, & 0 \le x \le \frac{1}{2} - \frac{1}{n} \\ nx + (1 - \frac{1}{2}n), & \frac{1}{2} - \frac{1}{n} \le x \le \frac{1}{2} \\ 1, & \frac{1}{2} \le x \le 1. \end{cases}$$

(1) Show that $(f_n)$ is a Cauchy sequence.
(2) Is $(X, d)$ complete?

**Exercise 7.3.15.** Let $X = C([0, 1])$. Show that $(X, d_\infty)$ is complete where $d_\infty$ is the supremum metric, i.e. the metric defined for all $f, g \in X$ by

$$d_\infty(f, g) = \sup_{x \in [0,1]} |f(x) - g(x)|$$

(hint: you need the fact that the uniform limit of a sequence of continuous functions is continuous. This should be known to the reader, if not, then this can be found in the next chapter).

**Exercise 7.3.16.** Let $X$ be the space of bounded real-valued functions on $[0, 1]$. Endow $X$ with the supremum metric which we denote by $d$. Show that $(X, d)$ is a complete metric space.

**Exercise 7.3.17.** Let $X$ be the set of all polynomials (of any degree) defined on $[0, 1]$ and let $d$ be the metric defined by

$$d(f, g) = \sup_{x \in [0,1]} |f(x) - g(x)|, \ f, g \in X.$$

(1) Show that the function $x \mapsto e^x$ is not a polynomial.
(2) Using the sequence $P_n(x) = \left(1 + \frac{x}{n}\right)^n$, $n \ge 1$, show that $(X, d)$ is not complete.
(3) Give another proof of the non-completeness of $(X, d)$ using the Weierstrass Theorem.

**Exercise 7.3.18.** Set $X = C^1([0,1], \mathbb{R})$ and endow it with the supremum metric defined by

$$d_\infty(f, g) = \sup_{0 \leq x \leq 1} |f(x) - g(x)|.$$

Is $(X, d_\infty)$ complete?

**Exercise 7.3.19.** Let $A = (0, \infty)$. Let $d$ be the usual metric and for any $x, y \in A$, define the function (cf. Exercise 2.5.12)

$$d'(x, y) = |\ln x - \ln y|.$$

(1) Show that $d'$ is a metric on $A$.
(2) Why is $(A, d)$ not complete?
(3) Show that $(A, d')$ is complete.
(4) Show that $(A, d)$ is homeomorphic to $(A, d')$.
(5) What can you deduce from the previous question? (cf. Exercise 2.5.12).

**Exercise 7.3.20.** Let $(X, d)$ be a metric space and let $A \subset X$.

(1) Show that if $A$ is complete, then it is closed in $X$.
(2) Show that if $A$ is closed and $X$ is complete, then $A$ is complete.

**Exercise 7.3.21.**

(1) Show that the finite union of complete metric spaces is complete.
(2) Need the previous result remain true for an arbitrary union?
(3) Show that the arbitrary intersection of complete spaces is complete.

**Exercise 7.3.22.** Let $(X, d)$ and $(Y, d')$ be two metric spaces. Endow the product space $Z = X \times Y$ with the *metric* $\delta$ defined by

$$\delta((x, y), (x', y')) = d(x, x') + d'(y, y').$$

Show that $(Z, \delta)$ is complete iff both $(X, d)$ and $(Y, d')$ are complete.

**Exercise 7.3.23.** Let $I$ be an interval and let $f : I \to \mathbb{R}$ be a continuous function which is also differentiable on $\overset{\circ}{I}$. Show that if $k \geq 0$, then $f$ satisfies the Lipschitz condition iff for all $x \in \overset{\circ}{I}$, we have $|f'(x)| \leq k$.

**Exercise 7.3.24.** Let $(X, d)$ be a complete metric space. Let $f : (X, d) \to (X, d)$ be a contracting mapping, i.e. it satisfies

$$d(f(x), f(y)) \leq k d(x, y)$$

for some $k \in [0, 1)$ and all $x, y \in X$.
Show that $f$ has a unique fixed point.

**Exercise 7.3.25.** Show, using the Fixed Point Theorem, that the equation

$$x^4 + 16x^3 - 32x + 8 = 0$$

has exactly one root in $[0, \frac{1}{2}]$.

**Exercise 7.3.26.** Let $X$ be a metric space.

(1) Let $X$ be a complete metric space. Let $f$ be a mapping defined from $X$ into $X$. We assume that for some integer $n \geq 1$, $f^{(n)}$ (the iterated map!) is a contraction. Show that $f$ has a unique fixed point.

(2) (a) Show that $\cos : \mathbb{R} \to \mathbb{R}$ is not a contraction.
   (b) Show that $x \mapsto \cos^{(2)} x$ is a contraction.
   (c) Deduce an approximate solution of the equation

$$\cos x = x.$$

**Exercise 7.3.27.** Let $X = \{x \in \mathbb{Q} : x \geq 1\}$ in the usual metric. Define a function on $X$ into $X$ by

$$f(x) = \frac{x}{2} + \frac{1}{x}.$$

(1) Show that

$$\forall x, y \in X : |f(x) - f(y)| \leq \frac{1}{2}|x - y|.$$

(2) Show that $f$ does not have a fixed point, that is,

$$\not\exists x \in X : f(x) = x.$$

(3) Why does not the result of the previous question contradict the Fixed Point Theorem?

**Exercise 7.3.28.** (cf. Exercise 5.3.36) Define a function $f$ in $X = [1, \infty)$ (with the usual metric) into the same interval by

$$f(x) = x + \frac{1}{x}.$$

(1) Show that

$$\forall x, y \geq 1 \ (x \neq y) : |f(x) - f(y)| < |x - y|.$$

(2) Show that $f$ does not have a fixed point, that is,

$$\forall x \in X : f(x) \neq x.$$

(3) Why does the result of the previous question not contradict the Fixed Point Theorem? Say why it does not contradict the result of Exercise 5.3.36 either.

**Exercise 7.3.29.** Let $(X, d)$ and $(X', d')$ be two isometric metric spaces. Show that $(X, d)$ is complete if and only if $(X', d')$ is complete.

**Exercise 7.3.30.** Let $a, b \in \mathbb{R}$ with $a \leq b$. Show that $[a, b]$ is compact using total boundedness.

**Exercise 7.3.31.** Let $(X, d)$ be a metric space. Show that $(X, d)$ is complete iff for each sequence $(A_n)$ of non-empty and closed subsets of $X$ verifying:

(1) $A_{n+1} \subset A_n$ for all $n$;
(2) $d(A_n) \to 0$ as $n$ goes to $\infty$;

we have

$$\bigcap_{n=1}^{\infty} A_n \neq \varnothing.$$

**Exercise 7.3.32.** Go back to Theorem 7.1.27 (the previous exercise!) and give examples showing that the hypotheses on the right hand side of the equivalence may not merely be dropped.

**Exercise 7.3.33.** Let $(X, d)$ be a complete metric space. Show that the countable intersection of open dense subsets of $X$ is dense in $X$.

**Exercise 7.3.34.** Let $(X, d)$ be a complete metric space. Show that if $(V_n)_n$ is a sequence of closed subsets of $X$ such that $\bigcup_{n=1}^{\infty} V_n = X$, then

$$\overline{\bigcup_{n=1}^{\infty} \overset{\circ}{V_n}} = X.$$

**Exercise 7.3.35.** Show that $\mathbb{R}$ is not countable (hint: use Corollary 7.1.30).

**Exercise 7.3.36.** Let $C$ be a non-empty closed and countable subset of usual $\mathbb{R}$. Show that $C$ has at least one isolated point.

**Exercise 7.3.37.** Assume that $(X, d)$ and $(Y, d')$ are two metric spaces. Assume also that $(X, d)$ is complete. Let $A \subset X$ be closed and let $f : A \to Y$ be a continuous function satisfying:

$$d'(f(x), f(x')) \geq d(x, x'), \ \forall x, x' \in A.$$

Prove that $f(A)$ is closed in $(Y, d')$.

**Exercise 7.3.38.** Consider the initial value problem:

$$\begin{cases} dy/dx = 2x(1 + y), \\ \quad\quad y(0) = 0. \end{cases}$$

Assume that the previous problem admits a unique solution (which it does under some conditions. See e.g. [**3**]). Use the method of successive approximations to find this solution.

**Exercise 7.3.39.** Let $X := C([a,b], \mathbb{R})$ be endowed with the supremum metric, denoted by $d_\infty$. Consider the Fredholm integral equation of the second type:

$$u(x) = g(x) + \lambda \int_a^b K(x,t)u(t)dt \cdots (E)$$

where $K : [a,b] \times [a,b] \to \mathbb{R}$ and $g : [a,b] \to \mathbb{R}$ are given continuous functions, and $\lambda$ is real. Define $T : X \to X$ by

$$T(u)(x) = g(x) + \lambda \int_a^b K(x,t)u(t)dt.$$

(1) Say briefly why $T$ is well defined.
(2) Let $M = \sup_{a \le x, t \le b} |K(x,t)|$.
    (a) Can $M$ be infinite?
    (b) Show that $T$ is a contraction whenever

$$|\lambda| < \frac{1}{M(b-a)}.$$

(3) Infer that $(E)$ has a unique solution $u \in C([a,b], \mathbb{R})$.
(4) Apply the previous result to solve the integral equation:

$$u(x) = \frac{5x}{6} + \frac{1}{2} \int_0^1 xtu(t)dt...(E')$$

**Exercise 7.3.40.** Let $X := C([a,b], \mathbb{R})$ be endowed with the supremum metric, denoted by $d_\infty$. Consider the Volterra integral equation of the second type:

$$u(x) = f(x) + \lambda \int_a^x K(x,t)u(t)dt...(E)$$

where $x \in [a,b]$.
    Set

$$T(u) = f(x) + \lambda \int_a^x K(x,t)u(t)dt.$$

Then as in Exercise 7.3.39, $T$ maps $X$ into $X$.

(1) Prove that $T^n$ is a contraction for a sufficiently large $n$.
(2) Deduce that $(E)$ has a unique solution.

(3) Using the method of successive approximations, and applying the previous result, find the only continuous solution (in
$(X, d_\infty)$) of:

$$u(x) = x - \int_0^x (x - t)u(t)dt...(E').$$

**Exercise 7.3.41.** Show that the (non-linear) system
$$\begin{cases} x_1 = \frac{1}{5}(2\sin x_1 + \cos x_2), \\ x_2 = \frac{1}{5}(\cos x_1 + 3\sin x_2) \end{cases}$$
has a unique solution $(x_1, x_2)$ in $\mathbb{R}^2$ (with respect to the metric $d_\infty$).

## 7.4. Tests

**Test 55.** Let $X$ be a metric space and let $A \subsetneq X$ be dense in $X$.
Must $A$ be complete?

**Test 56.** Show that $\mathbb{R}$ endowed with the "arctan metric" already
defined in Exercise 2.3.29 is not complete.

**Test 57.** Show that $[0, 1)$ is not countable.

**Test 58.** Show that the Cantor set is uncountable.

**Test 59.** On $\ell^2$, let
$$e_1 = (1, 0, 0, \cdots), \ e_2 = (0, 1, 0, \cdots), ...$$

(1) Show that $(e_n)_n$ is not a Cauchy sequence in $(\ell^2, d_2)$ and neither is any subsequence of it.
(2) What can deduce from the previous question?

## 7.5. More Exercises

**Exercise 7.5.1.** Set
$$x_n = 1 + \frac{1}{2} + \cdots + \frac{1}{n} = \sum_{k=1}^n \frac{1}{k}.$$

(1) Is $(x_n)$ Cauchy in $\mathbb{R}$?
(2) Does the series $\sum_{n \geq 1} \frac{1}{n}$ converge?

**Exercise 7.5.2.** Is the Cantor set complete?

**Exercise 7.5.3.** Let $(X, d)$ be a *ultrametric* space (see Exercise
2.3.22). Let $(x_n)$ be a sequence $(x_n)$ in $X$. Show that
$$(x_n) \text{ is Cauchy} \iff \lim_{n \to \infty} d(x_n, x_{n+1}) = 0.$$

**Exercise 7.5.4.** For all $(x, y) \in \mathbb{R}^2$, define the *metric d* by

$$d(x, y) = |x^3 - y^3|.$$

(1) Show that $\mathbb{R}$ is complete with respect to $d$.
(2) Answer the previous question with the metric $d$ defined by

$$d(x, y) = \left| \frac{1}{x} - \frac{1}{y} \right|$$

and $\mathbb{R}^*$ in lieu of $\mathbb{R}$.

**Exercise 7.5.5.** Let $X = C([-1, 1], \mathbb{R})$. We endow $X$ with the *metric d* defined for all $f, g \in X$ by

$$d(f, g) = \left( \int_{-1}^{1} |f(x) - g(x)|^2 dx \right)^{\frac{1}{2}}$$

and let

$$f_n(x) = \begin{cases} -1, & -1 \leq x \leq -\frac{1}{n}, \\ nx, & -\frac{1}{n} \leq x \leq \frac{1}{n}, \\ 1, & \frac{1}{n} \leq x \leq 1. \end{cases}$$

(1) Show that $(f_n)$ is a Cauchy sequence in $(X, d)$.
(2) Is $(X, d)$ complete?

**Exercise 7.5.6.** We endow $X = [0, 1)$ with the following *metric*

$$d(x, y) = \left| \frac{1}{1 - x} - \frac{1}{1 - y} \right|.$$

Is $(X, d)$ complete?

**Exercise 7.5.7.** Let $f : (\mathbb{N}, d) \to (\mathbb{N}, d)$ be a function defined for all $x \in \mathbb{N}$ by $f(x) = x + 1$, where $d$ is the metric defined in Exercise 2.3.7 (cf. Exercise 7.3.9). Without calculation, can $f$ be a contraction? Why?

**Exercise 7.5.8.** Show that the image of a Cauchy sequence by a uniform continuous function remains Cauchy.

**Exercise 7.5.9.** Let $f : \mathbb{R}^2 \to \mathbb{R}^2$ be defined by

$$f(x, y) = \left( \frac{1}{2} \cos y, \frac{1}{2} \sin x + 1 \right)$$

where $\mathbb{R}^2$ is equipped with the usual metric. Show that $f$ has a unique fixed point.

**Exercise** 7.5.10. Using the Fixed Point Theorem, show that the equation

$$4x^5 - 2x^2 - 4x + 1 = 0$$

has a unique root in $[0, 1/2]$. Now using the method of successive approximations, find this root up to three decimal places.

**Exercise** 7.5.11. Define $f : \mathbb{R} \to \mathbb{R}$ (usual $\mathbb{R}$!) by $f(x) = e^{-x}$. Show that $f$ is not a contraction yet it has a unique fixed point (hint: consider $f^{(2)} = f \circ f$).

**Exercise** 7.5.12. Let $(f_n)$ the sequence defined in Exercise 4.3.34. Is $\{f_n : n \in \mathbb{N}\}$ totally bounded?

**Exercise** 7.5.13. Let $(X, d)$ be a metric space. Recall the following important result (see, for instance, [**29**]):

**THEOREM** 7.5.1. *(Hausdorff) There exist a complete metric space* $(\tilde{X}, \tilde{d})$ *and an isometry* $f : X \to \tilde{X}$ *such that* $f(X)$ *is dense in* $\tilde{X}$.

The space $(\tilde{X}, \tilde{d})$ is usually called a **completion** of $(X, d)$.

(1) Let $A \subset X$. If $X$ is complete, then what is the completion of $A$?
(2) What is the completion of $(-1, 1)$ and that of $\mathbb{Q}$ in usual $\mathbb{R}$?
(3) Explain why the completion of the space of polynomials on $[0, 1]$ with respect to the supremum metric is the space of continuous functions on $[0, 1]$ with respect to the supremum metric.

**Exercise** 7.5.14. ([**6**]) Use the method of successive approximations to solve:

$$u(x) = e^x - \frac{x}{3} + \frac{1}{3} \int_0^1 xtu(t)dt.$$

**Exercise** 7.5.15. Use the method of successive approximations to solve:

$$u(x) = x + \int_0^x u(t)dt.$$

**Exercise** 7.5.16. Let $X := C([a, b], \mathbb{R})$ be endowed with the supremum metric. Consider the **nonlinear Fredholm integral equation of the second kind**:

$$u(x) = g(x) + \lambda \int_a^b K(x, t; u(t))dt \cdots (E)$$

where the kernel $K$ and $g$ are given continuous functions (on e.g. $[a,b]^2 \times [c,d]$ and $[a,b]$ respectively), and $\lambda$ is real. Assume that $K$ satisfies

$$|K(x,y;z) - K(x,y;z')| \leq k|z-z'|,$$

i.e. $K$ satisfies a Lipschitz condition with respect to $z$. Show that if $M = \sup_{x,y,z} |K(x,y;z)|$, then $(E)$ has a unique solution in $X$ provided that:

$$|\lambda| < \frac{1}{M(b-a)}.$$

# CHAPTER 8

# Function Spaces

## 8.1. Essential Background

### 8.1.1. Types of Convergence.

**DEFINITION 8.1.1.** *Let $(X, d)$ and $(Y, d')$ be two metric spaces. Let $(f_n)$ be a sequence of functions defined from $X$ into $Y$. We say that $(f_n)$ converges **pointwise** to $f : X \to Y$ if:*

$$\lim_{n \to \infty} f_n(x) = f(x) \quad \text{(for each } x \in \mathbb{R}\text{)}.$$

**DEFINITION 8.1.2.** *Let $X$ be a topological space and let $(Y, d')$ be a metric space. Let $(f_n)$ be a sequence of functions defined from $X$ into $Y$. We say that $(f_n)$ converges **uniformly** to $f : X \to Y$ if:*

$$\forall \varepsilon > 0, \exists N \in \mathbb{N}, \forall x \in X, \forall n \in N : (n \geq N \implies d'(f_n(x), f(x)) < \varepsilon).$$

**REMARK.** If $f_n$ is real-valued, defined on $A \subset \mathbb{R}$ and converges pointwise to $f$, then saying that $(f_n)$ converges to $f$ uniformly amounts to saying that

$$\lim_{n \to \infty} \sup_{x \in A} |f_n(x) - f(x)| = 0.$$

**EXAMPLES 8.1.3.**

(1) *Let $f_n(x) = x^n$ where $x \in [0, 1]$. Then $(f_n)$ converges pointwise to the function $f$ defined by*

$$f(x) = \begin{cases} 0, & 0 \leq x < 1, \\ 1, & x = 1. \end{cases}$$

(2) *The same sequence does not converge uniformly to $f$ for:*

$$\lim_{n \to \infty} \sup_{0 \leq x \leq 1} |f_n(x) - f(x)| \geq \lim_{n \to \infty} \sup_{0 \leq x < 1} |x^n - 0| = 1$$

*and hence $\lim_{n \to \infty} \sup_{0 \leq x \leq 1} |f_n(x) - f(x)| \neq 0$.*

Here are some applications of uniform convergence. We content ourselves to results of sequences of functions defined on a subset of $\mathbb{R}$.

**THEOREM 8.1.4.** *Let $(f_n)$ be a sequence of real-valued functions defined on $[a, b]$. Assume that $(f_n)$ converges **uniformly** on $[a, b]$ to a function $f$. Then:*

153

(1) $(f_n)$ converges pointwise to $f$.

(2) If all $f_n$ are continuous on $[a, b]$, then $f$ too is continuous on $[a, b]$.

(3) If all $f_n$ are Riemann-integrable on $[a, b]$, then $f$ is Riemann-integrable on $[a, b]$, and

$$\lim_{n \to \infty} \int_a^b f_n(x) dx = \int_a^b f(x) dx.$$

An akin result on differentiability needs a little more care.

**THEOREM** 8.1.5. *Let $(f_n)$ be a sequence of differentiable real-valued functions defined on $[a, b]$ such that $(f_n(x_0))$ converges for some $x_0 \in [a, b]$. Assume that the sequence of derivatives $(f_n')$ converges **uniformly** on $[a, b]$. Then $(f_n)$ converges uniformly on $[a, b]$ to a function $f$. Besides,*

$$\lim_{n \to \infty} f_n'(x) = f'(x).$$

Pointwise convergence may imply uniform convergence under stronger hypotheses.

**THEOREM** 8.1.6 (**Dini**). *Let $(f_n)$ be an **increasing** sequence of real-valued continuous functions defined on (the compact!) $[a, b]$. Assume that $(f_n)$ converges pointwise on $[a, b]$ to a continuous function $f$. Then $(f_n)$ converges uniformly to $f$ on $[a, b]$.*

**REMARK.** By $(f_n)$ being increasing, we mean increasing with respect to $n$.

**REMARK.** The same conclusions of the Dini Theorem hold:

(1) if "increasing" is replaced by "decreasing";

(2) or if $[a, b]$ is replaced by another compact set.

**DEFINITION** 8.1.7. *Let $X$ be a topological space and let $(Y, d)$ be a metric space. Let $(f_n)$ be a sequence of functions from $X$ into $(Y, d)$. We say that $(f_n)$ **converges uniformly on compacta** (or **compactly**) to $f : X \to (Y, d)$ if:*

$$\forall A \text{ compact } \subset X, \forall \varepsilon > 0, \exists N \in \mathbb{N},$$
$$\forall n \in \mathbb{N} : (n \geq N \implies d(f_n(x), f(x)) < \varepsilon) \text{ for all } x \in A.$$

**REMARK.** The "$N$" in the foregoing definition may depend on both $A$ and $\varepsilon$.

**THEOREM** 8.1.8. *Let $X$ be a topological space and let $(Y, d)$ be a metric space. Let $(f_n)$ be a sequence of functions from $X$ into $(Y, d)$.*

(1) *If* $(f_n)$ *converges uniformly to* $f$, *then* $(f_n)$ *converges uniformly on compacta to* $f$.

(2) *If* $(f_n)$ *converges uniformly on compacta to* $f$, *then* $(f_n)$ *converges pointwise to* $f$.

(3) *Let* $X$ *be compact. If* $(f_n)$ *converges on compacta to* $f$, *then* $(f_n)$ *converges uniformly to* $f$.

**8.1.2. Weierstrass Approximation Theorem.** The Weierstrass theorem has a wide range of applications in different areas of mathematics. We have already given in the previous chapter some of its uses. We *rephrase* it here for convenience.

**THEOREM 8.1.9 (Weierstrass).** *The subspace of real-valued polynomials defined on* $[a, b]$ *is dense in the space of real-valued continuous functions defined on* $[a, b]$, *with respect to the supremum metric.*

**REMARK.** In other words, the Weierstrass Theorem tells us that any real-valued continuous function (on $[a, b]$) may be approximated uniformly by a sequence of real-valued polynomials (on $[a, b]$).

**REMARK.** There exist generalizations of the Weierstrass Theorem. The most famous one is the Stone-Weierstrass Theorem. For readers' convenience we shall state here, but without further use in this book. First we need to introduce two notions.

**DEFINITION 8.1.10.** *Let* $X$ *be a compact space equipped with the uniform metric. The subset* $A \subset C(X, \mathbb{R})$ *is called* **separating** *if*

$$\forall x, y \in X, x \neq y, \exists f \in A : f(x) \neq f(y).$$

**DEFINITION 8.1.11.** *A vector subspace* $A$ *of* $C(X, \mathbb{R})$ *is called a* **subalgebra** *if for all* $f, g \in A$: $fg$ *(the pointwise product) remains in* $A$.

**THEOREM 8.1.12 (Stone-Weierstrass).** *Let* $X$ *be a compact space equipped with the uniform metric. Then every separating subalgebra of* $C(X, \mathbb{R})$ *which contains the constant functions is dense in* $C(X, \mathbb{R})$.

**8.1.3. The Arzelà-Ascoli Theorem.** The epilogue of this chapter is devoted to the Arzelà-Ascoli Theorem. But first we have the following definition:

**DEFINITION 8.1.13.** *A subset* $A \subset C[0, 1]$ *is said to be* **equicontinuous** *at a point "a" if:*

$$\forall \varepsilon > 0, \exists \alpha > 0, \forall f \in A, \forall x \in [0, 1] : (|x - a| < \alpha \implies |f(x) - f(a)| < \varepsilon).$$

*A subset* $A \subset C[0, 1]$ *is* **equicontinuous** *if it is equicontinuous at each point of* $[0, 1]$.

EXAMPLES 8.1.14.

(1) *A finite subset of $C[0,1]$ is equicontinuous.*

(2) *The set of real-valued functions $f$ defined on $[0,1]$ and satisfying the **Hölder condition**:*

$$\exists k, \alpha > 0, \forall x, y \in [0,1]: \ |f(x) - f(y)| \leq k|x - y|^{\alpha}$$

*is equicontinuous.*

In an infinite dimensional space, compact spaces are rare. For instance, the closed unit ball in $C[0,1]$ is not compact with respect to the supremum metric (see Exercise 5.3.29 or Exercise 5.5.10). The following theorem characterizes compact subsets of $C[0,1]$.

THEOREM 8.1.15 (**Arzelà-Ascoli**). *Let $A \subset C[0,1]$ be closed and bounded. Then $A$ is compact iff it is equicontinuous.*

REMARK. The Arzelà-Ascoli Theorem is tremendously important in mathematics: It has applications in Ordinary Differential Equations, Distribution Theory and Holomorphic Functions, among others.

## 8.2. True or False: Questions

QUESTIONS. Comment on the following questions/statements and indicate those which are false and those which are true if it applies. Justify your answers.

(1) Let $(f_n)$ be a monotonic sequence of continuous real-valued functions on $\mathbb{R}$. If $(f_n)$ converges pointwise to $f$ and $f$ is continuous, then $(f_n)$ converges uniformly to $f$.

(2) Let $(f_n)$ be a sequence of continuous real-valued functions having $f$ as its pointwise limit. If each function $f_n$ is increasing, then so will be $f$.

(3) Let $(f_n)$ be a sequence of bounded real-valued functions. Then its pointwise limit $f$ is bounded. What about the uniform limit?

(4) Let $(f_n)$ be a sequence of continuous real-valued functions which converges pointwise to $f$ on $[0, 1)$. If $(f_n)$ converges uniformly on $[0, a]$ for each $0 < a < 1$, then the convergence is uniform on $[0, 1)$.

(5) Let $(f_n)$ be a sequence of continuous real-valued functions which converges pointwise to $f$ on $(0, 1]$. If $f$ is continuous, then the convergence becomes uniform.

(6) Let $(f_n)$ be a sequence of continuous real-valued functions defined on $[a, b]$ which converges pointwise to $f$. Then

$$\lim_{n\to\infty} \int_a^b f_n(x)dx = \int_a^b \lim_{n\to\infty} f_n(x)dx \left( = \int_a^b f(x)dx \right).$$

(7) Let $(f_n)$ be a sequence of continuous and differentiable real-valued functions that converges pointwise to $f$. Then

$$\lim_{n\to\infty} f_n'(x) = f'(x).$$

## 8.3. Exercises With Solutions

**Exercise 8.3.1.** Let $(f_n)$ be the sequence of real-valued functions defined by $f_n(x) = \frac{\sin nx}{\sqrt{n}}$ for all $x \in [0, 1]$.

(1) What is the pointwise limit of $(f_n)$?

(2) Does $(f_n)$ converge uniformly to the pointwise limit? Why?

**Exercise 8.3.2.** Let $(f_n)$ be the sequence of real-valued functions defined by $f_n(x) = \frac{e^{nx}}{\sqrt{n}}$ for all $x \in [-1, 0]$.

(1) Show that $(f_n)$ converges pointwise to some function $f$.

(2) Show that the convergence is also uniform.

(3) Investigate the uniform convergence of $(f_n')$.

**Exercise 8.3.3.** Let
$$f_n(x) = \left(1 + \frac{x}{n}\right)^n, \quad x \in \mathbb{R}, n \in \mathbb{N}.$$
Using the Dini Theorem, show that $(f_n)$ converges uniformly to $e^x$ on $[0,1]$.

**Exercise 8.3.4.** Let $(P_n)$ be a sequence of polynomials on $[-1,1]$ defined by
$$\begin{cases} P_0 = 0, \\ P_{n+1}(x) = P_n(x) + \frac{1}{2}(x^2 - P_n^2(x)), \quad \forall n \geq 0. \end{cases}$$
Show that $(P_n)$ converges uniformly to $f(x) = |x|$.

**Exercise 8.3.5.** Let $(f_n(x))$ be a sequence of functions defined from $[0, \frac{1}{2})$ into $\mathbb{R}$ by $f_n(x) = x^n$. Investigate the convergence of $(f_n)$:
  (1) pointwise;
  (2) uniformly;
  (3) compactly.
The same questions for $(g_n)$ defined from $\mathbb{R}$ into $\mathbb{R}$ by: $g_n(x) = \frac{n+1}{n}x$.

**Exercise 8.3.6.** Let $(X, d)$ and $(Y, d')$ be two metric spaces and let $(f_n)$ be a sequence of equicontinuous functions from $X$ into $Y$. Set
$$V = \{x \in X : (f_n(x)) \text{ is Cauchy in } Y\}.$$
Show that $V$ is closed in $Y$.

**Exercise 8.3.7.** Let $k$ be a continuous real-valued function on $[0,1] \times [0,1]$. Define a (linear) map $K$ for all $f \in C([0,1])$ by
$$Kf(x) = \int_0^1 k(x, y) f(y) dy.$$
($K$ is usually called an **integral operator** with **kernel** $k$). Let $(f_n)$ be bounded in $(C([0,1]), d_\infty)$.
  (1) Show that $(Kf_n)$ is equicontinuous.
  (2) Using the Arzelà-Ascoli Theorem, deduce that $(Kf_n)$ is relatively compact in $C([0,1])$ (we then say that $K$ is a **compact operator**).

**Exercise 8.3.8.** Let $f : [a, b] \to \mathbb{R}$ be a continuous function. The **moments** of $f$ are the numbers
$$m(f) = \int_a^b x^n f(x) dx, \quad \text{for all } n = 1, 2, 3, \cdots$$
Show that if all the $m(f)$ vanish, then so does $f$ (hint: Use the Weierstrass Approximation Theorem).

**Exercise** 8.3.9. Endow $C([0,1],\mathbb{R})$ with the supremum distance. Show that $C([0,1],\mathbb{R})$ is separable (hint: show that the set of polynomials with rational coefficients is dense in $C([0,1],\mathbb{R})$).

## 8.4. Tests

**Test** 60. What is the pointwise limit (denoted by $f$) of the sequence of functions $(f_n)$ defined by

$$f_n(x) = 1 + x + \cdots + x^n$$

on $(-1,1)$? Does $(f_n)$ converge uniformly to $f$?

**Test** 61. Let $(f_n)$ be a sequence of real-valued functions defined on $[0,1]$ by

$$f_n(x) = (x(1-x))^n + x.$$

Find the pointwise limit and investigate the uniform convergence of $(f_n)$.

**Test** 62. Let $(P_n)$ be a polynomial sequence which converges uniformly to $f$. Is $f$ continuous?

**Test** 63. Need a finite union of equicontinuous sets be equicontinuous?

## 8.5. More Exercises

**Exercise** 8.5.1. Define a sequence of functions $(f_n)$ as follows

$$f_n(x) = f\left(x + \frac{1}{n^2}\right)$$

where $f : \mathbb{R} \to \mathbb{R}$ is a continuous function.
  (1) Verify that $(f_n)$ converges pointwise to $f$.
  (2) Show that the convergence is also uniform.
  (3) Deduce the value of $\lim_{n\to\infty} \int_0^1 f\left(x + \frac{1}{n^2}\right) dx.$

**Exercise** 8.5.2. The same questions as those of Exercise 8.3.5 for $(f_n)$ defined from $\mathbb{R}$ into $\mathbb{R}$ by:

$$f_n(x) = \begin{cases} \frac{1}{n}\sqrt{n^2 - x^2}, & |x| < n, \\ 0, & \text{otherwise.} \end{cases}$$

**Exercise** 8.5.3. Let $(X,d)$ and $(Y,d')$ be two metric spaces and let $(f_n)$ be a sequence of functions from $X$ into $Y$ assumed to be equicontinuous at $a \in X$. Let $(x_n)$ be a sequence in $X$ which converges to $a$. Show that if $(f_n(a))$ converges to $b$, then so does $(f_n(x_n))$.

**Exercise** 8.5.4. Let $f$ be a uniformly continuous real-valued function on $\mathbb{R}$. Let $a \in \mathbb{R}$ and set $f_a(x) = f(x - a)$. Show that $\{f_a : a \in \mathbb{R}\}$ is equicontinuous on $\mathbb{R}$.

**Exercise** 8.5.5.

(1) Let $X$ be a metric space and let $(f_n)$ be a sequence in $C(X)$. Show that if $(f_n)$ is an equicontinuous family at a point $a$, then for all sequences $(x_n)$ converging to $a$, $(f_n(x) - f_n(x_n))$ converges to 0.

(2) What about the converse? (hint: consider $f_n(x) = \sin nx$ and $x_n = a + \frac{\pi}{2n}$).

**Exercise** 8.5.6. Let $f_n(x) = \sin(\sqrt{4(n\pi)^2 + x})$ be defined for all $x \geq 0$.

(1) What is the pointwise limit of $(f_n)$?

(2) Show that $(f_n)$ is equicontinuous.

(3) Show that $(f_n)$ is not relatively compact in $C([0, \infty), \mathbb{R})$ equipped with the supremum metric.

(4) Why does the result of the previous question not contradict the Arzelà-Ascoli Theorem?

# Part 2

# Solutions

# CHAPTER 1

# General Notions: Sets, Functions et al

## 1.2. Solutions to Exercises

**SOLUTION** 1.2.1. We use a proof by contradiction. So assume that $\sqrt{2}$ is rational, that is,

$$\sqrt{2} = \frac{a}{b}, \ a, b \in \mathbb{N}.$$

We may also assume that $a/b$ is simplified to lowest terms, i.e. the greatest common divisor of $a$ and $b$ is 1. Hence

$$2 = \frac{a^2}{b^2} \text{ or } a^2 = 2b^2.$$

So, $a^2$ is even and since $a \in \mathbb{N}$, $a$ too is even. Write $a = 2c$ ($c \in \mathbb{N}$). Then, this gives

$$4c^2 = 2b^2 \text{ or } 2c^2 = b^2.$$

As before, $b$ too is even. Thus, we have ended up with the greatest common divisor of $a$ and $b$ being at least equal to 2 (and remaining even!). Accordingly, $\sqrt{2} \notin \mathbb{Q}$.

**SOLUTION** 1.2.2. Suppose that $\frac{\ln 2}{\ln 3}$ is rational, i.e.

$$\frac{\ln 2}{\ln 3} = \frac{p}{q}, \ p, q \in \mathbb{N}.$$

Hence

$$\frac{\ln 2}{\ln 3} = \frac{p}{q} \Longrightarrow q \ln 2 = p \ln 3 \Longrightarrow \ln 2^q = \ln 3^p \Longrightarrow 2^q = 3^p$$

which may occur only if $p = q = 0$. Thus, we arrived at a contradiction. Therefore $\frac{\ln 2}{\ln 3} \notin \mathbb{Q}$.

**SOLUTION** 1.2.3. The sequence $(x_n)_n$ is strictly increasing and the sequence $(y_n)_n$ is strictly decreasing and it is well-known that

$$\lim_{n \to \infty} x_n = \lim_{n \to \infty} y_n = e$$

and we have $x_n < e < y_n$. Now, assume that $e = a/b$ where $a \in \mathbb{N}$ and $b \in \mathbb{N}$. Since $x_n < e < y_n$ for all $n$, taking $n = b$ gives us

$$x_b < e = \frac{a}{b} < y_b, \text{ i.e. } x_b < e = \frac{a}{b} < x_b + \frac{1}{b!b}.$$

Hence we have been led to

$$b!bx_b < b!a < b!bx_b + 1.$$

But $b!bx_b$ is an integer which we denote by $p$. Hence we have

$$p < b!a < p + 1$$

where $p \in \mathbb{N}$. But this is impossible since $b!a \in \mathbb{N}$. Thus $e \in \mathbb{R} \setminus \mathbb{Q}$.

**REMARK**. In fact, $e$ is not even an algebraic number (see Exercise 1.3.10 for the definition and straightforward properties of algebraic numbers).

**SOLUTION** 1.2.4.

(1) "(1) $\Rightarrow$ (2)": Clearly $B \subset A \cup B$. Next, let $x \in A \cup B$, i.e. $x \in A$ or $x \in B$. Since by assumption $A \subset B$, we see that in either case we have $x \in B$. Therefore, $B \supset A \cup B$ and thus

$$A \cup B = B.$$

(2) "(2) $\Rightarrow$ (3)": Clearly, $A \cap B \subset A$. Now, let $x \in A$. We know that $A \subset A \cup B$ and since by hypothesis, $A \cup B = B$, it follows that $A \subset B$. Hence $x \in B$ and so $x \in A \cap B$. Accordingly

$$A = A \cap B.$$

(3) "(3) $\Rightarrow$ (1)": This easily follows from

$$A = A \cap B \subset B.$$

**SOLUTION** 1.2.5. All three inclusions are true. We can check this in a traditional way. Alternatively, we may exploit Exercise 1.2.4. To show that $A \cap C \subset B \cap C$, we use the third statement of Exercise 1.2.4, given that $A \subset B$ (which also means that $A \cap B = A$). We then have

$$(A \cap C) \cap (B \cap C) = A \cap B \cap C = A \cap C.$$

Thus,

$$A \cap C \subset B \cap C.$$

To show the other inclusion, we use (2) of Exercise 1.2.4 (remembering that $A \subset B$ is equivalent to $A \cup B = B$). We have

$$(A \cup C) \cup (B \cup C) = A \cup B \cup C = B \cup C.$$

Consequently,

$$A \cup C \subset B \cup C.$$

To see why the last inclusion holds, apply the first one (by recalling that $A \setminus C = A \cap C^c$ and changing $C$ by $C^c$!)...

**SOLUTION** 1.2.6. First, recall that

$$x \in A \cup B \Longleftrightarrow x \in A \text{ or } x \in B,$$

$$x \in A \cap B \Longleftrightarrow x \in A \text{ and } x \in B.$$

Hence

$$x \notin A \cup B \Longleftrightarrow x \notin A \text{ and } x \notin B,$$

and

$$x \notin A \cap B \Longleftrightarrow x \notin A \text{ or } x \notin B.$$

(1) Suppose that $A \subset B$ and let us show that $B^c \subset A^c$. So, let $x \in B^c$, then $x \notin B$. But $A \subset B$ and so $x \notin A$, i.e. $x \in A^c$.

A similar way may be used to establish the other implication. Alternatively, we may exploit what we have just shown. Indeed,

$$B^c \subset A^c \Longrightarrow (A^c)^c \subset (B^c)^c \Longrightarrow A \subset B,$$

as desired.

(2) To show the equality we may show that both inclusions

$$(A \cap B)^c \subset A^c \cup B^c \text{ and } (A \cap B)^c \supset A^c \cup B^c$$

hold.

  (a) $(A \cap B)^c \subset A^c \cup B^c$: Let $x \in (A \cap B)^c$. Then $x \notin A \cap B$ and so $x \notin A$ **or** $x \notin B$. Hence $x \in A^c$ **or** $x \in B^c$. Thus, $x \in A^c \cup B^c$.

  (b) $(A \cap B)^c \supset A^c \cup B^c$: Let $x \in A^c \cup B^c$. Then $x \in A^c$ **or** $x \in B^c$ and so $x \notin A$ **or** $x \notin B$. Hence $x \notin A \cap B$. Therefore, $x \in (A \cap B)^c$.

(3) We can show as before that the two inclusions "$(A \cup B)^c \subset A^c \cap B^c$" and "$(A \cup B)^c \supset A^c \cap B^c$" hold. We can also prove the equivalence directly (something we could have done in the previous question anyway).

$$x \in (A \cup B)^c \Longleftrightarrow x \notin A \cup B$$
$$\Longleftrightarrow x \notin A \text{ and } x \notin B$$
$$\Longleftrightarrow x \in A^c \text{ and } x \in B^c$$
$$\Longleftrightarrow x \in A^c \cap B^c.$$

(4) Assume that $A \subset B$ and let's show that $f(A) \subset f(B)$. Let $y \in f(A)$. By definition, there exists at least an $x$ in $A$ such that $y = f(x)$. Since, $A \subset B$, we get $y = f(x) \in f(B)$, and this finishes the proof.

(5) Suppose that $A \subset B$ and let's show that $f^{-1}(A) \subset f^{-1}(B)$. Let $x \in f^{-1}(A)$. This means that $f(x) \in A$. Whence, $f(x) \in B$ (as assumed). Consequently, $x \in f^{-1}(B)$.

**SOLUTION** 1.2.7. We have to show that

$$(A \cap B) \cap (A \setminus B) = \varnothing \text{ and } A = (A \cap B) \cup (A \setminus B).$$

We know that $A \setminus B = A \cap B^c$. Hence

$$(A \cap B) \cap (A \setminus B) = A \cap B \cap A \cap B^c = A \cap B \cap B^c = A \cap \varnothing = \varnothing.$$

Now,

$$(A \cap B) \cup (A \setminus B) = (A \cap B) \cup (A \cap B^c) = A \cap (B \cup B^c) = A \cap X = A,$$

as required.

**SOLUTION** 1.2.8.

(1) $\mathbb{Q}$ is not an interval for if it were, then we would have

$$\forall x, y \in \mathbb{Q}, \forall z \in \mathbb{R} : (x < z < y \implies z \in \mathbb{Q}).$$

But this fails if for instance, $x = 2$, $y = 3$ and $z = e$ because

$$2 < e < 3 \text{ and } e \notin \mathbb{Q}.$$

(2) A somehow similar argument for $\mathbb{R} \setminus \mathbb{Q}$ applies! Take for instance, $x = e$, $y = \pi$ and $z = 3 \notin \mathbb{R} \setminus \mathbb{Q}$.

(3) As for $\mathbb{R}^*$, take $x = -1$, $y = 1$ and $z = 0 \notin \mathbb{R}^*$.

(4) What would you do here?...

**SOLUTION** 1.2.9. Let $x \in A$. Then by assumption there is a subset $U_x$ containing $x$. Hence $x \in \bigcup_{a \in A} U_a$ and we have proved that $A \subset \bigcup_{a \in A} U_a$.

Conversely, let $x \in \bigcup_{a \in A} U_a$. Then for some $a$ in $A$, $x \in U_a$. Since $U_a \subset A$ (by hypothesis), $x \in A$. Therefore, $\bigcup_{a \in A} U_a \subset A$. Thus

$$A = \bigcup_{a \in A} U_a.$$

**SOLUTION** 1.2.10. We give two ways of showing this.

(1) We use a proof by induction. The statement is true for $n = 1$. Indeed, if $X$ is constituted of one element $X = \{x\}$, say, then evidently

$$\mathcal{P}(X) = \{\varnothing, X\}$$

and hence

$$\text{card} \mathcal{P}(X) = 2 = 2^1.$$

Assume now that a set $X$ of $n$ elements has a power set of $2^n$ elements, and let us show that a set of $n+1$ elements (call it $Y$) has a power set of $2^{n+1}$ elements.

Let $Y = \{x_1, x_2, \cdots, x_n, x_{n+1}\}$. It is clear that the power set of $Y$ is constituted of all elements of $\mathcal{P}(X)$, to which we add unions of each element of $\mathcal{P}(X)$ with $\{x_{n+1}\}$. By assumption, $\mathrm{card}\mathcal{P}(X) = 2^n$. This, combined with what we have just said, yields

$$\mathrm{card}\mathcal{P}(Y) = 2 \times 2^n = 2^{n+1},$$

as required.

(2) Assume that $\mathrm{card}X = n$. The number of subsets of $X$ having $k$ elements is $\binom{n}{k}$ where

$$\binom{n}{k} = \frac{n!}{k!(n-k)!}.$$

We also know that $\mathcal{P}(X)$ contains parts having: 0 elements (precisely one set), 1 element, 2 elements, $\cdots$, $n$ elements. Hence

$$\mathrm{card}\mathcal{P}(X) = \binom{n}{0} + \binom{n}{1} + \cdots + \binom{n}{n}$$

But this reminds of Newton Binomial Theorem of $(1+1)^n = 2^n$. Accordingly,

$$\mathrm{card}\mathcal{P}(X) = (1+1)^n = 2^n.$$

**SOLUTION 1.2.11.**

(1) We have

$$y \in f(\bigcup_{i \in I} A_i) \Longleftrightarrow \exists x \in \bigcup_{i \in I} A_i : \ y = f(x)$$
$$\Longleftrightarrow \exists i \in I, \ x \in A_i \text{ and } y = f(x)$$
$$\Longleftrightarrow \exists i \in I : \ y \in f(A_i)$$
$$\Longleftrightarrow y \in f(\bigcup_{i \in I} A_i).$$

(2) We have

$$\bigcap_{i \in I} A_i \subset A_j, \ \forall j \in I \Longrightarrow f(\bigcap_{i \in I} A_i) \subset f(A_j), \ \forall j \in I.$$

Thus

$$f(\bigcap_{i \in I} A_i) \subset \bigcap_{i \in I} f(A_i).$$

**REMARK.** Observe that the equality does not hold even for finite intersections. For instance, take the real-valued function $f$ defined on $\mathbb{R}$ by $f(x) = x^2$. Then consider for example $A = (-\infty, 0]$ and $B = [0, \infty)$ and one can see easily that

$$f(A \cap B) = f(\{0\}) = \{0\} \neq [0, \infty) = f(A) \cap f(B).$$

(3) We have

$$x \in f^{-1}\left(\bigcup_{i \in I} B_i\right) \iff f(x) \in \bigcup_{i \in I} B_i$$

$$\iff \exists i \in I, \ f(x) \in B_i$$

$$\iff \exists i \in I, \ x \in f^{-1}(B_i)$$

$$\iff x \in \bigcup_{i \in I} f^{-1}(B_i).$$

(4) The same idea of proof as in the previous question.

(5) We have

$$x \in f^{-1}(B^c) \iff f(x) \in B^c$$

$$\iff f(x) \notin B$$

$$\iff x \notin f^{-1}(B)$$

$$\iff x \in [f^{-1}(B)]^c.$$

(6) We have

$$x \in (g \circ f)^{-1}(C) \iff (g \circ f)(x) \in C$$

$$\iff g(f(x)) \in C$$

$$\iff f(x) \in g^{-1}(C)$$

$$\iff x \in f^{-1}(g^{-1}(C)).$$

(7) We have

$$y \in f(f^{-1}(B)) \implies \exists x \in f^{-1}(B): \ y = f(x) \implies y = f(x) \in B.$$

The reverse inclusion is not always true. For a counterexample, take $f : \{0, 1\} \to \{0, 1\}$ defined by

$$f(0) = f(1) = 0.$$

So if $B = \{0, 1\}$, then $f(B) = \{0\}$ and so

$$f^{-1}(B) = \{x \in \{0, 1\}: \ f(x) = 0 \in \{0, 1\}\} = B.$$

Hence

$$f(f^{-1}(B)) = \{0\} \neq B.$$

(8) A similar argument to that of the previous question may be applied. The reverse inclusion here does not hold either. Consider again $f : \{0,1\} \to \{0,1\}$ defined by

$$f(0) = f(1) = 0.$$

Then if we take $A = \{0\}$, then we get $f(A) = \{0\}$ and so

$$f^{-1}(f(A)) = f^{-1}(\{0\}) = \{0,1\} \neq A.$$

SOLUTION 1.2.12. Let $x \in A \cap f^{-1}(U)$. Then $x \in A$ and $x \in f^{-1}(U)$ or $f(x) \in U$. Since $x \in A$, we get $f_A(x) \in U$, i.e. $x \in f_A^{-1}(U)$.

Conversely, let $x \in f_A^{-1}(U)$ which, by definition, means that $x \in A$ and $f_A(x) \in U$. Hence $x \in A$ and $f(x) \in U$. Thus $x \in A$ and $x \in f^{-1}(U)$ or $x \in A \cap f^{-1}(U)$.

SOLUTION 1.2.13.

(1) Assume that $A \subset B$. Let $x \in X$. If $x \in A$, then $x \in B$ (assumption!). Hence $\mathbb{1}_A(x) = 1 = \mathbb{1}_B(x)$.

If $x \notin A$ (hence $\mathbb{1}_A(x) = 0$), then either $x \in B$ (hence $\mathbb{1}_B(x) = 1$) or $x \notin B$ (hence $\mathbb{1}_B(x) = 0$). Therefore, in all cases we will always have $\mathbb{1}_A \leq \mathbb{1}_B$.

Assume now that $\mathbb{1}_A \leq \mathbb{1}_B$ and let $x \in A$. Then $\mathbb{1}_A(x) = 1$. By assumption, we must have $\mathbb{1}_B(x) = 1$ and this means that $x \in B$.

(2) Just apply the previous result...

(3) Let $x \in X$. Then:

(a) If $x \in A \cap B$, then clearly

$$\mathbb{1}_{A \cap B}(x) = 1 = 1 \times 1 = \mathbb{1}_A(x)\mathbb{1}_B(x).$$

(b) If $x \notin A$ and $x \notin B$, then

$$\mathbb{1}_{A \cap B}(x) = 0 = 0 \times 0 = \mathbb{1}_A(x)\mathbb{1}_B(x).$$

(c) If $x \in A$ and $x \notin B$, then

$$\mathbb{1}_{A \cap B}(x) = 0 = 1 \times 0 = \mathbb{1}_A(x)\mathbb{1}_B(x).$$

(d) If $x \notin A$ and $x \in B$, then

$$\mathbb{1}_{A \cap B}(x) = 0 = 0 \times 1 = \mathbb{1}_A(x)\mathbb{1}_B(x).$$

(4) Just apply the previous result in the particular case $A = B$...

(5) Let $x \in X$.

(a) $x \in A^c$: Then $\mathbb{1}_{A^c}(x) = 1$. At the same time, $x \notin A$ so that $\mathbb{1}_A(x) = 0$. Hence we clearly have

$$\mathbb{1}_{A^c}(x) = 1 - \mathbb{1}_A(x).$$

(b) $x \in A$: Then $\mathbb{1}_A(x) = 1$ and $\mathbb{1}_{A^c}(x) = 0$ so that we equally have

$$\mathbb{1}_{A^c}(x) = 1 - \mathbb{1}_A(x).$$

(6) We may prove this result as before by investigating all possible cases. Alternatively, we may do the following:

$$
\begin{aligned}
\mathbb{1}_{A \cup B} &= 1 - \mathbb{1}_{(A \cup B)^c} \text{ (by (5))} \\
&= 1 - \mathbb{1}_{A^c \cap B^c} \\
&= 1 - \mathbb{1}_{A^c} \mathbb{1}_{B^c} \text{ (by (3))} \\
&= 1 - (1 - \mathbb{1}_A)(1 - \mathbb{1}_B) \text{ (by (5) again)} \\
&= \mathbb{1}_A + \mathbb{1}_A - \mathbb{1}_{A \cap B},
\end{aligned}
$$

as desired.

**SOLUTION** 1.2.14. Let $x \in X$. Recall that $\mathbb{1}_{\left(\bigcup_{n=1}^{\infty} A_n\right)}$ can be either one or zero (depending on whether $x \in \bigcup_{n=1}^{\infty} A_n$ or not).

(1) If $x \notin \bigcup_{n=1}^{\infty} A_n$, then $x \notin A_n$ for all $n$, i.e. $\mathbb{1}_{A_n}(x) = 0$ for all $n$. Hence

$$\mathbb{1}_{\left(\bigcup_{n=1}^{\infty} A_n\right)}(x) = 0 = \sum_{n=1}^{\infty} 0 = \sum_{n=1}^{\infty} \mathbb{1}_{A_n}(x).$$

(2) If $x \in \bigcup_{n=1}^{\infty} A_n$, then $x$ is in *one and only one* $A_n$ (as the sets are disjoint!) which we denote by $A_m$ (hence $\mathbb{1}_{A_n}(x) = 0$ for all $n \neq m$). Then obviously

$$\mathbb{1}_{\left(\bigcup_{n=1}^{\infty} A_n\right)}(x) = 1 = \mathbb{1}_{A_m}(x) = \sum_{n=1}^{\infty} \mathbb{1}_{A_n}(x).$$

**SOLUTION** 1.2.15.

(1) Assume that $g \circ f$ is injective. Let $x, x' \in X$ be such that $f(x) = f(x')$. Hence

$$g(f(x)) = g(f(x')), \text{ that is, } (g \circ f)(x) = (g \circ f)(x').$$

But $g \circ f$ is injective and so $x = x'$, showing that $f$ is injective.

(2) Suppose that $g \circ f$ is surjective. To show that $g$ is surjective, let $z \in Z$. Since $g \circ f$ is onto,

$$\exists x \in X, \ z = (g \circ f)(x) = g(f(x)).$$

Take $y = f(x)$. Then $y$ is in $Y$ and verifies $g(y) = z$. Thus, $g$ is surjective.

(3) Just apply the two previous results...

**SOLUTION 1.2.16.**

(1) Suppose that $f$ and $g$ are injective. To show that $g \circ f$ is injective, let $x, y \in X$ be such that $(g \circ f)(x) = (g \circ f)(y)$. Since $g$ is injective, it follows that $f(x) = f(y)$. Now, as $f$ is injective, then $x = y$. Therefore, $g \circ f$ is injective.

(2) Assume that $f$ and $g$ are surjective, i.e.

$$f(X) = Y \text{ and } g(Y) = Z.$$

To show that $g \circ f$ is surjective, we are required to show that $(g \circ f)(X) = Z$. But clearly,

$$(g \circ f)(X) = g(f(X)) = g(Y) = Z,$$

and this finishes the proof.

**SOLUTION 1.2.17.** The verification of the statements is obvious from the the following facts:

(1) $0 \in \mathbb{Q}$,

(2) if $r \in \mathbb{Q}$, then $-r \in \mathbb{Q}$,

(3) if $r, r' \in \mathbb{Q}$, then $r + r' \in \mathbb{Q}$...

**SOLUTION 1.2.18.** We use Theorem 1.1.49.

(1) We have for all $x > 0$,

$$f'(x) = -\frac{1}{x} \text{ and } f''(x) = +\frac{1}{x^2} > 0.$$

Hence $f$ is convex.

(2) The function $f : x \mapsto f(x) = x^p$ is twice differentiable on $\mathbb{R}_+^*$. We have for all $x > 0$,

$$f''(x) = p(p-1)x^{p-2},$$

which is positive iff $p \geq 1$. Therefore, $f$ is convex iff $p \geq 1$.

(3) Let $a \in \mathbb{R}$. Then

$$f'(x) = ae^{ax} \text{ and so } f''(x) = a^2 e^{ax}.$$

Thus $f''(x) \geq 0$ for all $x$ (and all $a$), i.e. $f$ is convex for all $a$.

**SOLUTION 1.2.19.** Let $X$ be a countable set and let $A \subset X$. Since $X$ is countable, by Theorem 1.1.74, there exists an injective map $f$ from $X$ into $\mathbb{N}$. Since $A \subset X$, the restriction of $f$ to $A$ remains injective. By Theorem 1.1.74 again, $A$ is countable.

**SOLUTION 1.2.20.**

(1) We need only show that $f$ is one-to-one. Let $(n,m), (n',m') \in \mathbb{N} \times \mathbb{N}$ such that $f(n,m) = f(n',m')$, i.e. $2^n 3^m = 2^{n'} 3^{m'}$. We need to show that $(n,m) = (n',m')$. We have

$$2^n 3^m = 2^{n'} 3^{m'} \implies \ln(2^n 3^m) = \ln(2^{n'} 3^{m'})$$
$$\implies n \ln 2 + m \ln 3 = n' \ln 2 + m' \ln 3.$$

Hence if $n \neq n'$, then

$$\frac{m - m'}{n' - n} = \frac{\ln 2}{\ln 3}.$$

But $(m-m')/(n'-n) \in \mathbb{Q}$ whereas $\ln 2 / \ln 3 \notin \mathbb{Q}$ (by Exercise 1.2.2). Therefore, $n = n'$ and so $3^m = 3^{m'}$. Now, just take the logarithm again to obtain $m = m'$. Accordingly, $f$ is one-to-one and so $\mathbb{N} \times \mathbb{N}$ is countable.

**REMARK.** We could have proved the countability as follows: Write

$$\mathbb{N} \times \mathbb{N} = \bigcup_{n=1}^{\infty} (\{n\} \times \mathbb{N}).$$

Then we can easily see that all $\{1\} \times \mathbb{N}$, $\{2\} \times \mathbb{N}$, $\cdots$ are countable (find a bijection between each of these sets and $\mathbb{N}$!). In the end, $\mathbb{N} \times \mathbb{N}$, being a countable union of countable sets, is itself countable.

(2) Since $A$ and $B$ are countable, there are *surjections* $f : \mathbb{N} \to A$ and $g : \mathbb{N} \to B$. Hence the map $h : \mathbb{N} \times \mathbb{N} \to A \times B$ defined, for all $(n,m) \in \mathbb{N} \times \mathbb{N}$ by

$$h(n,m) = (f(n), g(m))$$

is a surjection. This leads to the countability of $A \times B$ since $\mathbb{N} \times \mathbb{N}$ is countable.

**SOLUTION 1.2.21.** Let $m \in \mathbb{N}$ and set

$$A_m = \left\{ \frac{n}{m} : n \in \mathbb{Z} \right\}.$$

Then each $A_m$ is countable. Since we can obviously write

$$\mathbb{Q} = \bigcup_{m=1}^{\infty} A_m,$$

we obtain by Theorem 1.1.77 that $\bigcup_{m=1}^{\infty} A_m$ or $\mathbb{Q}$ is countable.

**SOLUTION 1.2.22.**

(1) Before giving the proof (due to Cantor), we recall that the decimal expansion of a number may not be unique. For example,

$$\frac{1}{10} = 0.1000\cdots \text{ and } \frac{1}{10} = 0.0999\cdots$$

To show that $[0,1]$ is not countable, we can show instead that any countable subset $[0,1]$ is proper. So, let

$$A = \{x_0, x_1, x_2, \cdots x_n, \cdots\}$$

be a countable subset of $[0,1]$. Hence each $x_n$ has a decimal expansion of the form

$$x_n = 0, x_{n,1}x_{n,2}x_{n,3}\cdots$$

where for every $k \in \mathbb{N}$, $x_{n,k} \in \{0, 1, \cdots, 9\}$. These numbers look like

$$0, x_{1,1}x_{1,2}x_{1,3}\cdots$$

$$0, x_{2,1}x_{2,2}x_{2,3}\cdots$$

$$0, x_{3,1}x_{3,2}x_{3,3}\cdots$$

$$\cdots\cdots\cdots\cdots$$

Next, define another number

$$y = 0, y_1y_2y_3\cdots$$

as follows:
   (a) If $x_{nn} \leq 5$, then we set $y_n = 6$.
   (b) If $x_{nn} \geq 6$, then we set $y_n = 3$, say.
   Then $y \in [0,1]$ and since $y_n \neq 0$ and $y_n \neq 9$, the decimal expansion of $y$ is unique (that is, we are not in the situation of $1/10$ discussed above!).
   Finally, as $\forall n \in \mathbb{N}$, $y_n \neq x_{nn}$, then $y \neq x_n$ for any $n$. Consequently, $y \notin A$ showing that $A$ is a proper subset of $[0,1]$. As alluded to above, this means that $[0,1]$ is uncountable.

(2) Since $[0,1] \subset \mathbb{R}$, $\mathbb{R}$ is not countable either. Indeed, if $\mathbb{R}$ were countable, every subset of it, in particular $[0,1]$, would too be countable!

(3) If $\mathbb{R} \setminus \mathbb{Q}$ were countable, then since $\mathbb{Q}$ is already countable, $\mathbb{Q} \cup (\mathbb{R} \setminus \mathbb{Q}) = \mathbb{R}$ would be countable too, which is absurd! Therefore, $\mathbb{R} \setminus \mathbb{Q}$ is not countable.

**SOLUTION** 1.2.23. Let $f : \bigcup_{n \geq 0} \mathbb{Q}^{n+1} \to \mathbb{Q}[X]$ be the map defined by

$$f(a_0, a_1, \cdots, a_n) = a_n x^n + \cdots + a_1 x + a_0.$$

By construction, $f$ is onto. Since $\bigcup_{n \geq 0} \mathbb{Q}^{n+1}$ is countable (why?), so is $\mathbb{Q}[X]$.

**SOLUTION** 1.2.24. One way of proving that $X$ is uncountable is to show that all countable subsets of $X$ are proper, i.e. whenever a subset of $X$ is countable, it will be strictly contained in $X$. Let $A \subset X$ be countable. Then there is some sequence of functions $(f_n)$ (whose image consists of zeroes and ones) such that $A = \{f_n : n \in \mathbb{N}\}$. Define $f : \mathbb{N} \to \{0, 1\}$ by

$$f(n) = 1 - f_n(n), \ \forall n \in \mathbb{N}.$$

Now, if $f_n(n) = 1$, then $f(n) = 0$; and if $f_n(n) = 0$, then $f(n) = 1$. Hence $f \in A$. However, $f \notin A$ since if it were, then we would have

$$f(n) = f_n(n) = 1 - f_n(n),$$

which is absurd.

**SOLUTION** 1.2.25. Since the intervals are nested, we can write (can't we?) for all $n$:

$$a_1 \leq a_2 \leq \cdots \leq a_n \leq \cdots \leq b_n \leq \cdots \leq b_2 \leq b_1.$$

Then $(a_n)$ is bounded above by $b_1$, say. Hence $\sup_{n \in \mathbb{N}} a_n$ exists and is finite. Call it $\alpha$. This implies that

$$\alpha \leq b_n, \ \forall n \in \mathbb{N}.$$

On the other hand, it is obvious that $a_n \leq \alpha$ for all $n$. Accordingly,

$$a_n \leq \alpha \leq b_n, \ \forall n \in \mathbb{N}.$$

Thus,

$$\bigcap_{n \in \mathbb{N}} I_n \neq \varnothing.$$

**SOLUTION** 1.2.26. The method of proofs is standard and it mainly uses the Archimedian Property.

(1) Obviously $0 \in \left(\frac{-1}{n}, \frac{1}{n}\right)$ for all $n \geq 1$ and hence

$$\{0\} \subset \bigcap_{n \in \mathbb{N}} \left(\frac{-1}{n}, \frac{1}{n}\right).$$

Let $x \neq 0$. There exists $N$ such that $|x| \geq \frac{1}{N} > 0$ (why?). Hence $x \notin \left(\frac{-1}{N}, \frac{1}{N}\right)$ and $x$ is not in the intersection. This proves that

$$\bigcap_{n \in \mathbb{N}} \left(\frac{-1}{n}, \frac{1}{n}\right) \subset \{0\},$$

establishing the other inclusion and hence the equality.

(2) We always have $\bigcup_{n \in \mathbb{N}} [-n, n] \subset \mathbb{R}$. Now let $x \in \mathbb{R}$. Again by the Archimedian Property, we know that there exists $n \in \mathbb{N}$ such that $|x| \leq n$, i.e. $x \in [-n, n]$. Hence $x \in \bigcup_{n \in \mathbb{N}} [-n, n]$. Thus,

$$\bigcup_{n \in \mathbb{N}} [-n, n] = \mathbb{R}.$$

(3) If there were an $x \in \bigcap_{n \in \mathbb{N}} [n, \infty)$, then this would imply that $\mathbb{N}$ is bounded and this is absurd. Therefore, no real is in the previous intersection and so

$$\bigcap_{n \in \mathbb{N}} [n, \infty) = \varnothing.$$

(4) To be shown as the preceding questions...

**SOLUTION** 1.2.27.

(1) Consider the sequence of closed and unbounded intervals $I_n = [n, \infty)$, where $n \in \mathbb{N}$. Then $(I_n)$ is decreasing. By Exercise 1.2.26, we know that

$$\bigcap_{n \in \mathbb{N}} [n, \infty) = \varnothing.$$

(2) Consider the sequence of bounded and not closed intervals $I_n = (0, \frac{1}{n}]$, where $n \in \mathbb{N}$. Then by the same method applied in Exercise 1.2.26 we may establish that

$$\bigcap_{n \in \mathbb{N}} (0, 1/n] = \varnothing.$$

**SOLUTION** 1.2.28. The reader may try to find and prove the following results:

(1) $\bigcap_{n \in \mathbb{N}} A_n = \{1\}$ and $\bigcup_{n \in \mathbb{N}} A_n = \mathbb{N}$.

(2) $\bigcap_{n \in \mathbb{N}} A_n = (-1, 1)$ and $\bigcup_{n \in \mathbb{N}} A_n = \mathbb{R}$.

(3) $\bigcap_{n \in \mathbb{N}} A_n = [0, 1]$ and $\bigcup_{n \in \mathbb{N}} A_n = (-1, 2)$.

(4) $\bigcap_{n \in \mathbb{N}} A_n = \{0\}$ and $\bigcup_{n \in \mathbb{N}} A_n = [0, 1)$.

(5) $\bigcap_{n \in \mathbb{N}} A_n = [0, 1)$ and $\bigcup_{n \in \mathbb{N}} A_n = (-1, 1)$.

**SOLUTION** 1.2.29. Assume that $[0, 1]$ is countable, i.e. assume that

$$[0, 1] = \{x_1, x_2, \cdots, x_n, \cdots\}.$$

To use the Nested Interval Property, we first need to find a decreasing sequence of subintervals of $[0, 1]$. Consider the following subintervals of $[0, 1]$ of equal length (here $\frac{1}{3}$):

$$\left[0, \frac{1}{3}\right], \quad \left[\frac{1}{3}, \frac{2}{3}\right] \text{ and } \left[\frac{2}{3}, 1\right].$$

Then $x_1$ cannot be in all intervals at the same time, so choose one of the three intervals which does not contain $x_1$. Call it $I_1$, so that $x_1 \notin I_1$. Denote $I_1$ by $[a_1, b_1]$, then divide it into three equal subintervals of equal length ($\frac{1}{9}$ this time) as:

$$\left[a_1, a_1 + \frac{1}{9}\right], \quad \left[a_1 + \frac{1}{9}, a_1 + \frac{2}{9}\right] \text{ and } \left[a_1 + \frac{2}{9}, b_1\right].$$

As before, at least one of these intervals does not contain $x_2$. Denote it by $I_2$. Carrying on in this way, we obtain a sequence of decreasing and closed intervals $(I_n)$, i.e.

$$\cdots \subset I_n \subset \cdots \subset I_2 \subset I_1,$$

and where by construction $x_n \notin I_n$ for each $n \in \mathbb{N}$. By the Nested Interval Property, we know that there is at least one point $a \in [0, 1]$ common to every interval $I_n$. Since we have assumed that $[0, 1] = \{x_1, x_2, \cdots, x_n, \cdots\}$, it follows that $a$ is one of the $x_n$ for some $n$. This means that $a = x_n \notin I_n$ which contradicts the fact that $a$ must be in *all* intervals $I_n$. Accordingly, $[0, 1]$ cannot be countable.

**SOLUTION** 1.2.30. Let $a, b \geq 0$ and let $p, q > 1$. If $a = 0$ or $b = 0$, then the inequality is clearly satisfied. So, assume that $a, b > 0$ and let $f : \mathbb{R}^+ \to \mathbb{R}$ be the function defined by

$$f(x) = \frac{x^p}{p} + \frac{1}{q} - x.$$

It can easily be established that $f(x) \geq 0$ for all $x \geq 0$. In particular, for $x = ab^{\frac{1}{1-p}}$ we have

$$ab^{\frac{1}{1-p}} \leq \frac{(ab^{\frac{1}{1-p}})^p}{p} + \frac{1}{q} = \frac{a^p b^{\frac{p}{1-p}}}{p} + \frac{1}{q}$$

and hence

$$ab^{\frac{1}{1-p}} b^q \leq \frac{a^p b^{\frac{p}{1-p}} b^q}{p} + \frac{b^q}{q}.$$

But $\frac{1}{p} + \frac{1}{q} = 1$ implies that

$$b^{\frac{1}{1-p}} b^q = b \quad \text{and} \quad b^{\frac{p}{1-p}} b^q = 1.$$

In the end, we obtain

$$ab \leq \frac{a^p}{p} + \frac{b^q}{q},$$

as expected.

**SOLUTION** 1.2.31. Let $p > 1$ and $q > 1$ be such that $\frac{1}{p} + \frac{1}{q} = 1$. We want to show that

$$ab \leq \frac{a^p}{p} + \frac{b^q}{q}$$

for all positive $a, b$.

By Proposition 1.1.51, we know that $x \mapsto e^x$ is convex on $\mathbb{R}$. In other words, for all $x, y \in \mathbb{R}$ and all $t \in [0, 1]$:

$$e^{tx + (1-t)y} \leq te^x + (1-t)e^y.$$

In particular, setting $x = \ln a^p$, $y = \ln b^q$ and $t = \frac{1}{p}$ gives

$$e^{\frac{1}{p} \ln a^p + (1 - \frac{1}{p}) \ln b^q} \leq \frac{1}{p} e^{\ln a^p} + \left(1 - \frac{1}{p}\right) e^{\ln b^q},$$

that is,

$$e^{\frac{1}{p} \ln a^p} e^{(\frac{1}{q}) \ln b^q} \leq \frac{1}{p} e^{\ln a^p} + \frac{1}{q} e^{\ln b^q}.$$

Hence

$$ab = e^{\ln a} e^{\ln b} \leq \frac{a^p}{p} + \frac{b^q}{q},$$

as required.

**SOLUTION** 1.2.32.

(1) Let $x, y \in \mathbb{R}$. We may assume WLOG that $x < y$. We are required to find an $r \in \mathbb{Q}$ such that $x < r < y$. Set $a = y - x > 0$. By the Archimedean property, for some $p \in \mathbb{N}$, $pa > 1$ or $\frac{1}{p} < a$. Set $q = [px] + 1$ where $[\cdot]$ denotes the greatest integer function. Then $q - 1 \leq px < q$. Hence

$$x < \frac{q}{p} \leq x + \frac{1}{p} < x + a = y$$

and the proof will be complete by taking $r = \frac{q}{p}$ (in $\mathbb{Q}$).

(2) Let $x, y \in \mathbb{R}$. Then $x - \sqrt{2}$ and $y - \sqrt{2}$ remain in $\mathbb{R}$. By the previous question, we know that there is a rational $r$ such that

$$x - \sqrt{2} < r < y - \sqrt{2} \text{ or } x < r + \sqrt{2} < y.$$

Since $r + \sqrt{2}$ is clearly irrational (cf. Exercise 1.3.5), we have just shown that between any two reals, there an irrational, as desired.

**SOLUTION** 1.2.33. To prove the inequality, we first note that

$$\forall x \in [0,1]: \ |f(x)| \leq \sup_{0 \leq x \leq 1} |f(x)|$$

and hence

$$\forall x \in [0,1]: \ |f(x)||g(x)| \leq \sup_{0 \leq x \leq 1} |f(x)||g(x)|.$$

Integrating with respect to $x$ on $[0,1]$, and taking into account that the final value of $\sup_{0 \leq x \leq 1} |f(x)|$ is a number, yield

$$\int_0^1 |f(x)g(x)|dx = \int_0^1 |f(x)||g(x)|dx$$

$$\leq \int_0^1 \sup_{0 \leq x \leq 1} |f(x)||g(x)|dx$$

$$= \sup_{0 \leq x \leq 1} |f(x)| \int_0^1 |g(x)|dx.$$

**SOLUTION** 1.2.34. No, this limit does not exist. To see this, we will show that there are two sequences both converging to $a$ but $\lim_{x \to a} f(x)$ has two different outcomes.

Notice first that as $\mathbb{Q}$ is dense in $\mathbb{R}$, then there is a sequence $(a_n)$ in $\mathbb{Q}$ such that $a_n \to a$. Hence $f(a_n) = 0$ (by definition of $f$) and so $\lim_{n \to \infty} f(a_n) = 0$.

On the other hand, the set of irrationals is also dense in $\mathbb{R}$, and so there is a sequence of irrationals $(b_n)$ such that $b_n \to a$. By definition of $f$, we have $f(b_n) = b_n \to a$. But, we have supposed that $a \neq 0$. Therefore, the limit $\lim_{x \to a} f(x)$ does not exist.

**SOLUTION** 1.2.35. We have already seen in Exercise 1.2.34 that the limit of $f$ at each point $a \neq 0$ does not exist. Hence $f$ is discontinuous at each point of $\mathbb{R}^*$. In fact, the function $f$ is only continuous at $x = 0$. Let's show that. We have $f(0) = 0$. Moreover, if $x \in \mathbb{Q}$ and $x \to 0$, then

$$\lim_{x \to 0} f(x) = \lim_{x \to 0} 0 = 0.$$

Now, if $x \in \mathbb{R} \setminus \mathbb{Q}$ and $x \to 0$, then

$$\lim_{x \to 0} f(x) = \lim_{x \to 0} x = 0.$$

Thus $f$ is continuous at 0.

As for the function $g$, we use a very similar argument to show that $g$ is everywhere discontinuous on $\mathbb{R}$.

**SOLUTION** 1.2.36. Let $g(x) = f(x) - x$ where $x \in [a, b]$. Then $g$, being a sum of two continuous functions, is itself continuous. Moreover,

$$g(a) = f(a) - a \geq 0 \text{ et } g(b) = f(b) - b \leq 0$$

for $a \leq f(x) \leq b$, $\forall x \in [a, b]$. We have to treat three cases, namely:

(1) If $f(a) = a$, then we are done! Indeed, in such case the fixed point is $a$.

(2) If $f(b) = b$, then $b$ is a fixed point for $f$.

(3) If $f(a) \neq a$ and $f(b) \neq b$, then $g(a)g(b) < 0$. Hence by the Intermediate Value Theorem we know that

$$\exists c \in (a, b) : g(c) = 0,$$

that is, $f(c) = c$ for some $c \in (a, b)$. This finishes the proof.

# CHAPTER 2

# Metric Spaces

## 2.2. True or False: Answers

**ANSWERS**.

(1) Yes, the discrete metric can be defined on $X$...

(2) The answer is yes. For a proof, see Test 1.

(3) In Topology, especially for beginners, this is something to avoid absolutely. We always have to emphasize the metric with respect to which we are working. Knowing this, we may consider this question as false or badly formulated. The given set is not open in $\mathbb{R}$ endowed with the usual metric. But e.g. with respect to the discrete metric, it becomes open as every subset is open in this metric space (see Exercise 2.3.5 below).

(4) No! It is not open, but certainly not because it is closed! It is not open for

$$\forall r > 0, \ (-r, r) \not\subseteq \{0\}.$$

(5) The answer is yes! This may come as a surprise to some of the readers but they have to be prepared to more "surprises" (interesting though) in Topology. We always remind the students that not every thing true in the classical $\mathbb{R}$, $\mathbb{C}$, $\mathbb{R}^n$,...etc will be true in an arbitrary metric (or topological) space.

Let us go back to our question. We give two counterexamples for the sake of diversity of examples.

(a) Let $\mathbb{R}$ be endowed with the *discrete metric* (see Exercise 2.3.5 below). Then

$$B(0, 2/3) = \{0\} \subseteq B(0, 1/2) = \{0\}$$

(this also shows that two balls of the same center with different radii may well be identical!).

(b) We give a less "exotic" example! Let $\mathbb{N}$ be endowed with the induced standard metric of $\mathbb{R}$. Then

$$B_c(3, 2) = \{n \in \mathbb{N} : |n - 3| \le 2\} = \{1, 2, 3, 4, 5\}$$

and

$$B_c(0,4) = \{n \in \mathbb{N} : |n| \leq 4\} = \{1,2,3,4\} \subsetneq B_c(3,2).$$

(6) The answer is no in general. We will show in Exercise 2.3.3 that

$$d(x,y) = \left| \frac{1}{x} - \frac{1}{y} \right|$$

defines a metric on $\mathbb{R}^*$. The given set, i.e. $(0,1)$ is not bounded with respect to this metric since $\left| \frac{1}{x} - \frac{1}{y} \right| \to +\infty$ as $x, y \to 0^+$.

**REMARK.** With respect to this metric, $\mathbb{N}$ or more generally, $[1, \infty)$ (which are obviously not bounded in usual $\mathbb{R}$) are bounded sets!

(7) The answer is no. In $(\mathbb{R}, |\cdot|)$ ($\mathbb{R}$ endowed with the usual metric), consider

$$A = [-1,1], \ B = [1,2] \text{ and } C = [2,4].$$

Then

$$d(A,C) = 1 > d(A,B) + d(B,C) = 0.$$

(8) The answer is yes. Let $(X,d)$ be a metric space. Let $B_c(x,r)$ be the closed ball of radius $r > 0$ and center $x$. We need only prove that the complement of $B_c(x,r)$ is open in $(X,d)$. There are trivial cases: $B_c(x,r) = X$ and $B_c(x,r) = \varnothing$. In such case, $B_c(x,r)$ is clearly closed. Next, we note that

$$[B_c(x,r)]^c = \{y \in X : d(x,y) > r\}.$$

Then take $s = d(x,y) - r > 0$ and let $z \in B(y,s)$, i.e.

$$d(y,z) < s = d(x,y) - r.$$

Hence

$$r < d(x,y) - d(y,z) \leq d(x,z) + d(z,y) - d(y,z) = d(x,z),$$

i.e. $z \in [B_c(x,r)]^c$ and so $B(y,s) \subset [B_c(x,r)]^c$ proving the openness of $[B_c(x,r)]^c$ or the closedness of $B_c(x,r)$.

The converse is not true in general. Consider the same idea for a counterexample as the one in Exercise 2.3.18, where it is also proved that every open ball in a metric space is an open set.

(9) We recall that, in a metric space $(X,d)$, the sphere of center $x$ and radius $r > 0$ is given by

$$S(x,r) = \{y \in X : d(x,y) = r\}$$

which can be written as

$$S(x,r) = B_c(x,r) \cap [B(x,r)]^c,$$

i.e. as a *finite* intersection of closed sets and hence $S(x,r)$ is closed (by Corollary 2.1.18).

(10) The answer is yes and this happens for example in a ultrametric space (see Exercise 2.3.22).

## 2.3. Solutions to Exercises

**SOLUTION** 2.3.1. We need only show the triangle inequality for $x$, $y$ and $z$ not all distinct. If $x = z$, then $d(x,z) = 0$ and hence

$$0 = d(x,z) \leq d(x,y) + d(y,z).$$

If $x = y$, then

$$d(x,z) = d(y,z) = d(x,y) + d(y,z),$$

i.e. we have equality in this case and so we will have in the case $y = z$. The proof is over.

**SOLUTION** 2.3.2. Let $x, y, z \in X$. We have

$$d(x,z) \leq d(x,y) + d(y,z) \text{ and hence } d(x,z) - d(y,z) \leq d(x,y).$$

Inverting the roles of $x$ and $y$ yields

$$d(y,z) - d(x,z) \leq d(y,x) = d(x,y).$$

Thus

$$|d(x,z) - d(y,z)| \leq d(x,y).$$

**SOLUTION** 2.3.3.

(1) No, $d$ is not a metric. For example, $1 \neq -1$ but $d(1,-1) = 0$.

(2) Yes, $d$ is a metric on $\mathbb{R}$. Let us show that. First, we note that the range of $d$ is $\mathbb{R}^+$. Let us check the remaining properties of a metric.

(a) Let $x, y \in \mathbb{R}$. Then

$$d(x,y) = 0 \iff |x^3 - y^3| = 0 \iff x^3 = y^3 \iff x = y.$$

(b) Let $x, y \in \mathbb{R}$. Then

$$d(x,y) = |x^3 - y^3| = |y^3 - x^3| = d(y,x).$$

(c) Let $x, y, z \in \mathbb{R}$. Then

$$d(x,z) = |x^3 - z^3| = |x^3 - y^3 + y^3 - z^3| \leq |x^3 - y^3| + |y^3 - z^3|$$

and hence

$$d(x,z) \leq d(x,y) + d(y,z).$$

(3) No, as $d$ never vanishes.

(4) Yes, $d$ is indeed a metric. Let us show this.

    (a) Let $x, y \in \mathbb{R}^*$. We have

$$x = y \iff \frac{1}{x} = \frac{1}{y} \iff \left| \frac{1}{x} - \frac{1}{y} \right| = 0 \iff d(x, y) = 0.$$

    (b) Let $x, y \in \mathbb{R}^*$. We obviously have

$$d(x, y) = d(y, x).$$

    (c) Let $x, y, z \in \mathbb{R}^*$. We have

$$d(x, z) = \left| \frac{1}{x} - \frac{1}{z} \right| = \left| \frac{1}{x} - \frac{1}{y} + \frac{1}{y} - \frac{1}{z} \right| \leq \left| \frac{1}{x} - \frac{1}{y} \right| + \left| \frac{1}{y} - \frac{1}{z} \right|$$

    and so

$$d(x, z) \leq d(x, y) + d(y, z).$$

(5) No, since $d(1, 1) = 2 \neq 0$.

**SOLUTION** 2.3.4. The open ball of center $x$ and radius $r > 0$ in the usual metric of $\mathbb{R}$ is given by

$$B(x, r) = \{y \in \mathbb{R} : |x - y| < r\} = (x - r, x + r).$$

So if $(x - r, x + r) = (0, 1)$, then

$$\begin{cases} x - r = 0, \\ x + r = 1 \end{cases} \text{ and whence } \begin{cases} x = \frac{1}{2}, \\ r = \frac{1}{2}. \end{cases}$$

Consequently,

$$B(1/2, 1/2) = (0, 1).$$

**SOLUTION** 2.3.5.

(1) Clearly $\forall x, y \in X$, $d(x, y) \geq 0$. Also, the first two properties of a metric are evidently satisfied.

    Now, let us show the triangle inequality. If $x = z$, then $d(x, z) = 0$ and hence

$$d(x, z) \leq d(x, y) + d(y, z).$$

If $x \neq z$, then either $x \neq y$ or $y \neq z$ (otherwise $x = y = z$) and hence

$$1 = d(x, z) \leq d(x, y) + d(y, z) = 1 \text{ or } 2.$$

(2) We show that

$$B(x, r) = \{x\} \text{ if } r \leq 1 \text{ and } B(x, r) = X \text{ if } r > 1.$$

If $r \leq 1$, then

$$B(x, r) = \{y \in X : d(x, y) < r \leq 1\} = \{x\}$$

since $d(x, y)$ equals either 1 or 0.

Since

$$d(x, y) \le 1 < r, \ \forall x, y \in X,$$

we see that $B(x, r) = X$ whenever $r > 1$.

Similarly, we can prove that

$$B_c(x, r) = \{x\} \text{ for } r < 1 \text{ and } B_c(x, r) = X \text{ for } r \ge 1.$$

(3) If $r \ne 1$, then obviously $S(x, r) = \varnothing$. If, however, $r = 1$, then

$$S(x, 1) = \{y \in X : \ d(x, y) = 1\} = X \setminus \{x\}.$$

(4) The way of proving $\varnothing$ and $X$ are open is the same as in Exercise 2.3.17 below. Now, let $U$ be any subset of $X$. Let $x \in U$. Then

$$B(x, 1/2) = \{x\} \subset U,$$

proving the openness of $U$.

(5) If $V$ is a subset of $X$, then so is its complement $V^c$. Hence $V^c$ is open and thus $V$ is closed.

(6) Denote the usual metric on $\mathbb{R}$ by $|\cdot|$. To show $d$ is not equivalent to $|\cdot|$, we can equivalently show that the function $(x, y) \mapsto \frac{d(x,y)}{|x-y|}$ is not bounded on $\mathbb{R}^2$. If it were, then it would have been so for $y = 0$ and $x = e^{-n}$. But

$$\frac{d(e^{-n}, 0)}{|e^{-n}|} = \frac{1}{e^{-n}} = e^n \longrightarrow \infty \text{ as } n \longrightarrow \infty.$$

Thus the two metrics are not equivalent.

**SOLUTION** 2.3.6. We first note that $d'$ is a positive and well-defined function on $X \times X$ since $d$ is a metric (the fact that $d$ is a metric will be used without further notice). Now we verify the other axioms.

(1) Let $x, y \in X$.
   (a) If $x = y$, then $d(x, y) = 0$. Hence $d'(x, y) = \sqrt{d(x, y)} = 0$.
   (b) If $d'(x, y) = 0$, then $\sqrt{d(x, y)} = 0$ or $d(x, y) = 0$. Thus $x = y$.

(2) Let $x, y \in X$. We obviously have

$$d'(x, y) = \sqrt{d(x, y)} = \sqrt{d(y, x)} = d'(y, x).$$

(3) Let $x, y, z \in X$. Since $d(x, z) \le d(x, y) + d(y, z)$, since $\sqrt{\cdot}$ is increasing on $\mathbb{R}^+$ and since $\sqrt{a + b} \le \sqrt{a} + \sqrt{b}$ (for all $a, b \ge 0$), one can write

$$d'(x, z) = \sqrt{d(x, z)} \le \sqrt{d(x, y) + d(y, z)} \le \sqrt{d(x, y)} + \sqrt{d(y, z)}.$$

Accordingly,

$$d'(x, z) \le d'(x, y) + d'(y, z).$$

**SOLUTION** 2.3.7.
(1) Let $x, y \in \mathbb{N}$. Then obviously $d(x, y) = 0 \Leftrightarrow x = y$.
(2) Also, for all $x, y \in \mathbb{N}$, $d(x, y) = d(y, x)$.
(3) Lastly, let $x, y, z \in \mathbb{N}$. Then

$$3 + \frac{x + z}{xz} = 3 + \frac{1}{x} + \frac{1}{z} \le 3 + \frac{1}{x} + \frac{1}{y} + 3 + \frac{1}{y} + \frac{1}{z}$$

and hence

$$d(x, z) \le d(x, y) + d(y, z), \ \forall x, y, z \in \mathbb{N}.$$

**SOLUTION** 2.3.8. First, observe that for all $x, y \in \mathbb{R}^n$, $d(x, y) \ge 0$. Let us show now the axioms of a metric.

(1) If $x = y$, then $|x_k - y_k| = 0$ (for all $k = 1, \cdots, n$) and hence $d(x, y) = 0$. Conversely, let $x, y \in \mathbb{R}^n$. We have

$$d(x, y) = 0 \implies |x_k - y_k| = 0,$$

$\forall k \in \{1, \cdots, n\} \implies x_k = y_k, \ \forall k \in \{1, \cdots, n\}$, i.e. $x = y$.

(2) Let $x, y \in \mathbb{R}^n$. We have

$$d(x, y) = \sum_{k=1}^{n} |x_k - y_k| = \sum_{k=1}^{n} |y_k - x_k| = d(y, x).$$

(3) Let $x, y, z \in \mathbb{R}^n$. Since

$$|x_k - z_k| \le |x_k - y_k| + |y_k - z_k|, \ \forall k \in \{1, \cdots, n\},$$

summing in $k$ yields

$$d(x, z) \le d(x, y) + d(y, z).$$

**SOLUTION** 2.3.9.
(1) Set for any $1 \le k \le n$,

$$A_k = \frac{a_k}{\left(\sum_{k=1}^{n} a_k^p\right)^{\frac{1}{p}}} \text{ and } B_k = \frac{b_k}{\left(\sum_{k=1}^{n} b_k^q\right)^{\frac{1}{q}}}.$$

Applying the Young Inequality (see Exercise 1.2.30) to $A_k$ and $B_k$ gives us

$$A_k B_k \le \frac{A_k^p}{p} + \frac{B_k^q}{q},$$

i.e.

$$\frac{a_k}{\left(\sum_{k=1}^{n} a_k^p\right)^{\frac{1}{p}}} \frac{b_k}{\left(\sum_{k=1}^{n} b_k^q\right)^{\frac{1}{q}}} \le \frac{1}{p} \frac{a_k^p}{\sum_{k=1}^{n} a_k^p} + \frac{1}{q} \frac{b_k^q}{\sum_{k=1}^{n} b_k^q}.$$

Summing in $k$ yields

$$\frac{\sum_{k=1}^{n} a_k b_k}{\left(\sum_{k=1}^{n} a_k^p\right)^{\frac{1}{p}} \left(\sum_{k=1}^{n} b_k^q\right)^{\frac{1}{q}}} \leq \frac{1}{p} + \frac{1}{q} = 1$$

and the desired result will then follows.

(2) To prove the Minkowski Inequality, we will use the Hölder Inequality twice. We have

$$\sum_{k=1}^{n} (a_k + b_k)^p = \sum_{k=1}^{n} (a_k + b_k)^{p-1} (a_k + b_k)$$

$$= \sum_{k=1}^{n} (a_k + b_k)^{p-1} a_k + \sum_{k=1}^{n} (a_k + b_k)^{p-1} b_k$$

$$\leq \left(\sum_{k=1}^{n} (a_k + b_k)^{(p-1)q}\right)^{\frac{1}{q}} \left(\sum_{k=1}^{n} a_k^p\right)^{\frac{1}{p}}$$

$$+ \left(\sum_{k=1}^{n} (a_k + b_k)^{(p-1)q}\right)^{\frac{1}{q}} \left(\sum_{k=1}^{n} b_k^p\right)^{\frac{1}{p}}.$$

But $\frac{1}{p} + \frac{1}{q} = 1 \Rightarrow (p-1)q = p$ and so

$$\sum_{k=1}^{n} (a_k + b_k)^p \leq \left(\sum_{k=1}^{n} (a_k + b_k)^p\right)^{\frac{1}{q}} \left(\sum_{k=1}^{n} a_k^p\right)^{\frac{1}{p}}$$

$$+ \left(\sum_{k=1}^{n} (a_k + b_k)^p\right)^{\frac{1}{q}} \left(\sum_{k=1}^{n} b_k^p\right)^{\frac{1}{p}}$$

$$= \left(\sum_{k=1}^{n} (a_k + b_k)^p\right)^{\frac{1}{q}} \left(\left(\sum_{k=1}^{n} b_k^p\right)^{\frac{1}{p}} + \left(\sum_{k=1}^{n} a_k^p\right)^{\frac{1}{p}}\right).$$

Hence (and since $\frac{1}{p} = 1 - \frac{1}{q}$)

$$\left(\sum_{k=1}^{n} (a_k + b_k)^p\right)^{1 - \frac{1}{q}} = \left(\sum_{k=1}^{n} (a_k + b_k)^p\right)^{\frac{1}{p}} \leq \left(\sum_{k=1}^{n} b_k^p\right)^{\frac{1}{p}} + \left(\sum_{k=1}^{n} a_k^p\right)^{\frac{1}{p}}$$

and the proof is over.

(3) (a) We first prove that $d_\infty$ is a metric and this does not require the previous questions.

    (i) Let $x, y \in \mathbb{R}^n$. If $x = y$, then obviously $d_\infty(x, y) = 0$.

Now, if $d_\infty(x, y) = 0$, then

$$\forall k = 1, \cdots, n : \ |x_k - y_k| = 0$$

and this leads to $x = y$.

(ii) We evidently have for all $x, y \in \mathbb{R}^n$

$$d_\infty(x, y) = d_\infty(y, x).$$

(iii) Let $x, y, z \in \mathbb{R}^n$. Then for all $1 \le k \le n$, we have

$$|x_k - z_k| \le |x_k - y_k| + |y_k - z_k| \le d_\infty(x, y) + d_\infty(y, z)$$

and hence

$$d_\infty(x, z) \le d_\infty(x, y) + d_\infty(y, z).$$

Now we show that each $d_p$ is a metric. Let $p \ge 1$.

(i) Let $x, y \in \mathbb{R}^n$. Then if $d_p(x, y) = 0$, one has

$$\sum_{k=1}^{n} |x_k - y_k|^p = 0 \implies |x_k - y_k|^p = 0, \ \forall k = 1, \cdots, n \implies x = y.$$

The other implication is obvious.

(ii) The second axiom is verified as for all $x, y \in \mathbb{R}^n$

$$d_p(x, y) = d_p(y, x).$$

(iii) The triangle inequality is a simple consequence of the Minkowski Inequality. Setting and substituting $x_k - y_k = a_k$ and $y_k - z_k = b_k$ (for all $k = 1, \cdots, n$) in the Minkowski Inequality give us the desired triangle inequality.

(b) First we show the equivalence of the metrics. We have for all $1 \le k \le n$,

$$|x_k - y_k|^p \le \sum_{k=1}^{n} |x_k - y_k|^p = (d_p(x, y))^p.$$

Hence, taking the $p$th root and taking the "max" over $k$ give

$$d_\infty(x, y) \le d_p(x, y)$$

and we are half-way through. We also have for all $1 \le k \le n$,

$$|x_k - y_k|^p \le \max_{1 \le k \le n} (|x_k - y_k|^p) = [d_\infty(x, y)]^p.$$

Summing in $k$ and the taking the $p$th root yield

$$d_p(x, y) \le n^{\frac{1}{p}} d_\infty(x, y),$$

showing the equivalence of the metrics.

To prove the last desired limit, we note that since for all $p \geq 1$,

$$d_\infty(x, y) \leq d_p(x, y) \leq n^{\frac{1}{p}} d_\infty(x, y),$$

taking the limit as $p \to \infty$ and observing that $n^{\frac{1}{p}} \to 1$, we easily obtain

$$\lim_{p \to \infty} d_p(x, y) = d_\infty(x, y).$$

(c) The answer is yes! Since $d_p$ is equivalent to $d_\infty$ and $d_\infty$ is equivalent to $d_q$, Proposition 2.1.32 gives the equivalence of $d_p$ and $d_q$.

**SOLUTION 2.3.10.**

(1) (a) Let us show that $\delta$ is in effect a metric on $X$.
  (i) Let $x, y \in X$. We have

$$\delta(x, y) = 0 \iff \min(1, d(x, y)) = 0 \iff d(x, y) = 0 \iff x = y$$

  since $d$ is a distance.
  (ii) Let $x, y \in X$. It is plain that $\delta(x, y) = \delta(y, x)$.
  (iii) Let $x, y, z \in X$.
    (A) If $d(x, y) \leq 1$ and $d(y, z) \leq 1$, then

$$\delta(x, z) \leq d(x, z) \leq d(x, y) + d(y, z) = \delta(x, y) + \delta(y, z).$$

    (B) If $d(x, y) > 1$, then

$$\delta(x, z) \leq 1 = \delta(x, y) \leq \delta(x, y) + \delta(y, z).$$

    (C) The same arguments apply if $d(y, z) > 1$.
  (b) Now we show that $\rho$ is a metric on $X$.
    (i) The first property of a metric is trivial.
    (ii) The second property is also trivial.
    (iii) Now, let $x, y, z \in \mathbb{R}$. Since $d(x, z) \leq d(x, y) + d(y, z)$ and using the hint, we easily see that

$$\rho(x, z) = \frac{d(x, z)}{1 + d(x, z)} \leq \frac{d(x, y) + d(y, z)}{1 + d(x, y) + d(y, z)} \leq \frac{d(x, y)}{1 + d(x, y)} + \frac{d(y, z)}{1 + d(y, z)}.$$

    Thus $\rho$ is a metric on $X$.
(2) The set $X$ is bounded (even if $X$ is $\mathbb{R}$, say) with respect to both metrics since the given metrics are bounded as

$$\forall x, y \in X : \ \delta(x, y) \leq 1 \text{ and } \rho(x, y) \leq 1.$$

**REMARK.** By doing some arithmetic one can prove the triangle inequality for $\rho$ directly without calling on the hint.

**SOLUTION 2.3.11.**

(1) Let $x = (x_n) = (1, 1, \cdots, 1, \cdots)$ and
$y = (y_n) = (1, \frac{1}{\sqrt{2}}, \frac{1}{\sqrt{3}}, \cdots, \frac{1}{\sqrt{n}}, \cdots)$. Then

$$\sum_{n=1}^{\infty} |x_n|^2 = 1^2 + 1^2 + \cdots + 1^2 + \cdots = +\infty$$

and

$$\sum_{n=1}^{\infty} |y_n|^2 = \sum_{n=1}^{\infty} \left(\frac{1}{\sqrt{n}}\right)^2 = \sum_{n=1}^{\infty} \frac{1}{n} = +\infty$$

since both series are divergent. This means that neither $x$ nor $y$ belongs to $\ell^2$.

Now let $x' = (x'_n) = (1, 1, \cdots, 1, 0, \cdots)$ and $y' = (y'_n) = (1, \frac{1}{2}, \frac{1}{3}, \cdots, \frac{1}{n}, \cdots)$. Then both $x'$ and $y'$ belong to $\ell^2$ as

$$\sum_{n=1}^{\infty} |x'_n|^2 = 1^2 + 1^2 + \cdots + 1^2 + 0 + \cdots < +\infty$$

and

$$\sum_{n=1}^{\infty} |y'_n|^2 = \sum_{n=1}^{\infty} \left(\frac{1}{n}\right)^2 = \sum_{n=1}^{\infty} \frac{1}{n^2} < +\infty,$$

because both series converge.

(2) We start by showing that $d_2$ is well-defined (i.e., the series involved in the definition of $d_2$ converges). We have for all $k$

$$|x_k - y_k|^2 = |x_k|^2 + 2\operatorname{Re}(x_k y_k) + |y_k|^2 \le |x_k|^2 + 2|\operatorname{Re}(x_k y_k)| + |y_k|^2$$

Hence

$$|x_k - y_k|^2 \le |x_k|^2 + 2|x_k y_k| + |y_k|^2$$

for every $k$. The Cauchy-Schwarz Inequality then gives us

$$\sum_{k=1}^{n} |x_k - y_k|^2 \le \sum_{k=1}^{n} |x_k|^2 + 2\sqrt{\sum_{k=1}^{n} |x_k|^2}\sqrt{\sum_{k=1}^{n} |y_k|^2} + \sum_{k=1}^{n} |y_k|^2.$$

Therefore

$$\sum_{k=1}^{n} |x_k - y_k|^2 \le \sum_{k=1}^{\infty} |x_k|^2 + 2\sqrt{\sum_{k=1}^{\infty} |x_k|^2}\sqrt{\sum_{k=1}^{\infty} |y_k|^2} + \sum_{k=1}^{\infty} |y_k|^2 < +\infty$$

as $x, y \in \ell^2$. Thus the partial sum $\sum_{k=1}^{n} |x_k - y_k|^2$ is bounded above and increasing since all terms are positive. Thus the series $\sum_{k \geq 1} |x_k - y_k|^2$ converges.

Let us now verify that $d$ satisfies the properties of a metric.

(a) Let $x, y \in \ell^2$.

(i) Obviously $x = y \Rightarrow d_2(x, y) = 0$.

(ii) We have

$$d_2(x, y) = 0 \iff \sum_{n=1}^{\infty} |x_n - y_n|^2 = 0$$
$$\implies |x_n - y_n|^2 = 0, \ \forall n$$
$$\implies x_n = y_n, \ \forall n,$$

that is, $x = y$.

(b) For any $x, y \in \ell^2$, $d_2(x, y) = d_2(y, x)$.

(c) Let $x, y, z \in \ell^2$. From the Minkowski Inequality with $p = 2$ (see Exercise 2.3.9) one has

$$\sqrt{\sum_{k=1}^{n} |x_k - z_k|^2} \leq \sqrt{\sum_{k=1}^{n} |x_k - y_k|^2} + \sqrt{\sum_{k=1}^{n} |y_k - z_k|^2}$$

and hence

$$\sqrt{\sum_{k=1}^{n} |x_k - z_k|^2} \leq \sqrt{\sum_{k=1}^{\infty} |x_k - y_k|^2} + \sqrt{\sum_{k=1}^{\infty} |y_k - z_k|^2}.$$

A similar argument to one used just above shows that

$$\sqrt{\sum_{k=1}^{\infty} |x_k - z_k|^2} \leq \sqrt{\sum_{k=1}^{\infty} |x_k - y_k|^2} + \sqrt{\sum_{k=1}^{\infty} |y_k - z_k|^2},$$

establishing the triangle inequality for $d_2$.

**SOLUTION** 2.3.12. To show that $d$ is a metric, we first need to check that the series involved in the definition of $d$ converges. For any $n$, one always have

$$\frac{1}{2^n} \times \frac{d_n(x_n, y_n)}{1 + d_n(x_n, y_n)} \leq \frac{1}{2^n} = u_n.$$

Since $\sum_{n=1}^{\infty} u_n = 1 < +\infty$, $d(x, y) < \infty$.

Now Exercise 2.3.10 can be used to show the properties of a metric for $d$. We leave details to readers.

**Solution** 2.3.13. Let $(X_1, d_1)$ and $(X_2, d_2)$ be metric spaces. Define the metric

$$d_\infty[(x_1, x_2), (y_1, y_2)] = \max(d_1(x_1, y_1), d_2(x_2, y_2))$$

in $(X_1 \times X_2, d_\infty)$.

(1) Let $(y_1, y_2) \in B((x_1, x_2), r)$. Then

$$\max(d_1(x_1, y_1), d_2(x_2, y_2)) < r.$$

Hence

$$d_1(x_1, y_1) < r \text{ and } d_2(x_2, y_2) < r$$

i.e. $y_1 \in B_1(x_1, r)$ and $y_2 \in B_2(x_2, r)$, that is, $(y_1, y_2) \in B_1(x_1, r) \times B_2(x_2, r)$, and this proves one inclusion.

(2) Now, let $(y_1, y_2) \in B_1(x_1, r) \times B_2(x_2, r)$, i.e. $y_1 \in B_1(x_1, r)$ and $y_2 \in B_2(x_2, r)$, i.e.

$$d_1(x_1, y_1) < r \text{ and } d_2(x_2, y_2) < r.$$

Therefore,

$$\max(d_1(x_1, y_1), d_2(x_2, y_2)) < r,$$

proving the remaining inclusion.

**Solution** 2.3.14. Before we start, observe that both $d$ and $d'$ are positive.

(1) Since $f$ and $g$ are continuous, so is $|f - g|$ and hence $d(f, g)$ is a well-defined quantity and so is $d'(f, g)$ as $|f - g|$ is continuous on the *closed* and *bounded* $[0, 1]$.

   First, we show that $d$ is a metric on $X$.

   (a) Let $f, g \in X$. We have

$$f = g \implies d(f, g) = 0 \text{ (this is obvious)} .$$

   Now since $|f - g|$ is a *continuous* and *positive* function, we have

$$\int_0^1 |f(x) - g(x)| dx = 0 \implies |f(x) - g(x)| = 0, \ \forall x \in [0, 1] \iff f = g.$$

   (b) It is plain that for all $f$ and $g$ in $X$ one has $d(f, g) = d(g, f)$.

(c) Let us show now the triangle inequality. Let $f, g, h \in X$. We have for all $x \in [0, 1]$

$$|f(x) - h(x)| = |f(x) - g(x) + g(x) - h(x)| \leq |f(x) - g(x)| + |g(x) - h(x)|.$$

Hence

$$\int_0^1 |f(x) - h(x)| dx \leq \int_0^1 |f(x) - g(x)| dx + \int_0^1 |g(x) - h(x)| dx,$$

i.e. $d(f, h) \leq d(f, g) + d(g, h)$.

Now we prove that $d'$ is a metric on $X$ too.

(a) If $f = g$, then obviously $d'(f, g)$, being the supremum of the zero function, must vanish as well. Conversely, if $d'(f, g) = 0$, then for all $x \in [0, 1]$ we have

$$0 \leq |f(x) - g(x)| \leq \sup_{x \in [0,1]} |f(x) - g(x)| = d'(f, g) = 0.$$

Thus $f = g$.

(b) For all $f, g \in X$, we have

$$d'(f, g) = \sup_{x \in [0,1]} |f(x) - g(x)| = \sup_{x \in [0,1]} |g(x) - f(x)| = d'(g, f).$$

(c) Let $f, g, h \in X$. Then we have for all $x \in [0, 1]$

$$|f(x) - h(x)| \leq |f(x) - g(x)| + |g(x) - h(x)| \leq d'(f, g) + d'(g, h).$$

Thus

$$d'(f, h) \leq d'(f, g) + d'(g, h).$$

(2) We recall that $d$ and $d'$ are said to be equivalent if

$$\exists a, b > 0, \ \forall f, g \in X : \ ad'(f, g) \leq d(f, g) \leq bd'(f, g).$$

We show that the LHS inequality does not hold (the RHS one does hold, cf Exercise 1.2.33) and in order to simplify the proof we may take $g = 0$. Now take $f(x) = x^n$, defined on $[0, 1]$, for a given $n$ in $\mathbb{N}$. Then $d(f, 0) = \frac{1}{n+1}$ and $d'(f, 0) = 1$. If $d$ were equivalent to $d'$ we would have

$$\exists a > 0, \forall n \in \mathbb{N} : \ a \leq \frac{1}{n+1}$$

which would imply that $\mathbb{N}$ is bounded, something impossible, i.e. we reached a contradiction. Thus $d$ and $d'$ are not equivalent.

(3) No, $d$ is no longer a metric in this case. For if one takes

$$f(x) = \begin{cases} 0, & x \in [0,1), \\ 2, & x = 1. \end{cases}$$

and $g$ is the zero function, then $f$ and $g$ are both Riemann-integrable. But, $f \neq g$ and yet

$$d(f,g) = \int_0^1 |f(x) - 0| dx = 0.$$

**SOLUTION** 2.3.15. The answer is no. The function $d$ does verify the second and third property of a metric and one implication of the first one (namely $f = g \Rightarrow d(f,g) = 0$). However,

$$d(f',g') = 0 \not\Rightarrow f = g.$$

For instance take any $f \in X$ and $g(x) = f(x) + 1$. Then $f \neq g$ but $d(f,g) = 0$ (because $f' = g'$).

**SOLUTION** 2.3.16.

(1) The answer is no. One simple example is to consider $\mathbb{R}$ equipped with the discrete metric (see Exercise 2.3.5). Take $r = 2$ and $s = 3$ and $x = 0$ and $y = -1$. Then we know that

$$B(0,2) = B(-1,3) = \mathbb{R}$$

and of course $r \neq s$ and $x \neq y$.
(2) The result is true for example in $\mathbb{R}$ endowed with its standard metric.

**SOLUTION** 2.3.17.

(1) First, we show that $X$ is open. Let $x \in X$. Then for some (and in this case all) $r > 0$

$$B(x,r) \subset X.$$

Thus $X$ is open in $(X, d)$. The openness of $\varnothing$ follows from the observation that if $x \in \varnothing$ ($x$ does not exist!), then for some (and here any) $r > 0$

$$B(x,r) = \varnothing \subset \varnothing.$$

(2) Let $\{U_i\}_{i \in I}$ be an arbitrary family of open sets in $(X, d)$. We have to show that their union remains open. To this end, let $x \in \bigcup_{i \in I} U_i$. Then there is some $j \in I$ such that $x \in U_j$. But,

$U_j$ is open and hence it contains an open ball, $B(x, r)$ $(r > 0)$ say. Hence

$$B(x, r) \subset U_j \subset \bigcup_{i \in I} U_i.$$

Thus the arbitrary union of open sets is open.

(3) We only prove it for two open sets. The proof for a finite intersection follows easily by induction. Let $U$ and $V$ be two open sets in $X$. Let $x \in U \cap V$. Hence $x \in U$ and $x \in V$. By the openness of both $U$ and $V$, we know that

$$\exists r > 0, \ B(x, r) \subset U \text{ and } \exists s > 0, \ B(x, s) \subset V$$

Taking $a = \min(r, s)$ leads to

$$B(x, a) \subset U \cap V,$$

showing the openness of $U \cap V$.

The last result does not extend to infinite intersections. For instance, all the intervals $(-\frac{1}{n}, \frac{1}{n})$ are open in usual $\mathbb{R}$ (and for all $n \in \mathbb{N}$), however,

$$\bigcap_{n \in \mathbb{N}} (-1/n, 1/n) = \{0\}$$

which is not open.

**SOLUTION** 2.3.18. Let $(X, d)$ be a metric space and let $B(x, r)$ be an open ball of center $x \in X$ and radius $r > 0$. Let $y \in B(x, r)$ (so $d(x, y) < r$). We must show that $B(x, r)$ contains another open ball centered at $y$. Let $s = r - d(x, y) > 0$. Then it may be easily shown that

$$B(y, s) \subset B(x, r),$$

completing the proof.

There are many counterexamples for the converse. For instance, in a discrete metric space $X$ (for convenience choose $X$ such that $\operatorname{card} X \geq 3$), open balls are only $X$ or singletons (nothing else). On the other hand, it is known that any subset is open, so $\{x, y\}$, where $x, y \in X$, is open too but as observed just above, it is not an open ball.

**SOLUTION** 2.3.19. Assume that a set $U$ is written as a union of open balls. Then, obviously $U$ is open since every open ball is an open set and the arbitrary union of open sets is open (see the foregoing exercise!).

Conversely, suppose $U$ is open. By definition of an open set in a metric space, for every $x$ in $U$ we can always find $r > 0$ for which

$B(x, r) \subset U$. By Proposition 1.1.24, we get

$$U = \bigcup_{x \in U} B(x, r),$$

and this completes the proof.

**SOLUTION** 2.3.20. None of the given intervals is open in $\mathbb{R}$. For instance $[a, b)$ is not open in $\mathbb{R}$ since $a \in [a, b)$ but

$$\forall r > 0, B(a, r) = (a - r, a + r) \not\subset [a, b).$$

**SOLUTION** 2.3.21.

(1) Straightforward verification for the first two properties...
    As for the triangle inequality, let $x, y, z \in \mathbb{R}^2$. Then we
    have

$$\begin{aligned} \delta(x, z) &= d(x, \mathbf{0}) + d(z, \mathbf{0}) \\ &\leq d(x, \mathbf{0}) + d(y, \mathbf{0}) + d(y, \mathbf{0}) + d(z, \mathbf{0}) \\ &= \delta(x, y) + \delta(y, z). \end{aligned}$$

(2) Let $a \neq \mathbf{0}$. We have

$$B(a, r) = \{x \in \mathbb{R}^2 : \ d(x, \mathbf{0}) + d(a, \mathbf{0}) < r\}$$

if $x \neq a$. Choosing $r < d(a, \mathbf{0})$, e.g. $r = \frac{1}{3} d(a, \mathbf{0})$ (which is legitimate since $a \neq \mathbf{0}$), we see that $d(x, \mathbf{0}) + d(a, \mathbf{0}) < r$ cannot be realized and hence we are left with $\delta(x, a) = 0$, i.e. $B(a, \frac{1}{3} d(a, \mathbf{0})) = \{a\}$. Therefore, each singleton is open.

(3) If $\{\mathbf{0}\}$ were open in $(\mathbb{R}^2, \delta)$, then there would exist a positive $r$ such that $B(\mathbf{0}, r) \subset \{\mathbf{0}\}$. But

$$\begin{aligned} B(\mathbf{0}, r) &= \{(x_1, x_2) \in \mathbb{R}^2 : \ \delta((x_1, x_2), \mathbf{0}) < r\} \\ &= \{(x_1, x_2) \in \mathbb{R}^2 : \ x_1^2 + x_2^2 < r^2\} \end{aligned}$$

cannot be a subset of $\{\mathbf{0}\}$ for any $r > 0$. Thus $\{\mathbf{0}\}$ is not open in $(\mathbb{R}^2, \delta)$.

(4) In $\mathbb{R}^2 \setminus \{\mathbf{0}\}$ every singleton is open. Hence every set is open since it can be written as a union (even arbitrary, see Exercise 2.3.19) of singletons. Thus every set in $\mathbb{R}^2 \setminus \{\mathbf{0}\}$ is closed too. Accordingly, every subset of $\mathbb{R}^2 \setminus \{\mathbf{0}\}$ is clopen and hence $\delta$ coincides with the discrete metric on $\mathbb{R}^2 \setminus \{\mathbf{0}\}$.

**SOLUTION** 2.3.22.

(1) The only property to prove is the triangle inequality. We may assume WLOG that $d(x, y) < d(y, z)$. Then

$$d(x, z) \leq \max(d(x, y), d(y, z)) = d(y, z) \leq d(x, y) + d(y, z).$$

(2) Assume $d(x,y) \neq d(y,z)$ or we could just assume that $d(x,y) < d(y,z)$ (couldn't we?). We can write by hypothesis

$$(d(x,y) <) \; d(y,z) \leq \max(d(x,y), d(x,z))$$

and a fortiori

$$\max(d(x,y), d(x,z)) = d(x,z), \text{ i.e. } d(y,z) \leq d(x,z).$$

The "ultrametricity" hypothesis also gives us

$$d(x,z) \leq \max(d(x,y), d(y,z)) = d(y,z)$$

and so $d(x,z) = d(y,z)$.

Geometrically, this means that in an ultrametric space every triangle is isosceles.

(3) Let $B(x,r)$ be the open ball of center $x$ and of radius $r > 0$. Let $y \in B(x,r)$ (hence $d(x,y) < r$). We need to show that $B(x,r) = B(y,r)$. Let $z \in B(x,r)$, i.e. $d(x,z) < r$. Hence

$$d(y,z) \leq \max(d(x,y), d(x,z)) < r \Longrightarrow z \in B(y,r).$$

The other inclusion can be dealt with similarly.

(4) Well, every closed ball is a closed set *in any metric space*. We show that every closed ball is open too. Let $B_c(x,r)$ be a closed ball of center $x$ and of radius $r > 0$. Let $y \in B_c(x,r)$, i.e. $d(x,y) \leq r$. We are done if we show that this closed ball contains the open ball $B(y,r)$. Let $z \in B(y,r)$. Then

$$d(x,z) \leq \max(d(x,y), d(y,z)) \leq r$$

and thus $z \in B_c(x,r)$.

We leave to the reader to show that every open ball is closed.

**SOLUTION** 2.3.23. Assume that $f$ is continuous and let $U$ be an open set in $X'$. We ought to show that $f^{-1}(U)$ is open in $X$. Let $x$ be in $f^{-1}(U)$, i.e. $f(x) \in U$. But $U$ is open and hence there exists some $\varepsilon > 0$, $B(f(x), \varepsilon) \subset U$. Since $f$ is continuous, the hypothesis implies that

$$f(B(x,\delta)) \subset B(f(x), \varepsilon) \subset U \text{ which leads to } B(x,\delta) \subset f^{-1}(U),$$

proving the openness of $f^{-1}(U)$.

Conversely, let us show that $f$ is continuous at $x \in X$ (assuming that the preimage by $f$ of every open set in $X'$ is open in $X$). Let $\varepsilon > 0$. Then it is clear that $B(f(x), \varepsilon)$ is open in $X'$. Then by hypothesis, $f^{-1}(B(f(x), \varepsilon))$ is also an open set but in $X$. This guarantees the existence of a strictly positive $\delta$ such that

$$B(x,\delta) \subset f^{-1}(B(f(x), \varepsilon))$$

or

$$f(B(x, \delta)) \subset B(f(x), \varepsilon),$$

establishing the continuity of $f$.

**SOLUTION 2.3.24.**

(1) Let $f : (X, d) \to (\mathbb{R}, |\cdot|)$ be a function such that $f(x) = d(x, a)$. Then $f$ is continuous on $X$ (it is in fact uniformly continuous) because for any $x, y \in X$

$$|f(x) - f(y)| = |d(x, a) - d(y, a)| \leq d(x, y) \text{ (by Exercise 2.3.2)}.$$

(2) Let $x, y$ be in $X$. If $b \in B$, then

$$d(x, B) \leq d(x, b) \leq d(x, y) + d(y, b).$$

Passing to the infimum over $B$ we obtain

$$d(x, B) \leq d(x, y) + d(y, B).$$

Inverting the roles of $x$ and $y$ yields

$$d(y, B) \leq d(y, x) + d(x, B) = d(x, y) + d(x, B)$$

and hence

$$|d(x, B) - d(y, B)| \leq d(x, y),$$

from which we easily establish the uniform continuity of $g$.

**SOLUTION 2.3.25.**

(1) Recall that $f$ is uniformly continuous on $\mathbb{R}^+$ iff

$$\forall \varepsilon > 0, \exists \alpha > 0, \forall x, y \in \mathbb{R}^+ \ (|x - y| < \alpha \Rightarrow d(f(x), f(y)) < \varepsilon).$$

Let $\varepsilon > 0$. Since $f$ is the identity mapping, we can write

$$d(f(x), f(y)) = d(x, y) = |\sqrt{x} - \sqrt{y}|.$$

It is well-known that

$$\forall x, y \geq 0 : |\sqrt{x} - \sqrt{y}| \leq \sqrt{|x - y|}.$$

So it becomes clear that in order to establish the uniform continuity of $f$, it suffices to take $\alpha = \varepsilon^2$.

(2) We could say that the function $g : \mathbb{R}^+ \to \mathbb{R}^+$ (both spaces equipped with the usual metric), defined for all $x \geq 0$, by $g(x) = \sqrt{x}$ is uniformly continuous on $\mathbb{R}^+$.

**SOLUTION 2.3.26.** The needed tools to answer these questions are the density of both $\mathbb{Q}$ and $\mathbb{R} \setminus \mathbb{Q}$ in $\mathbb{R}$.

(1) Consider

$$f(x) = \begin{cases} x - 1, & x \in \mathbb{Q}, \\ x + 1, & x \in \mathbb{R} - \mathbb{Q}. \end{cases}$$

Then $f$ is not continuous at any point of $\mathbb{R}$. To show this, let $x_0 \in \mathbb{R}$. Then

$$\exists (x_n) \in \mathbb{Q}, (y_n) \in \mathbb{R} \setminus \mathbb{Q} : x_n \to x_0 \text{ and } y_n \to x_0.$$

If $f$ were continuous at $x_0$, then would have

$$f(x_n) = x_n - 1 \to x_0 - 1 = x_0 + 1 \leftarrow y_n + 1 = f(y_n)$$

and obviously no such $x_0$ would satisfy that equation. Hence $f$ is discontinuous on the whole of $\mathbb{R}$.

(2) Consider

$$f(x) = \begin{cases} 0, & x \in \mathbb{Q}, \\ x, & x \in \mathbb{R} - \mathbb{Q}. \end{cases}$$

Then $f$ is only continuous at one point, that is at $x_0 = 0$. The proof is very similar to the previous one.

(3) Consider

$$f(x) = \begin{cases} x^2, & x \in \mathbb{Q}, \\ 2 - x, & x \in \mathbb{R} - \mathbb{Q}. \end{cases}$$

Then $f$ is only continuous at two points, namely 1 and $-2$. The proof is also left to the reader for its resemblance to Answer 1.

**SOLUTION** 2.3.27. >From Exercise 2.3.24 we know that $f$ is continuous. Now we have

$$\begin{aligned} B(a, r) &= \{x \in X : d(x, a) < r\} \\ &= \{x \in X : f(x) < r\} \\ &= \{x \in X : f(x) \in (-\infty, r)\}. \end{aligned}$$

Hence

$$B(a, r) = f^{-1}((-\infty, r)).$$

Since $(-\infty, r)$ is open in $\mathbb{R}$ and $f$ is continuous on $X$, $B(a, r) = f^{-1}((-\infty, r))$ must be open in $X$, as expected.

**SOLUTION** 2.3.28. The proof is easy and we shall prove only one implication. The other implication, being very akin, is left to the interested reader. Let $U$ be an open set in $(X, d)$. Then, for each $x \in U$, we can find some $r > 0$ such that $B(x, r) \subset U$. Since $d$ and $d'$ are equivalent, there are some $\alpha, \beta > 0$ such that for all $x, y \in X$ one has

$$\alpha d'(x, y) \leq d(x, y) \leq \beta d'(x, y).$$

Then if $B'(x, r')$ is the open ball in $(X, d')$, where $r' = \frac{r}{\beta}$, and if $y \in B'(x, r')$, then

$$d(x, y) \leq \beta d'(x, y) < \beta \frac{r}{\beta} = r,$$

implying that $y \in B(x, r) \subset U$ and hence $B'(x, r') \subset U$. Therefore, $U$ is open in $(X', d')$.

**SOLUTION** 2.3.29.

(1) The proof that $\delta$ is indeed a metric is left to the reader.
   The metric space $(\mathbb{R}, \delta)$ is bounded since

$$\forall x, y \in \mathbb{R}: \ \delta(x, y) = |\arctan x - \arctan y| < \pi.$$

(2) The range of the "arctan" function is $(-\frac{\pi}{2}, \frac{\pi}{2})$. This observation leads to

$$B(0, 2) = B(0, 4) = B(1, 4) = \mathbb{R}.$$

Thus, two balls with different radii (and the same center) may coincide with respect to this metric. Also, two balls with the same radius (and different centers) may also be equal!

(3) The two metrics cannot be equivalent. Assume they were, then we would have for some $\beta > 0$ and all $x, y \in \mathbb{R}$:

$$\beta |x - y| \leq |\arctan x - \arctan y| < \pi.$$

This would imply then that (usual) $\mathbb{R}$ is bounded! This is impossible.

(4) Yes. This can easily be seen from the fact that the identities

$$\mathrm{id} : ((-\frac{\pi}{2}, \frac{\pi}{2}), \delta) \to (\mathbb{R}, d) \text{ and id} : (\mathbb{R}, d) \to ((-\frac{\pi}{2}, \frac{\pi}{2}), \delta)$$

are both continuous for the functions "arctan" and "tan" are both continuous in usual $\mathbb{R}$.

**SOLUTION** 2.3.30.

(1) It is clear that for all $x, y \in X$

$$\rho(x, y) \leq d(x, y).$$

On the other hand, the ratio $\rho/d$ is not a bounded function on $X^2$, i.e. *there is no positive constant $C$ such that for all $x, y \in X$*

$$\rho(x, y) \geq C d(x, y),$$

as this would lead to $C \leq 1/(1 + d(x, y))$! (consider large enough $d(x, y)$). Therefore, $d$ and $\rho$ are not equivalent metrics.

(2) Let us show that $d$ and $\rho$ are topologically equivalent. Denote an open ball in $(X, d)$ by $B_d$ and an open ball in $(X, \rho)$ by $B_\rho$. Let $U$ be an open set in $(X, d)$. Let $x \in U$. Then $\exists r > 0$ such that $B_d(x, r) \subset U$. But

$$\frac{d(x,y)}{1+d(x,y)} < \frac{r}{1+r} \implies rd(x,y) + d(x,y) < rd(x,y) + r$$
$$\implies d(x,y) < r$$

(for each $x, y \in X$ and $r > 0$). Hence

$$B_\rho\left(x, \frac{r}{1+r}\right) \subset B_d(x,r) \subset U,$$

proving the openness of $U$ in $(X, \rho)$.

Conversely, let $U$ be an open set in $(X, \rho)$. Let $x \in U$. Then there is an $r > 0$ such that $B_\rho(x, r) \subset U$. But, if $r < 1$, then

$$d(x,y) < \frac{r}{1-r} \implies \frac{d(x,y)}{1+d(x,y)} = \rho(x,y) < r$$

(for all $x, y$). Hence

$$B_d\left(x, \frac{r}{1-r}\right) \subset B_\rho(x,r) \subset U$$

and if $r \geq 1$, then for all $x, y \in X$

$$\frac{d(x,y)}{1+d(x,y)} \leq 1 \leq r.$$

Hence $B_\rho(x,r) = X$ which forces us to have $X = U$. Thus, in either case $U$ is open in $(X, d)$. The proof is complete.

## 2.4. Hints/Answers to Tests

**SOLUTION** 1. Let $x, y \in X$. We have

$$0 = d(x,x) \leq d(x,y) + d(y,x) = d(x,y) + d(x,y) = 2d(x,y)$$

where we have used the three properties of a metric...

**SOLUTION** 2. Straightforward calculations based on known properties of the Logarithm function...

**SOLUTION** 3. Proceed as in Exercise 2.3.14. Also use the Cauchy-Schwarz Inequality for continuous functions...

**SOLUTION** 4. It is obvious that $d$ is positive. The first two properties are evidently verified. For the triangle inequality, there is a finite number of cases that must be checked...

**SOLUTION** 5. The second and the third properties use the fact that $d$ is a metric but they do not use the injectivity of $f$...To prove the first property, utilize the injectivity of $f$...

**SOLUTION** 6. The discrete metric space is an ultrametric space whereas the usual metric on $\mathbb{R}$ is not...

# CHAPTER 3

# Topological Spaces

## 3.2. True or False: Answers

**ANSWERS.**

(1) The answer is yes. We may define at least two topologies on $X$, namely the discrete and the indiscrete ones.

   Observe that if $\operatorname{card} X = 1$, then these topologies manifestly coincide.

(2) If we come to show that the intersection of two elements of $T$ is in $T$, then a proof by induction will allow us to deduce that the finite intersection of elements in $T$ is in $T$ too. This is probably known to most readers, but we wanted to remind the students that the proof by induction is used here as we have *finite* intersections.

(3) False! For example, define on $X = \{a, b, c, d\}$

$$T = \{\varnothing, \{a\}, X\} \text{ and } T' = \{\varnothing, \{b, d\}, X\}.$$

Then it can easily be established that $T$ and $T'$ are two topologies on $X$. However,

$$T \cup T' = \{\varnothing, \{a\}, \{b, d\}, X\}$$

does not define a topology on $X$.

(4) True! In fact, a more general result holds. Namely, if $\{T_i\}_{i \in I}$ is an *arbitrary* collection of topologies on the *same* set $X$, then their intersection is a topology on $X$. Let us prove it. Since $T_i$ are all topologies, they must all contain $\varnothing$ and hence so must their intersection. The same reasoning applies for $X$. Let $A, B \in \bigcap_{i \in I} T_i$. Then

$$\forall i \in I : A, B \in T_i \text{ and thus for all } i, A \cap B \in T_i.$$

Therefore, $A \cap B \in \bigcap_{i \in I} T_i$.

Finally, let $A_j \in \bigcap_{i \in I} T_i$ where $j \in J$. Then each $A_j$ belongs to all $T_i$. Since $T_i$ are all topologies, it follows that $\bigcup_{j \in J} A_j \in T_i$, for all $i \in I$, and so

$$\bigcup_{j \in J} A_j \in \bigcap_{i \in I} T_i.$$

The proof is complete.

(5) The answer is yes and both mappings are "increasing" with respect to "$\subset$". For the known properties state that

$$A \subset B \Longrightarrow \begin{cases} \overline{A} \subset \overline{B}, \\ \overset{\circ}{A} \subset \overset{\circ}{B}. \end{cases}$$

(6) The answer is yes if and only if $A = X$ (as $X$ is open). And the answer is no and will always be so if $A \neq X$, i.e. $A \subsetneq X$ as $\overset{\circ}{A} \subset A \subsetneq X$.

(7) The answer is no! Give $\mathbb{R}$ the usual topology and take $A = \mathbb{Q}$. Then

$$\overset{\circ}{\overline{A}} = \overset{\circ}{\mathbb{R}} = \mathbb{R} \neq \varnothing.$$

(8) The answer is yes if and only if $A = \varnothing$ (as $\varnothing$ is closed). And the answer is no and will ever be so if $A \neq \varnothing$ for the simple reason that by definition of the closure of a set, we have $\overline{A} \supset A$.

(9) True. For instance, endow $\mathbb{Z}$ with the induced usual metric, i.e.

$$d(n, m) = |n - m|, \ n, m \in \mathbb{Z}.$$

Then $d$ is not the discrete metric (is it?). But the topology here is discrete. To see this, we have to show that every singleton is open in $(\mathbb{Z}, d)$. This is, however, evident from the fact that for each $n \in \mathbb{Z}$

$$\{n\} = (n - 1, n + 1) \cap \mathbb{Z}$$

where $(n - 1, n + 1)$ is open in usual $\mathbb{R}$.

(10) Absolutely not! The two terminologies are totally different.

(11) True. To see this, assume that $X$ is a Hausdorff space and let $A \subset X$. Now, let $x, y \in A$ with $x \neq y$. By the Hausdorffness of $X$, there are two open sets $U$ and $V$ in $X$ containing $x$ and $y$ respectively such that $U \cap V = \varnothing$. The proof is now complete as $x \in A \cap U$, $y \in A \cap V$, $A \cap U$ and $A \cap V$ are open in $A$, and

$$(A \cap U) \cap (A \cap V) = A \cap (U \cap V) = A \cap \varnothing = \varnothing.$$

(12) The answer is yes. We provide a proof. Let $x, y \in X$ be two distinct points in $X$. Since $T$ is Hausdorff,

$$\exists U \in \mathcal{N}(x), \exists V \in \mathcal{N}(x) : \ U \cap V = \varnothing.$$

But, any open set in $T$ is open in $T'$ and hence $T'$ is Hausdorff too.

(13) The answer is no! See Exercise 3.3.34.

(14) Yes and the indiscrete topological space is a good example. For if $A$ is a nonvoid subset of a set $X$ endowed with the indiscrete topology, then the smallest closed set containing $A$ is $X$, i.e. $\overline{A} = X$.

(15) The answer is no. Endow $X = \mathbb{R}$ with the usual topology. Let $Y = (0, 2]$ be a topological subspace of $X$. Let $A = (1, 2]$. Then

$$\overset{\circ Y}{A} = (1, 2] \text{ and } \overset{\circ X}{A} = (1, 2)$$

and so

$$\overset{\circ Y}{A} = (1, 2] \neq (1, 2) = Y \cap \overset{\circ X}{A}.$$

(16) The answer is no. Apart from some obvious sets such as a finite set say, it is not clear how one can introduce such a definition in an arbitrary topological space. In metric spaces, this is possible thanks to balls which, even if they are subsets of an arbitrary set, their definition depends on a positive number and hence one can impose some constraint on a set to be bounded. If, however, the topological space is also given a structure of a vector space (more known as *topological vector spaces*), then one can introduce a definition of a bounded set. This will not be discussed in this book (see e.g. [21]).

(17) Firstly, this is a purely algebraic question but it is of interest especially in the product topology.

The answer is no! There are many counterexamples. Here is one: in $\mathbb{R} \times \mathbb{R}$, take

$$A = [0, \infty) \text{ and } B = [0, \infty).$$

Then

$$(A \times B)^c \neq A^c \times B^c! \text{ Check it out!}$$

What is *true* is the following:

$$(A \times B)^c = \{(x, y) \in X \times Y : \ x \notin A \text{ or } y \notin Y\} = (A^c \times Y) \cup (X \times B^c).$$

(18) The answer is no! Take $X = Y = \mathbb{R}$ both endowed with the usual topology. Then $U = [-1, 1)$ is neither closed nor open in $X$ and yet $U \times \varnothing = \varnothing$ is closed and open in $X \times Y$.

(19) The answer is yes! In euclidian $\mathbb{R}^2$, let $B((0,0),1)$ be the open ball (hence open) with center $(0,0)$ and radius 1, i.e.

$$B((0,0),1) = \{(x,y) \in \mathbb{R}^2 : x^2 + y^2 < 1\}.$$

Assume that this open ball could be written as

$$B((0,0),1) = U \times V.$$

Then $(0.8,0) \in B((0,0),1)$ giving $0.8 \in U$ and $(0,0.7) \in B((0,0),1)$ giving $0.7 \in V$. However, $(0.8,0.7) \notin B((0,0),1)$.

(20) The answer is no! First we recall that the exterior of $A$ is the interior of the complement of $A$. As a counterexample, take $A = \mathbb{Q}$ with respect to the usual topology of $\mathbb{R}$. Then

$$\overset{\circ}{\mathbb{Q}} \cup \overset{\circ}{\overbrace{\mathbb{R} \setminus \mathbb{Q}}} = \varnothing \cup \varnothing = \varnothing \neq X.$$

(21) The answer is no! In usual $X = \mathbb{R}$, take $A = \mathbb{Q}$. Then

$$\overset{\circ}{\mathbb{Q}} = \varnothing \neq \overset{\circ}{\overline{\mathbb{Q}}} = \overset{\circ}{\mathbb{R}} = \mathbb{R}.$$

(22) The answer is no! In usual $X = \mathbb{R}$, take $A = \mathbb{Q}$. Then

$$\overline{\mathbb{Q}} = \mathbb{R} \neq \overline{\overset{\circ}{\mathbb{Q}}} = \overline{\varnothing} = \varnothing.$$

(23) The answer is no! Let $A = \mathbb{N}$ in the usual topology of $\mathbb{R}$. Then $A$ is closed (why?) but $A' = \varnothing$. We show this last result. Let $x$ be any real number. Remember that

$$x \in \mathbb{N}' \iff \forall \varepsilon > 0, (x - \varepsilon, x + \varepsilon) \cap \mathbb{N} - \{x\} \neq \varnothing.$$

A quick observation tells us that the intersection intervening in the previous statement need not be non-empty for all $\varepsilon$.

The analogous question for open sets is true (see Exercise 3.3.17).

(24) False! Consider a non empty set $X$ with the indiscrete topology. Let $A = \{a\}$ where $a \in X$. The only non-empty open set is $X$ and hence

$$x \in A' \iff X \cap A - \{x\} \neq \varnothing.$$

We clearly see that all points but $a$ verifies the last equivalence and hence $A' = \{a\}^c$ which is not closed for $\{a\}$ is not open.

REMARK. In a metric space, every derived set is closed.

(25) The answer is yes. We provide simple examples with details below in some exercises. The topology here is the usual one.
  (a) Let $A = \{\frac{1}{n} : n \geq 1\}$. Then $A' = \{0\}$ and hence $A \cap A' = \varnothing$.

(b) For instance, take $A = \{0\}$. Then $A' = \varnothing$. Thus $A' \subsetneq A$.

(c) If $A = (0, 1)$, then $A' = [0, 1]$ and hence $A \subset A'$.

(d) If $A = \mathbb{R}$, then $A' = \mathbb{R}$ and so $A = A'$.

(26) Only the implication "$\Rightarrow$" holds. Let us show that. Let $x \in A'$. This means that

$$\forall U \in \mathcal{N}(x) : \ U \cap A - \{x\} \neq \varnothing.$$

Since $A \subset B$, we immediately get that $A - \{x\} \subset B - \{x\}$. Hence

$$\varnothing \neq U \cap A - \{x\} \subset U \cap B - \{x\}$$

yielding $U \cap B - \{x\} \neq \varnothing$ or $x \in B'$.

The other implication is false in general. One possible example is to take $A = [0, 1]$ and $B = (0, 1)$. Then $A' = B' = [0, 1]$ and hence

$$A' \subset B' \text{ whilst } A \not\subset B.$$

(27) The answer is no! In usual $\mathbb{R}$, let $A = \mathbb{Q}$. Then

$$\mathrm{Fr}(\overline{\mathbb{Q}}) = \mathrm{Fr}(\mathbb{R}) = \overline{\mathbb{R}} \setminus \overset{\circ}{\mathbb{R}} = \varnothing \neq \mathrm{Fr}(\mathbb{Q}) = \overline{\mathbb{Q}} \setminus \overset{\circ}{\mathbb{Q}} = \mathbb{R} \setminus \varnothing = \mathbb{R}.$$

(28) False! In usual $\mathbb{R}$, take again $A = \mathbb{Q}$. Then

$$\mathrm{Fr}(\overset{\circ}{\mathbb{Q}}) = \varnothing \neq \mathrm{Fr}(\mathbb{Q}) = \mathbb{R}.$$

What always holds is: $\mathrm{Fr}(\overset{\circ}{A}) \subset \mathrm{Fr}(A)$. The proof is a mere consequence of $\overline{\overset{\circ}{A}} \subset \overline{A}$.

(29) False! In usual $\mathbb{R}$, take $A = [0, 1]$ and $B = [1, 2]$. Then

$$\mathrm{Fr}(A \cup B) = \mathrm{Fr}([0, 2]) = \{0, 2\} \neq \mathrm{Fr}([0, 1]) \cup \mathrm{Fr}([1, 2]) = \{0, 1, 2\}.$$

Even for disjoint $A$ and $B$, the assertion has a negative answer. However, if $\overline{A}$ and $\overline{B}$ are disjoint, then

$$\mathrm{Fr}(A \cup B) = \mathrm{Fr}(A) \cup \mathrm{Fr}(B).$$

What always holds is: $\mathrm{Fr}(A \cup B) \subset \mathrm{Fr}(A) \cup \mathrm{Fr}(B)$. Here is a simple proof

$$\mathrm{Fr}(A \cup B) = (\overline{A \cup B}) \setminus (\overset{\circ}{\overbrace{A \cup B}})$$

$$= (\overline{A} \cup \overline{B}) \cap (\overset{\circ}{\overbrace{A \cup B}})^c$$

$$= (\overline{A} \cup \overline{B}) \cap (\overset{\circ}{\overbrace{A \cup B}})^c$$

$$\subset (\overline{A} \cup \overline{B}) \cap (\overset{\circ}{A} \cup \overset{\circ}{B})^c$$

$$\subset \mathrm{Fr}(A) \cup \mathrm{Fr}(B).$$

(30) The answer is again no! For in the usual topology of $\mathbb{R}$, $\mathbb{Q}$ verifies

$$\mathbb{Q} \subset \mathrm{Fr}\,\mathbb{Q} = \mathbb{R}.$$

(31) The answer is yes and this a simple consequence of the definition of the frontier, namely

$$\mathrm{Fr}(A) = \overline{A} \setminus \overset{\circ}{A} = \overline{A} \cap (\overset{\circ}{A})^c \subset \overline{A}.$$

(32) The answer is no! As the last but one answer, take $A = \mathbb{Q}$ and $B = \mathbb{R}$. Then $A \subset B$ whilst

$$\mathrm{Fr}\,A = \mathbb{R} \not\subset \mathrm{Fr}\,B = \varnothing.$$

**REMARK.** The hypothesis $A \subset B$ does not imply $\mathrm{Fr}\,B \subset \mathrm{Fr}\,A$ either. In the usual topology of $\mathbb{R}$, take $A = \mathbb{Q}^+$ and $B = \mathbb{Q}$. Then

$$A \subset B \text{ but } \mathrm{Fr}\,A = \mathbb{R}^+ \not\supset \mathrm{Fr}\,B = \mathbb{R}.$$

(33) None of the assertions holds! Let $X$ be a set endowed with the discrete metric such that $\mathrm{card}X \geq 2$. Let $r = 1$ and let $x \in X$. Then we know from Exercise 2.3.5 that $S(x, 1) = X \setminus \{x\}$. Now since every set is simultaneously open and closed in a discrete metric space, we have

$$\mathrm{Fr}(B(x, 1)) = \varnothing \text{ and } \mathrm{Fr}(B_c(x, 1)) = \varnothing$$

and thus

$$\mathrm{Fr}(B(x, r)) \neq S(x, r) \text{ and } \mathrm{Fr}(B_c(x, r)) \neq S(x, r).$$

(34) We have already defined the concept of a neighborhood in a topological space. As alluded to in the "Essential Background" Section, there is another common definition of this notion. We state here: *In a topological space $X$, a set $U \subset X$ is said to be*

a ***neighborhood*** *of* $a \in X$ *if there is an open set* $V$ *containing the point* "a" *such that* $V \subset U$. If $X$ is a metric space, then we may change $V$ by an open ball centered at $a$.

So, according to this definition, $(-1, 1]$ is a neighborhood of 0 since $0 \in (-1, 1) \subset (-1, 1]$ and $(-1, 1)$ is an open set in $\mathbb{R}$. $(-2, 2)$ is also a neighborhood of 0. The other sets are not for:

(a) $(0, 2]$ does not even contain 0;

(b) $[-1, 0]$ cannot contain an open set which contains 0;

(c) $[0, 1]$ cannot contain an open set which contains 0.

But, according to our definition, only $(-2, 2)$ is a neighborhood of 0.

(35) True and the reason is simple. If $V$ is closed in $T$, then $V^c$ is open in $T$ and by assumption $V^c$ is open in $T'$ too. Therefore, $V$ is closed in $T'$, as required.

(36) False! For a counterexample see Exercise 3.3.51.

## 3.3. Solutions to Exercises

SOLUTION 3.3.1.

- Possible topologies on $X$: The only possible topology here is $T = \{\varnothing, \{1\}\}$.
- Possible topologies on $Y$: There are four possible topologies in this case. They are

$$T_1 = \{\varnothing, Y\}; \ T_2 = \{\varnothing, \{1\}, Y\}; \ T_3 = \{\varnothing, \{2\}, Y\} \text{ and } T_4 = \mathcal{P}(Y).$$

SOLUTION 3.3.2.

(1) Obviously $X$ and $\varnothing$ are both in $T$ and the reader may easily check that finite intersections of elements of $T$ is again in $T$ and that the arbitrary union (finite in this case) of elements of $T$ is in $T$ too.

(2) The closed sets in this case are easy to find. They are the complements of the open sets, i.e. the complements of the elements of $T$. Hence they are

$$X, \ \{b, c, d, e\}, \ \{a, b, e\}, \ \{b, e\}, \ \{a\}, \ \varnothing.$$

(3) The closure of $\{a\}$ is the smallest closed set containing $\{a\}$ (which is also the intersection of all closed sets containing $\{a\}$). Hence $\overline{\{a\}} = \{a\}$ (or here simply since $\{a\}$ is closed!). In a similar way we find that $\overline{\{b\}} = \{b, e\}$.

The interior of $\{a\}$ is the biggest open set contained in $\{a\}$.

Hence $\overset{\circ}{\overbrace{\{a\}}} = \{a\}$ (or just because $\{a\}$ is open). Similarly $\overset{\circ}{\overbrace{\{b\}}} = \varnothing$. Hence

$$\mathrm{Fr}(\{a\}) = \overline{\{a\}} \setminus \overset{\circ}{\overbrace{\{a\}}} = \varnothing$$

and

$$\mathrm{Fr}(\{b\}) = \overline{\{b\}} \setminus \overset{\circ}{\overbrace{\{b\}}} = \{b, e\}.$$

(4) The smallest closed set containing $\{a, c\}$ is $X$. Hence, $\overline{\{a, c\}} = X$, i.e. $\{a, c\}$ is dense in $X$.

(5) A neighborhood of $c$ is any open set containing $c$ (remember that this is the definition adopted in this book). Hence

$$\mathcal{N}(c) = \{\{c, d\}, \{a, c, d\}, \{b, c, d, e\}, X\}.$$

Also,

$$\mathcal{N}(d) = \{\{c, d\}, \{a, c, d\}, \{b, c, d, e\}, X\}.$$

(6) No $T$ is not Hausdorff since

$$\exists c, d \in X \ (c \neq d), \forall (U, V) \in \mathcal{N}(c) \times \mathcal{N}(d) : U \cap V \neq \varnothing.$$

**SOLUTION 3.3.3.**

(1) $\varnothing \in T$ (by definition) while $\mathbb{N} \in T$ for $n = 1$. The other axioms of a topological are the matter of easy unions and intersections and hence left to the reader.

(2) No, the given set is not open (why?).

(3) The only two open sets containing 2 are $\{1, 2, 3, \cdots\}$ and $\{2, 3, 4, \cdots\}$. Hence

$$\mathcal{N}(2) = \{\{1, 2, 3, \cdots\}, \{2, 3, 4, \cdots\}\}.$$

Similarly

$$\mathcal{N}(3) = \{\{1, 2, 3, \cdots\}, \{2, 3, 4, \cdots\}, \{3, 4, 5 \cdots\}\}.$$

(4) No, $T$ is not Hausdorff. The previous question provides us with a counterexample (doesn't it?).

(5) The closed sets are $\varnothing$, $\mathbb{N}$ and any set of the form $\{1, 2, 3, \cdots, p\}$ where $p \in \mathbb{N}$.

(6) The interior of $\{4\}$ is the biggest open set contained in it, that is the empty set! The smallest closed set containing $\{4\}$, i.e. its closure, is $\{1, 2, 3, 4\}$.

The interior and the closure of $\{2, 4, 6, 8, \cdots\}$ are respectively $\varnothing$ and $\mathbb{N}$.

(7) We claim that a set $A$ is dense in $\mathbb{N}$ iff it is infinite. We first note that if $A$ is finite, then its closure is a finite set and hence $\overline{A} \neq \mathbb{N}$. Therefore, we have proved the implication "$\Rightarrow$". Now, if $A$ is infinite, then the smallest closed superset is $\mathbb{N}$ and this proves the implication "$\Leftarrow$".

**SOLUTION** 3.3.4. In order to show that $T$ is the discrete topology, it suffices to check that every subset of $X$ is open in $T$. Let $A$ be a subset of $X$. Then $A$ can always be written as a union (finite in this case) of those singletons. We have

$$\{a, b\} = \{a\} \cup \{b\}, \{a, c\} = \{a\} \cup \{c\}, \ldots\ldots, \{a, b, c\} = \{a\} \cup \{b\} \cup \{c\}.$$

Thus, any subset is open (and closed).

**SOLUTION** 3.3.5. Recall that by definition, the interior of a set $A$ is the largest open set contained in $A$ (which may be taken to be the union of all open sets contained in $A$).

Similarly, the closure of a set $A$ is the smallest closed set containing $A$ (which may be taken to be the intersection of all closed sets containing $A$).

We also note that if $U$ is an open set *contained* in $A$, then $U^c := V$ is a closed set *containing* $A^c$. These remarks allow us to write for open sets $U$

$$(\overset{\circ}{A})^c = \left(\bigcup_{U \subset A} U\right)^c = \bigcap_{U \subset A} U^c = \bigcap_{A^c \subset V} V = \overline{A^c}.$$

The other property can be easily established by replacing $A$ by $A^c$.

**SOLUTION** 3.3.6. Well, the right-to-left implication is trivial (is it not?). To show the other implication, assume that $\overline{A} \cap U \neq \varnothing$. Then there is some $x$ in both $U$ and $\overline{A}$. Hence for any open set $V$ containing $x$, $V \cap A \neq \varnothing$. In particular, taking $V = U$ (which is legitimate as $U$ is open and it contains $x$) gives $A \cap U \neq \varnothing$.

Another (simpler) proof goes as follows. If $A \cap U = \varnothing$, then $A \subset U^c$. But $U^c$ is closed and hence $\overline{A} \subset U^c$ or $\overline{A} \cap U = \varnothing$.

**SOLUTION** 3.3.7.

(1) Since $\varnothing$ and $X$ are closed, it immediately follows that:

$$\overline{\varnothing} = \varnothing \text{ and } \overline{X} = X.$$

(2) As before, $\overline{A}$ is closed and so $\overline{\overline{A}} = \overline{A}$.

(3) We know that $B \subset \overline{B}$. Since $A \subset B$, it follows that $A \subset \overline{B}$. This means that $\overline{B}$ is a closed set containing $A$. But $\overline{A}$ is the smallest closed set containing $A$. Consequently, $\overline{A} \subset \overline{B}$.

(4) By the previous property we can write:

$$\begin{cases} A \subset A \cup B \\ B \subset A \cup B \end{cases} \implies \begin{cases} \overline{A} \subset \overline{A \cup B} \\ \overline{B} \subset \overline{A \cup B} \end{cases} \implies \overline{A} \cup \overline{B} \subset \overline{A \cup B}.$$

Now, observe that $\overline{A} \cup \overline{B}$ is a closed set (being a finite union of closed sets!) containing $A \cup B$. By definition, $\overline{A \cup B}$ is the smallest closed set containing $A \cup B$. Therefore,

$$\overline{A} \cup \overline{B} \supset \overline{A \cup B},$$

and the equality holds.

(5) If $I$ is finite, apply the previous method. If $I$ is infinite, then for all $i$,

$$A_i \subset \bigcup_{i \in I} A_i \implies \overline{A_i} \subset \overline{\bigcup_{i \in I} A_i}$$

and so (only)

$$\bigcup_{i \in I} \overline{A_i} \subset \overline{\bigcup_{i \in I} A_i}.$$

(6) For all $i$ we may write:

$$\bigcap_{i \in I} A_i \subset A_i \implies \overline{\bigcap_{i \in I} A_i} \subset \overline{A_i}.$$

Accordingly,

$$\overline{\bigcap_{i \in I} A_i} \subset \bigcap_{i \in I} \overline{A_i}.$$

**SOLUTION** 3.3.8. We only prove a few properties. We leave the others to interested readers. For example, since $\varnothing$, $X$ and $\overset{\circ}{A}$ are all open, we have

$$\overset{\circ}{\varnothing} = \varnothing, \ \overset{\circ}{X} = X \text{ and } \overset{\circ}{\overset{\circ}{A}} = \overset{\circ}{A},$$

and this proves (1) and (2).

For (3), we give two ways of showing it:

- Assume that $A \subset B$. Since $\overset{\circ}{A} \subset A$, we get $\overset{\circ}{A} \subset B$. This shows that $B$ contains the open set $\overset{\circ}{A}$. However, $\overset{\circ}{B}$ is the largest open set contained in $B$. Therefore,

$$\overset{\circ}{A} \subset \overset{\circ}{B}.$$

- Alternatively, we use Proposition 3.1.17. We may write:

$$A \subset B \Longrightarrow B^c \subset A^c$$
$$\Longrightarrow \overline{B^c} \subset \overline{A^c} \text{ (by Theorem 3.1.21)}$$
$$\Longrightarrow (\overset{\circ}{B})^c \subset (\overset{\circ}{A})^c \text{ (by Proposition 3.1.17)}$$
$$\Longrightarrow \overset{\circ}{A} \subset \overset{\circ}{B},$$

as required.

In fact, we can prove the remaining statements using either of the previous ways. Let's show (1) using Proposition 3.1.17 and Exercise 3.3.7 (and then we stop!).

$$(\overset{\circ}{\overbrace{A \cap B}})^c = \overline{(A \cap B)^c} = \overline{A^c \cup B^c} = \overline{A^c} \cup \overline{B^c} = (\overset{\circ}{A})^c \cup (\overset{\circ}{B})^c = (\overset{\circ}{A} \cap \overset{\circ}{B})^c.$$

Therefore,

$$\overset{\circ}{\overbrace{A \cap B}} = \overset{\circ}{A} \cap \overset{\circ}{B}.$$

**SOLUTION** 3.3.9. The idea of proof is the same for each class (i.e. closures and interiors). There are trivial properties e.g.:

$$\overline{[a, b]} = [a, b]$$

as $[a, b]$ is closed. Similarly,

$$\overset{\circ}{(a, b)} = (a, b)$$

as $(a, b)$ is open. The same applies for $\overline{\mathbb{R}} = \mathbb{R}$ and $\overset{\circ}{\mathbb{R}} = \mathbb{R}$.

Now, we show two more properties. For example, we show that $\overline{(a, b]} = [a, b]$. We can write

$$(a, b] \subset [a, b] \Longrightarrow \overline{(a, b]} \subset \overline{[a, b]} = [a, b].$$

Next, we say why $a \in \overline{(a, b]}$. Recall that

$$a \in \overline{(a, b]} \Longleftrightarrow \forall \varepsilon > 0, \ (a - \varepsilon, a + \varepsilon) \cap (a, b] \neq \varnothing.$$

But it is clear that for any $\varepsilon > 0$ no matter how small (let alone the case when it is big!), we will always have $(a - \varepsilon, a + \varepsilon) \cap (a, b] \neq \varnothing$. This shows that $a \in \overline{(a, b]}$. Hence, we may write

$$(a, b] \subset \overline{(a, b]} \Longrightarrow [a, b] = \{a\} \cup (a, b] \subset \{a\} \cup \overline{(a, b]} = \overline{(a, b]}.$$

In the end, we have shown that:

$$\overline{(a, b]} = [a, b].$$

Let's show that $\overset{\circ}{[a, b)} = (a, b)$. We give two methods:

(1) It is plain that

$$(a, b) \subset [a, b) \text{ and so } (a, b) = \overset{\circ}{(a, b)} \subset \overset{\circ}{[a, b)}.$$

This proves one inclusion. For the reverse one, we first show that $a \notin \overset{\circ}{[a, b)}$. If this were to be true, then we would have

$$\exists r > 0, \ (a - r, a + r) \subset [a, b)$$

and is not true as

$$\forall r > 0, \ (a - r, a + r) \not\subset [a, b).$$

Therefore, $a \notin \overset{\circ}{[a, b)}$. Hence we can write

$$\overset{\circ}{[a, b)} \subset [a, b) \implies \overset{\circ}{[a, b)} = \overset{\circ}{[a, b)} \setminus \{a\} \subset [a, b) \setminus \{a\} = (a, b).$$

Thus

$$\overset{\circ}{[a, b)} = (a, b).$$

(2) After we prove the properties concerning closure for infinite intervals, we may exploit them to give a different proof to $\overset{\circ}{[a, b)} = (a, b)$. Indeed, utilizing Proposition 3.1.17 and Theorem 3.1.21, we can do the following:

$$(\overset{\circ}{[a, b)})^c = \overline{[a, b)^c} = \overline{(-\infty, a) \cup [b, \infty)} = \overline{(-\infty, a)} \cup \overline{[b, \infty)} = (-\infty, a] \cup [b, \infty)$$

and so

$$\overset{\circ}{[a, b)} = ((-\infty, a] \cup [b, \infty))^c = (-\infty, a]^c \cap [b, \infty)^c = (a, \infty) \cap (-\infty, b) = (a, b),$$

as required.

The remaining examples are left to the interested readers...

**SOLUTION 3.3.10.**

(1) Let us find the closure of $\mathbb{Q}$ in $\mathbb{R}$. We know that $\mathbb{Q} \subset \mathbb{R}$ and hence $\overline{\mathbb{Q}} \subset \overline{\mathbb{R}} = \mathbb{R}$. Let us show that $\mathbb{R} \subset \overline{\mathbb{Q}}$. Let $x \in \mathbb{R}$. By definition

$$x \in \overline{\mathbb{Q}} \iff \forall \varepsilon > 0, \ (x - \varepsilon, x + \varepsilon) \cap \mathbb{Q} \neq \varnothing.$$

But, any (open) interval contains always rational numbers and hence

$$\forall \varepsilon > 0, \ (x - \varepsilon, x + \varepsilon) \cap \mathbb{Q} \neq \varnothing.$$

Therefore, $\mathbb{R} \subset \overline{\mathbb{Q}}$ and thus $\overline{\mathbb{Q}} = \mathbb{R}$.

Since between any two real numbers there is always an irrational one, we can also prove that $\overline{\mathbb{R} \setminus \mathbb{Q}} = \mathbb{R}$...

We know that $\overline{(2,3]} = [2,3]$. Besides since $\{1\}$ is closed in $\mathbb{R}$ (because its complement in $\mathbb{R}$, being $(-\infty, 1) \cup (1, +\infty)$, is open since it is a union of open sets), one has $\overline{\{1\}} = \{1\}$. Consequently,

$$\overline{\{1\} \cup (2,3]} = \overline{\{1\}} \cup \overline{(2,3]} = \{1\} \cup [2,3]$$

(2) The interiors of $\mathbb{Q}$ and $\mathbb{R} \setminus \mathbb{Q}$ are both empty for a similar reason. We know that

$$x \in \overset{\circ}{\mathbb{Q}} \iff \exists r > 0 : \ (x - r, x + r) \subset \mathbb{Q}.$$

But, since an interval contains irrational numbers,

$$\forall r > 0 : \ (x - r, x + r) \not\subset \mathbb{Q}.$$

Thus $\overset{\circ}{\mathbb{Q}} = \varnothing$. Also, since an interval contains rational numbers, we obtain

$$\overset{\circ}{\widetilde{\mathbb{R} \setminus \mathbb{Q}}} = \varnothing.$$

(3) In fact, we can use the same counterexample for both cases. Let $A = \mathbb{Q}$ and let $B = \mathbb{R} \setminus \mathbb{Q}$. Then

$$\overline{A} \cap \overline{B} = \mathbb{R} \cap \mathbb{R} = \mathbb{R} \not\subset \varnothing = \overline{\varnothing} = \overline{\mathbb{Q} \cap \mathbb{R} \setminus \mathbb{Q}} = \overline{A \cap B}.$$

Also,

$$(A \cup B)^\circ = \overset{\circ}{\mathbb{R}} = \mathbb{R} \not\subset \varnothing = \varnothing \cup \varnothing = \overset{\circ}{A} \cup \overset{\circ}{B}.$$

(4) The answer is no for both cases. We give counterexamples. Consider $A_n = \left(-\frac{1}{n}, \frac{1}{n}\right)$, $n \geq 1$. Then $\overset{\circ}{A_n} = \left(-\frac{1}{n}, \frac{1}{n}\right)$. By Exercise 1.2.26, we know that

$$\bigcap_{n \geq 1} A_n = \bigcap_{n \geq 1} \left(-\frac{1}{n}, \frac{1}{n}\right) = \{0\}$$

and so

$$\left(\bigcap_{n \geq 1} A_n\right)^\circ = \overset{\circ}{\widetilde{\{0\}}} = \varnothing \neq \bigcap_{n \geq 1} \overset{\circ}{A_n} = \{0\}.$$

For the other equality, consider $A_n = \left[\frac{1}{n}, 1\right]$, $n \geq 1$. Hence $\overline{A_n} = A_n$ (for each $n$). Then by Exercise 1.2.26 again, we have

$$\bigcup_{n \geq 1} A_n = \bigcup_{n \geq 1} [1/n, 1] = (0, 1]$$

and so

$$\overline{\bigcup_{n\geq 1} A_n} = \overline{(0,1]} = [0,1] \neq \bigcup_{n\geq 1} \overline{A_n} = (0,1].$$

**SOLUTION 3.3.11.**

(1) Contrapositively, we show that if $A$ is countable, then it must have empty interior. So, assume that $A$ is countable and let $x \in \overset{\circ}{A}$. This means that

$$\exists r > 0, \ (x-r, x+r) \subset A.$$

Since $A$ is countable, by Proposition 1.1.75, $(x-r, x+r)$ must be countable too. But $(x-r, x+r)$ has the power of the continuum and so it cannot be countable. Therefore, $\overset{\circ}{A}$ contains no element!

(2) No! For instance, in the discrete topology of $\mathbb{R}$, $\mathbb{Q}$ is open and hence $\overset{\circ}{\mathbb{Q}} = \mathbb{Q} \neq \varnothing$, and yet $\mathbb{Q}$ is countable.

**SOLUTION 3.3.12.** Let $(X, d)$ be a metric space and let $x, y \in X$ such that $x \neq y$. Hence $d(x, y) > 0$. Set $r = \frac{d(x,y)}{3}$. Then $B(x, r)$ and $B(y, r)$ are two open balls containing $x$ and $y$ respectively. We are done as soon as we show that these two balls are also disjoint. Assume they were not and let $z \in B(x, r) \cap B(y, r)$. Then

$$d(x, z) < r \text{ and } d(y, z) < r.$$

By the triangle inequality, we know that

$$d(x, y) \leq d(x, z) + d(z, y) = d(x, z) + d(y, z) < 2r.$$

Since $r = d(x, y)/3$, we obtain

$$d(x, y) \leq 2d(x, y)/3 \text{ or merely } 1 \leq 2/3,$$

which is absurd. Therefore, the two balls are disjoint and so $(X, d)$ is Hausdorff.

**SOLUTION 3.3.13.** Let $x \in U$. Let $I_x$ be the largest open interval containing $x$ and contained in $U$. Such an interval exists because as $U$ is open, then for some $r > 0$, we have that

$$(x - r, x + r) \subset U.$$

Now, let

$$a_x = \inf\{a < x : \ (a, x) \subset U\} \text{ and } b_x = \sup\{b > x : \ (x, b) \subset U\}.$$

Hence $-\infty \leq a_x < x < b_x \leq \infty$ and then let $I_x = (a_x, b_x)$. It is clear that

$$x \in I_x \text{ and } I_x \subset U.$$

By Proposition 1.1.24, we know that

$$U = \bigcup_{x \in U} I_x.$$

It remains to show that $\{I_x\}_{x \in U}$ is a disjoint countable collection.

(1) Assume that $I_x \cap I_y \neq \varnothing$. Then $I_x \cup I_y$ too is an interval (Theorem 1.1.14). Since both $I_x$ and $I_y$ are open, so is their union. In other words, $I_x \cup I_y$ is an open interval which clearly contains $x$. But $I_x$ being the largest possible (containing $x$) yields:

$$I_x \cup I_y \subset I_x \text{ or simply } I_x \cup I_y = I_x.$$

A similar argument may be used to obtain

$$I_x \cup I_y \subset I_y \text{ or simply } I_x \cup I_y = I_y.$$

Accordingly,

$$I_x = I_y.$$

In other words, the collection of intervals $\{I_x\}_{x \in U}$ is disjoint.

(2) To see that $\{I_x\}_{x \in U}$ is countable, we know that each of these open intervals contains a rational number (remember that $\overline{\mathbb{Q}} = \mathbb{R}$). Since the intervals are disjoint, each one of them contains a *different* rational number. Hence, the collection can be put in a one-to-one correspondence with a countable set (here a subset of $\mathbb{Q}$).

Therefore, we may re-write the result as: *Any open set $U$ in usual $\mathbb{R}$ may be written as:*

$$U = \bigcup_{n \in \mathbb{N}} I_n$$

*where all $I_n$ are disjoint open intervals.*

**SOLUTION** 3.3.14. Suppose for the sake of contradiction that $D$ could be expressed as a disjoint union of open rectangles. Take some open rectangle $R$ from this union and let $x \in \text{Fr}R$, i.e. $x \in \overline{R} \setminus \overset{\circ}{R}$, that is, $x \in \overline{R}$ and $x \notin \overset{\circ}{R}$ or merely $x \notin R$ as $R$ is open. But

(1)
$$x \in \overline{R} \Longleftrightarrow \forall \varepsilon > 0, \ B(x, \varepsilon) \cap R \neq \varnothing$$

and

(2) $x \notin R$ forces $x$ to be in some other open rectangle $R'$ (as the union is assumed to be disjoint).

Since $R'$ is open and $x \in R'$, we know that

$$\exists r > 0, \ B(x, r) \subset R'...(*)$$

But for all $\varepsilon > 0$, $B(x, \varepsilon) \cap R \neq \varnothing$, so that in particular

$$B(x, r) \cap R \neq \varnothing.$$

This means that there is some $y$ in both $B(x, r)$ and $R$. By if $y$ is in $B(x, r)$, it is also in $R'$ by $(*)$. Therefore, $R \cap R' \neq \varnothing$, and this is a contradiction. Thus the union cannot be disjoint.

**SOLUTION 3.3.15.**

(1) The proof relies on Proposition 3.1.17. Let $A \subset X$. We take its closure, then the complement, then the closure etc... Also, we may start with $A^c$, then take its closure, then the complement etc...We can define two sequences of sets as follows:

$$A_1 = A, \ A_{2n} = \overline{A_{2n-1}} \ \text{and} \ A_{2n+1} = A_{2n}^c$$

and

$$B_1 = A^c, \ B_{2n} = \overline{B_{2n-1}} \ \text{and} \ B_{2n+1} = B_{2n}^c.$$

It is then clear that $A_2$, $A_4$ etc... are all closed. Hence $A_3$, $A_5$, etc... are all open. Thanks to the definition of the sequence $(A_n)$ (combined with Proposition 3.1.17), we may write

$$A_7 = A_6^c = \overline{A_5}^c = \overline{\overline{A_4^c}}^c = \overset{\circ}{A_4} = \overset{\circ}{\overline{A_3}}.$$

Hence clearly,

$$A_3 = \overset{\circ}{A_3} \subset \overset{\circ}{\overline{A_3}} = A_7 = \overset{\circ}{\overline{A_3}} \subset \overline{A_3}.$$

Passing to the closure, we obtain $\overline{A_3} = \overline{A_7}$ and so, $A_4 = A_8$. We may therefore prove by induction that:

$$\forall n \geq 4 : \ A_{n+4} = A_n.$$

We may argue similarly to show that

$$\forall n \geq 4 : \ B_{n+4} = B_n.$$

Therefore, with the periodicity of $(A_n)$ and $(B_n)$, we can observe that any $A_n$ or $B_n$ will be one of the *fourteen* sets:

$$A_1, A_2, \cdots, A_7; B_1, B_2, \cdots, B_7.$$

This finishes the proof.

(2) Let
$$A = (0,1) \cup (1,2) \cup \{3\} \cup ([4,5] \cap \mathbb{Q})$$

with respect to the standard topology of $\mathbb{R}$. Then the associated 14 different sets are:

$A_1 = A = (0,1) \cup (1,2) \cup \{3\} \cup ([4,5] \cap \mathbb{Q})$

$A_2 = \overline{A_1} = [0,2] \cup \{3\} \cup [4,5]$

$A_3 = A_2^c = (-\infty,0) \cup (2,3) \cup (3,4) \cup (5,\infty)$

$A_4 = \overline{A_3} = (-\infty,0] \cup [2,4] \cup [5,\infty)$

$A_5 = A_4^c = (0,2) \cup (4,5)$

$A_6 = \overline{A_5} = [0,2] \cup [4,5]$

$A_7 = A_6^c = (-\infty,0) \cup (2,4) \cup (5,\infty)$

$B_1 = A^c = (-\infty,0] \cup \{1\} \cup [2,3) \cup (3,4) \cup ([4,5] \setminus \mathbb{Q}) \cup (5,\infty)$

$B_2 = \overline{B_1} = (-\infty,0] \cup \{1\} \cup [2,\infty)$

$B_3 = B_2^c = (0,1) \cup (1,2)$

$B_4 = \overline{B_3} = [0,2]$

$B_5 = B_4^c = (-\infty,0) \cup (2,\infty)$

$B_6 = \overline{B_5} = (-\infty,0] \cup [2,\infty)$

$B_7 = B_6^c = (0,2),$

and as the French say: Voilà!

**SOLUTION** 3.3.16. Assume that $A$ is open. We are required to show that $A + B$ is open. It is clear that we can write

$$A + B = \bigcup_{b \in B} (A + \{b\}).$$

Since the arbitrary union of open sets is open, to show that $A + B$ is open, it then suffices to show that each $A + \{b\}$ is open where $b \in B$.

So let $x \in A + \{b\}$. Then there exists $a \in A$ such that $x = a + b$. Since $A$ is open, we have

$$\exists r > 0 : \ B(a,r) \subset A.$$

Hence (by an appeal to Exercise 2.5.14)

$$B(x,r) = B(a+b,r) = B(a,r) + \{b\} \subset A + \{b\}.$$

Thus $A + \{b\}$ is open.

**SOLUTION** 3.3.17. Let $a$ be a point in $A$. Since $A$ is open,

$$\exists r > 0, \ B(a, r) \subset A.$$

We need to show that $a$ is also a limit point of $A$, i.e. $a \in A'$, i.e.

$$\forall \varepsilon > 0, \ B(a, \varepsilon) \cap A \setminus \{a\} \neq \varnothing.$$

Let $\varepsilon > 0$. Choose a point $b$ such that $d(a, b) = \frac{1}{3} \min(r, \varepsilon)$. Then $b \in B(a, \varepsilon)$ and also $b \in B(a, r) \subset A$. Since $a \neq b$, $b \in A \setminus \{a\}$. Thus $B(a, \varepsilon) \cap A \setminus \{a\}$ is non-empty. The solution is complete.

**SOLUTION** 3.3.18.

(1) First, $\overset{\circ}{A} = \varnothing$. One way of seeing this is the following

$$A \subset \mathbb{Q} \Longrightarrow \overset{\circ}{A} \subset \overset{\circ}{\mathbb{Q}} = \varnothing \Longrightarrow \overset{\circ}{A} = \varnothing.$$

All points of $A$ are isolated. For instance 1 is an isolated point. For we can easily choose an $r > 0$ (for instance $r = \frac{1}{3}$) such that

$$(1 - r, 1 + r) \cap A \setminus \{1\} = \varnothing.$$

The only limit point is 0. By the Archimedian Property, for all $\varepsilon > 0$ there exists $n \in \mathbb{N}$ such that $\frac{1}{n} < \varepsilon$. Hence

$$\forall \varepsilon > 0 : \ (-\varepsilon, \varepsilon) \cap A \setminus \{0\} \neq \varnothing.$$

(2) Since 0 is the only limit point, we have

$$\overline{A} = A \cup A' = \left\{ \frac{1}{n} : \ n \geq 1 \right\} \cup \{0\}.$$

Thus $A$ is not closed. Finally, since $\overset{\circ}{A} = \varnothing$, $\mathrm{Fr}(A) = \overline{A}$.

(3) We need to verify that $\overset{\circ}{\overline{A}} = \varnothing$. If $x$ were a point in $\overset{\circ}{\overline{A}}$, then there would exist an $r > 0$ such that $(x - r, x + r) \subset \overline{A}$ which obviously does not hold. Therefore, $A$ is nowhere dense.

**SOLUTION** 3.3.19. In the usual topology of $\mathbb{R}$, the following set

$$A = (0, 1) \cup (1, 2] \cup \{3\}$$

will do as the reader can easily check that

$$\overset{\circ}{A} = (0, 1) \cup (1, 2),$$

$$\overline{A} = [0, 1] \cup [1, 2] \cup \{3\} = [0, 2] \cup \{3\},$$

$$\overset{\circ}{\overline{A}} = (0, 2),$$

and

$$\overline{\overset{\circ}{A}} = [0, 1] \cup [1, 2] = [0, 2],$$

i.e. the five sets are pairwise different.

**SOLUTION** 3.3.20. We show that $A' = \varnothing$. Since $X$ is discrete, $\{x\}$ is a (an open) neighborhood of $x \in X$. So

$$\exists U = \{x\} \in \mathcal{N}(x), \ U \cap A - \{x\} = \varnothing.$$

Since this is true for any $x \in X$, it follows that $A' = \varnothing$.

**SOLUTION** 3.3.21.

(1) First, since $A$ is a non-empty and bounded set of $\mathbb{R}$, both $\inf A$ and $\sup A$ exist. We know that

$$a = \inf A \Longleftrightarrow \begin{cases} \forall x \in A : \ x \geq a, \\ \forall \varepsilon > 0, \exists x_\varepsilon \in A : \ a \leq x_\varepsilon < a + \varepsilon. \end{cases}$$

Hence $x_\varepsilon \in [a, a + \varepsilon) \subset (a - \varepsilon, a + \varepsilon)$. We also know that in $\mathbb{R}$

$$a \in \overline{A} \Longleftrightarrow \forall \varepsilon > 0, \ (a - \varepsilon, a + \varepsilon) \cap A \neq \varnothing.$$

So let $\varepsilon > 0$. Then there is an $x_\varepsilon \in A$ s.t. $x_\varepsilon \in (a - \varepsilon, a + \varepsilon)$. Therefore

$$(a - \varepsilon, a + \varepsilon) \cap A \neq \varnothing.$$

Thus $a \in \overline{A}$.

The proof for $\sup A$ can be dealt with similarly...

(2) The answer is no! Take $A = (0, 1)$ in the discrete topology of $\mathbb{R}$, say. Then $A$ is closed, i.e. $\overline{A} = A$. However,

$$\inf A = 0 \notin \overline{A} = (0, 1) \text{ and } \sup A = 1 \notin \overline{A} = (0, 1).$$

(3) No! In usual $\mathbb{R}$, let $A = \{0\}$. Then $\overset{\circ}{A} = \varnothing$ and hence

$$\sup A = \inf A = 0 \notin \overset{\circ}{A}.$$

**SOLUTION** 3.3.22.

(1) First, we justify the existence of $\sup \overline{A}$. Since $A$ is bounded above, for some $m, M$:

$$A \subset (m, M]$$

($M$ being real and $m$ any number small than $M$ and it may even be $-\infty$). Depending on $m$, we then have $\overline{A} \subset (m, M]$ or $\overline{A} \subset [m, M]$. So, in either case, $\overline{A}$ is bounded above. Since it is also non empty (why?), $\sup \overline{A}$ exists. Obviously, we have

$$A \subset \overline{A} \Longrightarrow \sup A \leq \sup \overline{A}.$$

Let us prove the other inequality. Let $M = \sup A$ and let $x > M$. Setting $r = \frac{x - M}{2}$ (then $r > 0$), we claim that

$$A \cap (x - r, x + r) = \varnothing.$$

To see this, assume there is some $a$ in $A$ such that $a \in (x - r, x + r)$. Hence $a > x - r > M$ and then $a$ would be bigger than $M = \sup A$, a clear contradiction. Thus $x \notin \overline{A}$ (that is, we have shown that if $x \in \overline{A}$, then $x \leq M$) and hence $M$ is an upper bound for $\overline{A}$. This certainly leads to

$$\sup \overline{A} \leq M = \sup A,$$

completing the proof.

(2) The main point is that the existence of $\sup A$ does not imply any more that of $\sup \overline{A}$ in another topological space (besides what does "bounded" mean in an arbitrary topological space?).

In $\mathbb{R}$ equipped with the *indiscrete topology*, take $A = [0, 1]$. Then $\sup A = 1$. Since $A$ is dense in $\mathbb{R}$, $\sup \overline{A} = +\infty$ (or it does not exist as some prefer to say).

(3) First, and on the contrary to the "closure case", $\sup \overset{\circ}{A}$ may not even exist even if $\sup A$ exists and in the usual topology setting! For example, take $A = \{1\}$. Then $\overset{\circ}{A} = \varnothing$ and so

$$\sup \overset{\circ}{A} = -\infty \neq \sup A = 1.$$

Now even when $\sup \overset{\circ}{A}$ exists, then it need not be equal to $\sup A$. Indeed, in usual $\mathbb{R}$, let $A = (-1, 1) \cup \{2\}$. Then

$$\sup A = 2 \neq \sup \overset{\circ}{A} = \sup(-1, 1) = 1.$$

**REMARK.** We say a few words about $\sup \varnothing = -\infty$. It is known that if $A$ is a bounded subset, then $\inf A \leq \sup A$ iff $A$ is *non-empty*. So, since $\varnothing \subset \mathbb{R}$, every element of $\mathbb{R}$ is an upper bound for $\varnothing$ and the least upper bound is then $-\infty$. Similarly, every element of $\mathbb{R}$ is a lower bounded for $\varnothing$ and biggest among them is "$+\infty$", i.e. $\inf \varnothing = +\infty$. This, thankfully, agrees with what we have just recalled above, that is, $A$ is empty iff $\sup A < \inf A$.

**SOLUTION 3.3.23.**

(1) The inclusion $\overline{B(x, r)} \subset B_c(x, r)$ is easily proved. We know that

$$B(x, r) \subset B_c(x, r).$$

Passing to the closure, we obtain

$$\overline{B(x, r)} \subset \overline{B_c(x, r)} = B_c(x, r)$$

as $B_c(x, r)$ is closed. The proof is complete.

(2) Let $X = [-1, 0] \cup [1, 2]$ considered as a metric *subspace* of usual $\mathbb{R}$. We have

$$B(1, 1) = \{x \in X : \ |x - 1| < 1\} = ([-1, 0] \cup [1, 2]) \cap (0, 2) = [1, 2).$$

Hence $\overline{B(1, 1)} = \overline{[1, 2)} = [1, 2]$ while

$$B_c(1, 1) = \{x \in X : \ |x - 1| \le 1\} = ([-1, 0] \cup [1, 2]) \cap [0, 2] = \{0\} \cup [1, 2].$$

Thus $B_c(1, 1) \not\subset \overline{B(1, 1)}$.

We give another example. Let $X$ be a set with $\operatorname{card} X \ge 2$. Let us associate with $X$ the discrete metric. Let $x \in X$. We know that $B(x, 1) = \{x\}$ and since every subset in a discrete metric space is closed (and open!), we get

$$\overline{B(x, 1)} = \overline{\{x\}} = \{x\}.$$

This is on the one hand. On the other hand, we have

$$B_c(x, 1) = \{y \in X : \ d(x, y) \le 1\} = X \not\subset \overline{B(x, 1)}.$$

**SOLUTION 3.3.24.** Since $\varnothing, Y \in T$, $f^{-1}(\varnothing) = \varnothing$ and $f^{-1}(Y) = X$, we get that $\varnothing, X \in T'$.

Now, let $(V_i)_{i \in I}$ be in $T'$. Then for each $i \in I$, there is some $U_i$ in $T$ such that $V_i = f^{-1}(U_i)$. Hence

$$\bigcup_{i \in I} V_i = \bigcup_{i \in I} f^{-1}(U_i) = f^{-1}\left(\bigcup_{i \in I} U_i\right)$$

belongs to $T'$ because $\bigcup_{i \in I} U_i \in T$.

In the end, let $V_1$ and $V_2$ be in $T'$. So there are $U_1$ and $U_2$ in $T$ such that $f^{-1}(U_1) = V_1$ and $f^{-1}(U_2) = V_2$. Whence

$$V_1 \cap V_2 = f^{-1}(U_1) \cap f^{-1}(U_2) = f^{-1}(U_1 \cap U_2) \in T'$$

as $U_1 \cap U_2 \in T$. The proof is therefore complete.

**SOLUTION 3.3.25.** The proof is essentially very similar to the one just before. First, $\varnothing$ and $A$ both belongs to $T_A$ as

$$\varnothing = A \cap \varnothing \text{ and } A = A \cap X.$$

Second, let $\{V_i\}_{i \in I}$ be a collection in $T_A$. Then each $V_i$ may be written as $V_i = A \cap U_i$ for some $U_i \in T$. Hence

$$\bigcup_{i \in I} V_i = \bigcup_{i \in I} (A \cap U_i) = A \cap \left(\bigcup_{i \in I} U_i\right) \in T_A$$

because $\bigcup_{i \in I} U_i \in T$.

Finally, let $V_1$ and $V_2$ be in $T_A$. Then there are $U_1$ and $U_2$ (both in $T$) such that

$$V_1 = A \cap U_1 \text{ and } V_2 = A \cap U_2.$$

Therefore,

$$V_1 \cap V_2 = (A \cap U_1) \cap (A \cap U_2) = A \cap (U_1 \cap U_2)$$

is in $T_A$ as $U_1 \cap U_2 \in T$.

**SOLUTION 3.3.26.**

(1) Yes the set $\{3\}$ is open in $A = [0, 1) \cup \{3\}$. To show this we need to write it as an intersection of $A$ with an open set in $\mathbb{R}$. One possible choice is the following

$$\{3\} = ([0, 1) \cup \{3\}) \cap (2, 5).$$

(2) The answer is again yes as one can do the following

$$[0, 1) = [0, 1] \cap \underbrace{(-1, 1)}_{\text{open in } \mathbb{R}} \text{ and } (0, 1) = [0, 1] \cap \underbrace{(0, 1)}_{\text{open in } \mathbb{R}}.$$

(3) Yes indeed. Write

$$\{n\} = \mathbb{N} \cap (n - 1, n + 1).$$

(4) Both $[0, 1]$ and $(2, 3)$ are open in $A$ since

$$[0, 1] = A \cap (-1, 2) \text{ and } (2, 3) = A \cap (2, 3).$$

We can deduce that $[0, 1]$ and $(2, 3)$ are also closed (in $A$!). This is simply because the complement of $[0, 1]$ in $A$ is $(2, 3)$ and vice versa.

(5) The closure of $\left(0, \frac{1}{2}\right)$ in $\mathbb{R}$ is $\left[0, \frac{1}{2}\right]$. Hence the closure of $\left(0, \frac{1}{2}\right)$ in $A$ is

$$\overline{(0, 1/2)}^A = [0, 1/2] \cap A = (0, 1/2].$$

(6) Let us denote the subspace topology on $A$ by $T_A$. It is defined as

$$T_A = \{A \cap U : U \in T\}.$$

Hence we can find $T_A$ explicitly and we have

$$T_A = \{\varnothing, \{c, d\}, A\}.$$

We observe that the smallest closed set containing $\{b, d\}$ is $A$. So $\overline{\{b, d\}}^A = A$ (i.e. $\{b, d\}$ is dense in $A$). We can also obtain the same result using the relativity of closures as follows

$$\overline{\{b, d\}}^A = \overline{\{b, d\}}^X \cap A = \{b, c, d, e\} \cap \{b, c, d\} = \{b, c, d\} = A.$$

**SOLUTION 3.3.27.**

(1) Yes $X$ is clopen in $A$. Since $[\sqrt{2}, \pi]$ is closed in $\mathbb{R}$, $X$ is closed in $A$. Also, since $\sqrt{2}, \pi \notin \mathbb{Q}$, one can write

$$X = A \cap [\sqrt{2}, \pi] = A \cap (\sqrt{2}, \pi).$$

As $(\sqrt{2}, \pi)$ is open in $\mathbb{R}$, then $X$ is open in $A$, establishing the "clopenness" of $X$ in $A$.

(2) Yes $Y$ is clopen in $B$. Since $[0, 2]$ is closed in $\mathbb{R}$, so is $Y$ in $B$. Now since $0, 2 \notin \mathbb{R} \setminus \mathbb{Q}$, we can write

$$Y = B \cap [0, 2] - B \cap (0, 2),$$

meaning that $Y$ is open in $B$ as well.

(3) Yes $Z$ is clopen in $A$. This is easily seen from

$$Z = A \cap [\sqrt{2}, \pi) = A \cap (\sqrt{2}, \pi) = A \cap [\sqrt{2}, \pi].$$

(4) $Z'$ is not clopen, more precisely, it is neither closed nor open. For example, if it were open we would have: for all $x \in Z'$, there is some $r > 0$ such that $B(x, r)$ (the open ball in $B$) is contained in $Z'$. In particular, for $x = \sqrt{2}$ there corresponds an $r > 0$ such that

$$B(\sqrt{2}, r) = \{y \in B : |\sqrt{2} - y| < r\} = B \cap (\sqrt{2} - r, \sqrt{2} + r) \subset Z',$$

which does not hold as

$$\forall r > 0 : B \cap (\sqrt{2} - r, \sqrt{2} + r) \not\subset Z'.$$

**SOLUTION 3.3.28.** Since $[0, 1] \cap \mathbb{Q} \subset [0, 1]$, it follows that

$$\overline{[0, 1] \cap \mathbb{Q}} \subset \overline{[0, 1]} = [0, 1].$$

Now let $x \in [0, 1]$, then obviously

$$\forall \varepsilon > 0 : [0, 1] \cap \mathbb{Q} \cap (x - \varepsilon, x + \varepsilon) \neq \varnothing.$$

Thus $x \in \overline{[0, 1] \cap \mathbb{Q}}$.

**SOLUTION 3.3.29.**

(1) The elements of $X$ are real numbers in $[0, 1]$ whose digits are constituted of the numbers 3 and/or 5.

(2) The following observation

$$X \cap (0.3\overline{5}, 0.5\overline{3}) = \varnothing \text{ (why?)}$$

shows that $X$ cannot be dense in $[0, 1]$.

**SOLUTION 3.3.30.** We give three ways of answering this question.

(1) Since $X$ is not Hausdorff, it cannot be metrizable.

(2) This is somehow similar to the previous method. Let $B(x, r)$ be the open ball of center $x$ and radius $r = \frac{d(x,y)}{3} > 0$ ($x \neq y$). It is an open set in $X$. However, $B(x, r)$ contains $x$, i.e. $B(x, r) \neq \varnothing$ and $B(x, r)$ does not contain $y$, i.e. $B(x, r) \neq X$. Thus $B(x, r)$ is another open set in $X$. This clearly leads to a contradiction. Therefore, $X$ is not metrizable.

(3) If there is a metric that induces the topology of $X$, then $X$ and $\varnothing$ are not the only closed sets in $X$ since every finite set is closed in a metric space. Accordingly, $X$ cannot be metrizable.

**SOLUTION 3.3.31.**

(1) Let us show that $T$ is in effect a topology on $\mathbb{R}$. First $\varnothing$ belongs to $T$ by definition while $\mathbb{R}$ belongs to $T$ since its complement is finite (it is the empty set!).

Let $U$ and $V$ be two elements of $T$, i.e. $U^c$ and $V^c$ are both finite (we assume both $U$ and $V$ are not empty, otherwise this is obvious). We need to show that $U \cap V$ belongs to $T$, i.e. $(U \cap V)^c$ is finite. But $(U \cap V)^c = U^c \cup V^c$ and it is finite. Hence $U \cap V$ does belong to $T$.

Now let $\{U_i\}_{i \in I}$ be an arbitrary collection of elements of $T$. If $U_i = \varnothing$ for all $i \in I$, then their union is the empty set and hence it belongs to $T$. If at least one of these elements is non empty, call it $U_j$ (hence $U_j^c$ is necessarily finite), then we can write

$$\left( \bigcup_{i \in I} U_i \right)^c = \bigcap_{i \in I} U_i^c \subset U_j^c$$

and hence $\left( \bigcup_{i \in I} U_i \right)^c$ is finite. This finishes the proof.

(2) Obviously $\varnothing$ and $\mathbb{R}$ are closed. The only other closed sets are the finite ones. For if $A$ is finite, then it is closed as $A^c$ is open because its complement (which is $A$) is finite. And if $A$ is closed, then $A^c$ is open and hence $(A^c)^c = A$ is finite.

(3) Every finite subset is closed in $T$ and also in standard $\mathbb{R}$. There are many infinite subsets which are closed in the standard $\mathbb{R}$ (for example every interval of the type $[a, b]$). Thus the standard topology has more closed sets, so it is finer than $T$.

(4) It is not Hausdorff since if it were, then for any $x, y \in \mathbb{R}$ such that $x \neq y$ there would exist two neighborhoods $U \in \mathcal{N}(x)$ and $V \in \mathcal{N}(y)$ (belonging to $T$) such that $U \cap V = \varnothing$. Hence

$$U^c \cup V^c = \varnothing^c = \mathbb{R}$$

and this contradicts the finiteness of both $U^c$ and $V^c$.

(5) No, since a metrizable space has to be Hausdorff.

(6) (a) If $A$ is finite, then it is closed (by Question 2) and hence $\overline{A} = A$.

As for $\overset{\circ}{A}$ we will show that it equals the empty set. Let $B$ be an open set contained in $A$. Then $A^c$ is contained in $B^c$ and since $B^c$ is finite (why?), so is $A^c$ and hence $\mathbb{R} = A \cup A^c$ would be finite! This is impossible and so $\overset{\circ}{A} = \varnothing$.

(b) If $A$ is infinite, then two cases must be looked at.

(i) If $A^c$ is finite, then $A$ is open and hence $\overset{\circ}{A} = A$. Also since $A^c$ is finite, then from the previous question $A^c$ has an empty interior. Therefore, Proposition 3.1.17 yields

$$\left(\overline{A}\right)^c = \left(\overset{\circ}{A^c}\right) = \varnothing \implies \overline{A} = \mathbb{R},$$

i.e. $A$ is dense in $\mathbb{R}$ in this case.

(ii) Now the case $A^c$ being infinite. If $B \subset A$ where $B$ is open, then $A^c \subset B^c$ which is impossible since $B^c$ is finite. This means that $\overset{\circ}{A} = \varnothing$.

Now, since the only closed set which can contain the infinite set $A$ is $X$, we deduce immediately that $\overline{A} = X$.

(7) Yes, $\mathbb{R}$ is separable since $\mathbb{Q}$ is a countable subset of $\mathbb{R}$ which is dense in $\mathbb{R}$ (as it satisfies the hypotheses of Question (6-b-i)).

(8) If $X$ is finite, then $T$ becomes the discrete topology for a simple reason. That is, since $X$ is finite, any subset $A$ of $X$ will have a finite complement. Hence every subset is open and thus every subset is closed too.

**Solution 3.3.32.** No, $T$ is not a topology on $\mathbb{R}$. Assume it is and take the two sets $U = (-\infty, 0)$ and $V = (0, \infty)$. They are both open in $T$ since their complements are infinite. However,

$$(U \cup V)^c = U^c \cap V^c = \{0\} \text{ is finite and hence } U \cup V \notin T.$$

Thus $T$ is a not a topological space.

**Solution 3.3.33.**

(1) Yes $T$ is indeed a topology on $X$. For a change and also to facilitate the proof we will prove that $T$ is a topology using closed sets and hence we need to show that $\varnothing$ and $X$ are elements of $T$ (which is obvious here), that the finite union of

closed sets and the arbitrary intersection of closed sets are all in $T$. If we want a set to be closed in $T$, then it has to be countable. We are done as we know that the finite union and the arbitrary intersection of countable sets remain countable.

(2) No. If it were, $X$, which is uncountable, would be a union of two countable sets!! (cf. Exercise 3.3.31).

(3) Remember that the closed sets are the countable ones (together with $X$ and $\varnothing$). Now if $A$ is a closed set different from $X$, then it is countable and so will be any subset of $A$.

(4) Let $\{U_n\}_n$ be a countable family of open sets in $T$. We may assume that they are all non-empty, otherwise this will be trivial. To show $\displaystyle\bigcap_{n=1}^{\infty} U_n$ is open, i.e. $\displaystyle\left(\bigcap_{n=1}^{\infty} U_n\right)^c$ is closed, i.e. countable. But

$$\left(\bigcap_{n=1}^{\infty} U_n\right)^c = \bigcup_{n=1}^{\infty} U_n^c$$

which is countable as a countable union of countable sets.

This result may fail to hold in usual $\mathbb{R}$ as shown by the classical example $U_n = (-\frac{1}{n}, \frac{1}{n})$, $n \in \mathbb{N}$.

(5) The proof is very similar to that of the non-Hausdorffness of the space. Assume $U_1 \cap U_2 \cap \cdots \cap U_n = \varnothing$ where the $U_i$ ($i = 1, 2, \cdots, n$) are all open. Then

$$U_1^c \cup U_2^c \cup \cdots \cup U_n^c = \mathbb{R}$$

which is impossible since the left hand side is countable while the right hand side is not. Thus the finite intersection of non empty open sets is non-empty.

This result is not true in usual $\mathbb{R}$ in general. Consider for instance $(0, 1)$ and $(2, 3)$.

(6) No! Since $\mathbb{Q}$ is countable, it is closed. Thus $\overline{\mathbb{Q}} = \mathbb{Q} \neq \mathbb{R}$.

We claim that every uncountable set is dense in $\mathbb{R}$: To see this, let $A$ be an uncountable subset of $\mathbb{R}$. Then $A$ is not closed. Besides, $A$ cannot be a subset of any closed set apart from $X$ (why?). Thus $A$ is dense in $\mathbb{R}$ equipped with this topology and hence so are the other two given sets.

(7) Any *countable* set will make the topology $T$ discrete. For if $X$ is countable, $\{x\}$ ($x \in X$) will be clopen as $\{x\}$ and $X - \{x\}$ are both closed since they are both countable.

**SOLUTION 3.3.34.**

(1) It is a routine by now and it is left to the interested reader.

(2) No $T$ is not Hausdorff. Let $x, y \in X$ such that $x \neq y$ ($x$ or $y$ may equal $a$). Let $U$ and $V$ be two open sets containing $x$ and $y$ respectively. Then they contain $a$ too so that

$$\{a\} \subset U \cap V, \text{ i.e. } U \cap V \neq \varnothing.$$

Consequently, $X$ cannot be Hausdorff.

(3) Remember that

$$x \in \{a\}' \iff \forall U \in \mathcal{N}(x): \ U \cap \{a\} - \{x\} \neq \varnothing.$$

Since $U$ contains $a$, we see immediately that only $a$ does not belong to $\{a\}'$ and hence $\{a\}' = X - \{a\}$.

(4) Let $U \neq \varnothing$ be an open set in $T$. Then $a \in U$ and hence

$$\{a\} \subset U \implies \{a\}' \subset U' \implies X - \{a\} \subset U' \subset X.$$

Two cases are to be discussed:

(a) If $U' = X$, then

$$\overline{U} = U' \cup U = X \cup U = X.$$

(b) If $U' = X - \{a\}$, then

$$\overline{U} = U' \cup U = X - \{a\} \cup U = X$$

as $a \in U$.

Thus in either case open sets are dense in $X$.

(5) Let $V$ be a subset of $X$. It is clear that we only have one of the two possibilities: $a \in V$ or $a \notin V$.

(a) $a \in V$: In this case $V$ is *open* by definition.

(b) $a \notin V$: In this case, $a \in V^c$ and so by definition $V^c$ is open, that is, $(V^c)^c = V$ is *closed*.

(6) It is obvious that $\varnothing$ and $X$ are clopen. Now, let $U$ be clopen with $U \neq \varnothing$ and $U \neq X$. Then $U^c$ is also clopen. Since all (*non-empty*) open subsets are dense, We deduce that

$$\overline{U} = \overline{U^c} = X.$$

Hence

$$\text{Fr}\, U = \overline{U} \cap \overline{U^c} = X \cap X = X$$

and on the other hand (as $U$ and $U^c$ are closed)

$$\text{Fr}\, U = \overline{U} \cap \overline{U^c} = U \cap U^c = \varnothing$$

and this is a contradiction. Therefore, the only possible clopen subsets are $\varnothing$ and $X$.

(7) As in the case of $\{a\}'$, we find that $A' = X - \{a\}$ for any set $A$ which contains $a$.

(8) For any $X$, $X$ is separable because $\{a\}$ is a countable (finite!) subset of $X$ which is dense as $\{a\}$ is open in $T$.

   If $X$ is uncountable, then $X - \{a\}$ is not separable. To see this, assume that $A$ is a subset of $X - \{a\}$ which is countable and such that $\overline{A} = X - \{a\}$. Then clearly $a \notin A$ and so $A$ is closed. Therefore, $A = X - \{a\}$. But, this cannot occur since $A$ has to be countable whereas $X - \{a\}$ is uncountable.

(9) Let $A$ be a proper subset of $X$. We have to show that $\overset{\circ}{\overline{A}} = \varnothing$. Since $\overline{A}$ is closed, $a \notin \overline{A}$. The biggest *open* set contained in $\overline{A}$ must contain $\{a\}$. Thus $\overset{\circ}{\overline{A}} = \varnothing$.

(10) The induced topology is the discrete one. To see this, *any* $U \subset X - \{a\}$ is closed. This implies that *any* subset of $X - \{a\}$ is open too...Alternatively, we can show that every singleton is open. Indeed, every singleton $\{x\}$ ($x \in X - \{a\}$) is open in $X - \{a\}$ as it can be written as $X - \{a\} \cap \{x, a\}$...

   Finally, this induced topology is of course Hausdorff.

## SOLUTION 3.3.35.

(1) Let us show that $T$ is a topology on $[-a, a]$. First $\varnothing \in T$ since $\{0\} \not\subset \varnothing$ and $X = [-a, a] \in T$ since $(-a, a) \subset X = [-a, a]$.

   Now let $U$ and $V$ be both in $T$ and hence ($\{0\} \not\subset U$ or $(-a, a) \subset U$) and ($\{0\} \not\subset V$ or $(-a, a) \subset V$). In all possible cases we will have $U \cap V \in T$.

   Finally, let $\{U_i\}_{i \in I}$ be an arbitrary collection of elements of $T$. If $\{0\} \not\subset U_i$ for all $i \in I$, then $\{0\} \not\subset \bigcup_{i \in I} U_i$ and hence $\bigcup_{i \in I} U_i \in T$. If at least one $U_j$ does not contain $\{0\}$, then it will contain $(-a, a)$ and hence $(-a, a) \subset U_j \subset \bigcup_{i \in I} U_i$ and this also means that $\bigcup_{i \in I} U_i \in T$.

(2) The closed sets in this topology are $\{a\}$, $\{-a\}$, $\{-a, a\}$, $\varnothing$, $[-a, a]$ and any subset of $[-a, a]$ containing $\{0\}$.

(3) The set $A = \{\frac{a}{3}\}$ is not closed, but from the previous question, $\{0, \frac{a}{3}\}$ is a closed set and it is clearly the smallest set which contains $A$. Thus

$$\overline{A} = \left\{0, \frac{a}{3}\right\}.$$

(4) Let $B$ be any subset of $(-a, a) \subset X$. Let us show that 0 is a limit point for $B$, i.e. $0 \in B'$. We recall that

$$0 \in B' \iff \forall U \in \mathcal{N}(0): \ U \cap B - \{0\} \neq \varnothing.$$

Let $U$ be a (an open) neighborhood of 0. Then necessarily, $(-a, a) \subset U$. Therefore,

$$U \cap B - \{0\} \neq \varnothing,$$

as desired.

**SOLUTION** 3.3.36.

(1) First, $\varnothing \in T$ since $\exists a = 0 \in [0, 2]$ such that $\varnothing = [0, 0)$ and $X \in T$ since $\exists a = 2 \in [0, 2]$ such that $X = [0, 2)$. Now let $[0, a_i)_{i \in I}$ be an arbitrary collection of elements in $T$. Then

$$\bigcup_{i \in I} [0, a_i) = [0, a) \text{ where } a = \sup_{i \in I} a_i \in [0, 2].$$

This means that $\bigcup_{i \in I} [0, a_i) \in T$. Lastly, let $[0, a)$ and $[0, b)$ be two elements of $T$ where $0 \leq a \leq 2$ and $0 \leq b \leq 2$. Then one has

$$[0, a) \cap [0, b) = [0, c) \text{ where } c = \min(a, b) \in [0, 2]$$

and so $[0, a) \cap [0, b) \in T$.
(2) One example among many is the following: Take $U_n = [0, \frac{2}{n})$ for $n \geq 1$. It belongs to $T$ for all $n \geq 1$. However,

$$\bigcap_{n=1}^{\infty} [0, 2/n) = \{0\}$$

which cannot be an element of $T$ as it cannot be written in the form $[0, a)$ for any $a \in [0, 2]$.
(3) This topology cannot be separated as any two non-empty open sets will both contain zero and hence their intersection will never be empty.
(4) The closed sets are of the form $[b, 2)$ where $0 \leq b \leq 2$ (why?).
(5) There is somehow a technical way of answering this question but there is a more direct way of answering it. The closure of $A = [1, \frac{3}{2}]$ is by definition the smallest closed set containing $A$. But we have just seen that closed sets in $X$ are of the form $[b, 2)$ where $0 \leq b \leq 2$. Therefore the closure of $A$ is $[1, 2)$.

The interior of $A$, i.e. the largest open set contained in $A$, is empty as open sets are of the form $[0, a)$ where $0 \leq a \leq 2$ and none of them can be contained in $A$.

(6) Let $B = \mathbb{Q} \cap [0, 2)$. It is obviously countable. Besides, its closure, the smallest set containing it, is of the form $[b, 2)$ where $0 \leq b \leq 2$. Hence

$$\overline{\mathbb{Q} \cap [0, 2)} = [0, 2),$$

proving the density of $B$ in $(X, T)$. Thus $X$ is separable.

**SOLUTION** 3.3.37.

(1) Let $x \in X$. Since we always have $\{x\} \subset \overline{\{x\}}$, we only show that $\overline{\{x\}} \subset \{x\}$ or equivalently $\{x\}^c \subset \overline{\{x\}}^c$. Let $y \in \{x\}^c$, i.e. $y \notin \{x\}$ or $y \neq x$. Since $T$ is Hausdorff,

$$\exists (U, V) \in \mathcal{N}(x) \times \mathcal{N}(y): \ U \cap V = \varnothing.$$

This implies that $V \cap \{x\} = \varnothing$. Thus $y \notin \overline{\{x\}}$, i.e. $y \in \overline{\{x\}}^c$.

(2) The result is no longer true if $T$ is not assumed to be Hausdorff. Consider the indiscrete topology $T$ (which is not separated), on the set $X = \{1, 2\}$, defined by

$$T = \{\varnothing, \{1, 2\}\}.$$

It is plain that $\{2\}$ is not closed as the *only* closed sets are $\varnothing$ and $\{1, 2\}$.

(3) Well, an example of that is the co-finite topology (Exercise 3.3.31) in which all finite sets, and in particular singletons, are closed.

(4) Let $U_x$ be an open set containing $x$. We need to establish that $\bigcap_{x \in X} U_x = \{x\}$. Since for all $x \in X$, $x \in U_x$, we immediately see that $\{x\} \subset \bigcap_{x \in X} U_x$.

Conversely, let $y \in \bigcap_{x \in X} U_x$ with $y \neq x$. Since $X$ is Hausdorff, for some open set $U$ containing $x$ and for some open set $V$ containing $y$ we have $U \cap V = \varnothing$. But $y$ is in all open sets which contain $x$ and hence $U \cap V \neq \varnothing$ which is a contradiction and so $y = x$. The proof is complete.

(5) Let $X$ be $\mathbb{R}$ equipped with the co-finite topology. That $X$ is *not* Hausdorff was already established in Exercise 3.3.31. We show that the intersection of all open sets containing $x$ is $\{x\}$ itself. We only show $\bigcap_{x \in U} U \subset \{x\}$ where $U$ is open in $X$ and contains $x$. Let $y \in \bigcap_{x \in U} U$ and assume $y \neq x$. Then for all $U$ s.t. $x \in U$, we have $y \in U$. In particular, $\mathbb{R} - \{y\}$, being an open set (why?) which contains $x$, would have to contain $y$ too which is absurd! Thus $y = x$ and this completes the proof.

**SOLUTION 3.3.38.**

(1) $\mathbb{R}$ is obviously a union of elements of $\mathcal{B}$. An intersection of two elements of $\mathcal{B}$ is again an element of $\mathcal{B}$ for it is either the empty set or an interval of the same type as the elements of $\mathcal{B}$. Therefore, $\mathcal{B}$ is a base for $\mathbb{R}$.

(2) $\mathbb{R}_\ell$ is strictly finer than $\mathbb{R}$. We have to show that any open set in $\mathbb{R}$ is an open set in $\mathbb{R}_\ell$ or in terms of bases, any basis element in $\mathbb{R}$, i.e. an open interval, can be written as a union of members of the basis of $\mathbb{R}_\ell$. This is illustrated by

$$\forall a, b \in \mathbb{R}, \ (a,b) = \bigcup_{n \in \mathbb{N}} [a + 1/n, b).$$

To finish the proof, we need to exhibit an open set in $\mathbb{R}_\ell$ which is not one in $\mathbb{R}$. One choice is $[0, 1)$ and the proof is complete.

(3) First, from the previous question all open sets in $\mathbb{R}$ are open in $\mathbb{R}_\ell$. The same thing for closed sets. There are many other sets that are open and/or closed in $\mathbb{R}_\ell$. For example,

$$(-\infty, b) = \bigcup_{n \in \mathbb{N}} [-n, b), \quad [a, \infty) = \bigcup_{n \in \mathbb{N}} [a, n)$$

are open. They are also closed as

$$(-\infty, b)^c = [b, \infty) \text{ and } [a, \infty)^c = (-\infty, a)$$

are open. Also $[a, b)$ is open and it is also closed as

$$[a, b)^c = (-\infty, a) \cup [b, \infty)$$

is open.

There are of course non-clopen sets. For instance, $[a, b]$ is closed but not open (why?), and $(a, \infty)$ is open and not closed. Finally, there are sets which are neither open nor closed like $(a, b]$ (prove it!).

(4) Yes, $\mathbb{R}_\ell$ is separated since $\mathbb{R}_\ell$ is finer than $\mathbb{R}$. This was discussed and proved in the section "True or False" of this chapter.

We propose another method to show that $\mathbb{R}_\ell$ is Hausdorff. Let $x \neq y$ be two distinct reals. WLOG, take $x < y$ for example. Then it is clear that $[x, y)$ is an open set in $\mathbb{R}_\ell$ containing $x$ and $[y, y + 1)$ is an open set in $\mathbb{R}_\ell$ containing $y$. Besides, these two sets obey

$$[x, y) \cap [y, y + 1) = \varnothing,$$

completing the proof.

(5) Yes, $\mathbb{R}_\ell$ is separable. To see that, first note that $\mathbb{Q}$ is countable in $\mathbb{R}$ (this is independent of the topology of this exercise!). To prove the density of $\mathbb{Q}$ in $\mathbb{R}_\ell$, note that unions of intervals of the form $[a, b)$ always intersect $\mathbb{Q}$.

**SOLUTION 3.3.39.**

(1) $\mathbb{R}_K$ is finer than $\mathbb{R}$ since its basis contains the basis of $\mathbb{R}$. It is strictly finer since $\mathbb{R} - K$ is open in $\mathbb{R}_K$ (it is a union of sets of the form $(a, b) - K$) but it is not open in $\mathbb{R}$ for $K$ is not closed.

(2) As in the previous exercise, $\mathbb{R}_K$ is Hausdorff as it is finer than $\mathbb{R}$.

(3) No. The two topologies are not comparable and it is better to use bases elements.

(a) $\mathbb{R}_K \not\subset \mathbb{R}_\ell$: For $(-1, 1) - K$ is open in $\mathbb{R}_K$ (clear!) but not in $\mathbb{R}_\ell$.

(b) $\mathbb{R}_\ell \not\subset \mathbb{R}_K$: For $[-1, 0)$ is in $\mathbb{R}_\ell$ but not in $\mathbb{R}_K$.

(4) Yes, $K$ is closed in $\mathbb{R}_K$ for its complement is $\mathbb{R} - K$ which is open (already discussed above).

(5) We claim that $K' = \varnothing$. Any point in $K$ is not a limit point for $K$ (why?).

If $x \notin K$, then $U = \mathbb{R} - K$ is an open set containing $x$ and verifying $U \cap K - \{x\} = \varnothing$. Thus $K' = \varnothing$.

**SOLUTION 3.3.40.**

(1) Let $x \in \mathbb{R}$. Then there are always $a, b \in \mathbb{Q}$ such that $x \in (a, b)$ (why?). Now, let $(a, b)$ and $(c, d)$ be in $\mathcal{B}$ ($a$, $b$, $c$ and $d$ are tacitly assumed to be rationals). If $x$ is in the intersection of these sets, then $(a, b) \cap (c, d)$ is an interval of the same type. This proves that $\mathcal{B}$ is a basis.

Now, we prove that $\mathcal{B}$ actually generates the usual topology of $\mathbb{R}$. Let $U$ be an open set in $\mathbb{R}$. Let $x \in U$. Then

$$\exists y, z \in \mathbb{R}: \ x \in (y, z) \in U.$$

By the density of $\mathbb{Q}$ in $\mathbb{R}$,

$$\exists a, b \in \mathbb{Q}, \ a \in (y, x) \text{ and } b \in (x, z)$$

leading to

$$x \in (a, b) \subset U \ (a, b \in \mathbb{Q}).$$

The proof is over.

(2) For any real $x$, there are rationals $a$ and $b$ verifying $a \leq a < x < b$. Now, if the intersection of two basis elements of $\mathcal{B}'$ is

not empty, then it is necessarily of their form. Thus $\mathcal{B}'$ is a basis. Now, we have to show that $\mathbb{R}_{\mathcal{B}'} \neq \mathbb{R}_\ell$. It is clear that $[\pi, 4) \in \mathbb{R}_\ell$. If $[\pi, 4)$ were in $\mathbb{R}_{\mathcal{B}'}$, then there would be some rationals $a$ and $b$ such that

$$\pi \in [a, b) \subset [\pi, 4).$$

It then becomes clear that no rational $a$ would satisfy that condition. Therefore, $\mathcal{B}'$ does not generate the lower limit topology on $\mathbb{R}$.

**SOLUTION** 3.3.41. We know that (finite) open intervals constitute a basis for standard $\mathbb{R}$. We can also write (for any $a, b \in \mathbb{R}$, $a < b$)

$$(a, b) = (a, \infty) \cap (-\infty, b)$$

and this is sufficient to declare the collection of open infinite intervals a subbasis for $\mathbb{R}$.

**SOLUTION** 3.3.42. Remember that in usual $\mathbb{R}$,

$$d(A) = \sup_{x,y \in A} |x - y|$$

where $A \subset \mathbb{R}$. We then easily find that

$$d((0, 1) \cap \mathbb{Q}) = d((0, 1) \cap \mathbb{R} \setminus \mathbb{Q}) = 1.$$

**SOLUTION** 3.3.43.

(1) The set $\{d(x, a) : a \in A\}$ is obviously non-empty. It is also bounded from below as

$$d(x, a) \geq 0, \ \forall a \in A.$$

Thus $d(x, A)$ exists.

(2) Let $\varepsilon > 0$ and let $B(x, \varepsilon)$ be the open ball of center $x$ and radius $\varepsilon$. We then have

$$x \in \overline{A} \Longleftrightarrow \forall \varepsilon > 0, B(x, \varepsilon) \cap A \neq \varnothing$$
$$\Longleftrightarrow \forall \varepsilon > 0, \exists a_\varepsilon \in A : d(x, a_\varepsilon) < \varepsilon$$
$$\Longleftrightarrow \forall \varepsilon > 0, \exists a_\varepsilon \in A : 0 \leq d(x, a_\varepsilon) < \varepsilon + 0$$
$$\Longleftrightarrow d(x, A) = 0$$

by the greatest lower bound property.

Let $a \in X$. We have to show that $\{a\}$ is closed. We have

$$\overline{\{a\}} = \{x \in X : d(x, \{a\}) = 0\} = \{a\},$$

i.e. $\{a\}$ is closed.

(3) Since $A \subset \overline{A}$, we have

$$d(x, A) \geq d(x, \overline{A}).$$

Now for all $a \in A$ and for all $b \in \overline{A}$,

$$d(x, A) \leq d(x, a) \leq d(x, b) + d(b, a).$$

Hence

$$d(x, A) \leq d(x, b) + d(b, A).$$

But, $d(b, A) = 0$ as $b \in \overline{A}$ and so

$$d(x, A) \leq d(x, b), \ \forall b \in \overline{A}.$$

Taking the "inf" again over $b \in \overline{A}$ gives

$$d(x, A) \leq d(x, \overline{A})$$

leading to the wanted equality.

**SOLUTION** 3.3.44. Assume that $X$ is separable. By definition, $X$ has a dense countable subset, $\{x_n : n \in \mathbb{N}\}$, say. This means that

$$\overline{\{x_n : n \in \mathbb{N}\}} = X.$$

It is known that for any $x \in A$, $B(x, \frac{\varepsilon}{2})$ is a neighborhood of $x$. But since $\overline{\{x_n : n \in \mathbb{N}\}} = X$, we have in particular that

$$B\left(x, \varepsilon/2\right) \cap \{x_n : n \in \mathbb{N}\} \neq \varnothing.$$

In other words, there is at least an $x_n$ which is also in $B(x, \frac{\varepsilon}{2})$. Hence $d(x_n, x) < \frac{\varepsilon}{2}$. Now, set

$$n_x = \min\{n \in \mathbb{N} : \ d(x_n, x) < \varepsilon/2\}$$

(remember that $d(x_{n_x}, x) < \frac{\varepsilon}{2}$).

Now, define a function $f : A \to \mathbb{N}$ by $f(x) = n_x$. We will show that $f$ is injective. Let $x \neq y$ and assume that $f(x) = f(y)$ (for the sake of contradiction), that is, $n_x = n_y$. Hence

$$d(x_{n_x}, x) < \frac{\varepsilon}{2} \text{ and } d(x_{n_y}, y) < \frac{\varepsilon}{2}.$$

By the triangle inequality,

$$d(x, y) \leq d(x, x_{n_x}) + \underbrace{d(x_{n_x}, x_{n_y})}_{=0} + d(x_{n_y}, y)$$

and hence

$$d(x, y) \leq d(x, x_{n_x}) + d(x_{n_y}, y) < \frac{\varepsilon}{2} + \frac{\varepsilon}{2} = \varepsilon$$

and this contradicts the assumption that for some $\varepsilon > 0$ and all distinct $x, y \in A$: $d(x, y) \geq \varepsilon$.

This implies that $f$ is injective. Then Theorem 1.1.74 tells us that $A$ too would be countable, contradicting the hypothesis $A$ being uncountable. Therefore, the assumption that $X$ is separable has led to a contradiction. Thus, $X$ cannot be separable.

**SOLUTION** 3.3.45.

(1) Let $1 \leq p < \infty$. First we treat *real* $\ell^p$ spaces. We must show that there exists a countable set $X$ such that $\overline{X} = \ell^p$. Let

$$X_n = \{(q_1, q_2, \cdots, q_n, 0, 0, \cdots) : q_1, \cdots, q_n \in \mathbb{Q}\}.$$

Then each $X_n$ is countable and so is their union, which we denote by $X$. We are done as soon as we show that $X$ is dense in $\ell^p$, i.e. $\ell^p \subset \overline{X}$.

Let $x = (x_n) = (x_1, x_2, \cdots, x_n, x_{n+1}, \cdots) \in \ell^p$. We show that $x \in \overline{X}$. Let $q = (q_1, q_2, \cdots, q_N, 0, 0, \dots) \in X_N$, $N \in \mathbb{N}$. We ought to show that $d_p(x, q)$ is arbitrarily small. Since $\mathbb{Q}$ is dense in usual $\mathbb{R}$, we know that for all $x_1, \cdots, x_N$ and all $\varepsilon > 0$ there are $q_1, \cdots, q_N \in \mathbb{Q}$ such that

$$\sum_{i=1}^{N} |x_i - q_i|^p = |x_1 - q_1|^p + \cdots + |x_N - q_N|^p < \varepsilon$$

making $\sum_{i=1}^{N} |x_i - q_i|^p$ as small as we wish.

Now, for $i \geq N + 1$, we have

$$d_p(x, q)^p = \sum_{i=N+1}^{\infty} |x_i|^p \longrightarrow 0 \text{ as } N \longrightarrow \infty$$

for this is the tail of a convergent series! Thus (real!) $\ell^p$ is separable.

To see that complex $\ell^p$ is separable, just use Gaussian rational numbers $p + iq$, where $p, q \in \mathbb{Q}$, in the definition of $X_n$ and the latter remains countable. All the rest stays unchanged.

(2) A very similar method may be applied. Everything as before (mutatis mutandis), we will have in the end that ($x \in c_0$)

$$d_\infty(x, q) = \sup_{i \geq N+1} |x_i| \longrightarrow 0 \text{ as } N \longrightarrow \infty$$

(otherwise we would get a contradiction with $(x_n)$ tending to zero). Thus $(c_0, d_\infty)$ is separable.

**SOLUTION** 3.3.46. Let $X \subset \mathbb{N}$. Define a sequence $(x_n)$ as follows

$$x_n = \mathbb{1}_X(n) = \begin{cases} 1, & n \in X, \\ 0, & n \notin X. \end{cases}$$

Then the set all of all sequences of this type, denoted by $\{0,1\}^{\mathbb{N}}$, is uncountable (see e.g. Exercise 1.2.24).

It is clear that every $x_n$ is in $\ell^\infty$. Now, let $X, Y \subset \mathbb{N}$ such that $X \neq Y$. Then if $m \in \mathbb{N}$, then either $m \in X \setminus Y$ or $m \in Y \setminus X$. For example, assume that $m \in X \setminus Y$ (the other case can be dealt with similarly). Hence

$$d_\infty(\mathbb{1}_X, \mathbb{1}_Y) = \sup_{n \in \mathbb{N}} |\mathbb{1}_X(n) - \mathbb{1}_Y(n)| \geq |\mathbb{1}_X(m) - \mathbb{1}_Y(m)| = |1 - 0| = 1.$$

Therefore, by Proposition 3.1.37, we infer that $\ell^\infty$ is not separable.

**SOLUTION 3.3.47.**

(1) First, note that $A \times B \subset \overline{A} \times \overline{B}$. But $\overline{A} \times \overline{B}$ is closed (why?) and hence $\overline{A \times B} \subset \overline{A} \times \overline{B}$.

Now we prove the other inclusion. Let $(x, y) \in \overline{A} \times \overline{B}$. Then $x \in \overline{A}$ and $y \in \overline{B}$. Hence any neighborhood of $x$ intersect $A$ and so does any neighborhood of $y$ with $B$. Let $\Omega$ be a neighborhood of $(x, y)$. Then $\Omega$ is a union of elements of the form $U \times V$ where $U$ is open in $X$ and contains $x$, $V$ is open in $Y$ and contains $y$. We have

$$(U \times V) \cap (A \times B) = (U \cap A) \times (V \cap B) \neq \varnothing$$

as $x \in \overline{A}$ and $y \in \overline{B}$. Thus $\Omega$ too intersects $A \times B$ which leads to $(x, y) \in \overline{A \times B}$.

(2) There are different methods to prove this property. The one we use here is based on the known property $\overline{A^c} = (\overset{\circ}{A})^c$ and on the previous question. We have

$$\overline{(\overset{\circ}{A \times B})^c} = \overline{(A \times B)^c}$$
$$= \overline{(A^c \times Y) \cup (X \times B^c)}$$
$$= \overline{A^c \times Y} \cup \overline{X \times B^c}$$
$$= (\overline{A^c} \times \overline{Y}) \cup (\overline{X} \times \overline{B^c})$$
$$= (\overline{A^c} \times Y) \cup (X \times \overline{B^c})$$
$$= ((\overset{\circ}{A})^c \times Y) \cup (X \times (\overset{\circ}{B})^c)$$
$$= (\overset{\circ}{A} \times \overset{\circ}{B})^c.$$

Thus

$$\overset{\circ}{\overline{A \times B}} = \overset{\circ}{A} \times \overset{\circ}{B}.$$

**SOLUTION** 3.3.48. Using a proof by induction, it suffices to prove this result for two spaces. Let $X$ and $Y$ be two separable spaces, i.e. there are two *countable* subsets $A$ and $B$ of $X$ and $Y$ respectively such that

$$\overline{A} = X \text{ and } \overline{B} = Y.$$

Now, obviously $A \times B$ is countable. It is also dense in $X \times Y$ since

$$\overline{A \times B} = \overline{A} \times \overline{B} = X \times Y.$$

This completes the proof.

**SOLUTION** 3.3.49.

(1) $A$ has an empty interior since no open ball can be contained in $A$ (why?). Its closure is given by

$$\overline{A} = A \cup \{(0, y) : -1 \leq y \leq 1\}$$

(cf. Exercise 6.3.24). This comes from the fact any open ball centered at $(0, y)$ with $-1 \leq y \leq 1$ intersects $A$.

(2) The interior of $B$ is void. For no open ball can be contained in $B$ regardless of its radius. Also $B$ is closed. There are different ways of seeing this:

(a) Let $(x, y) \notin B$. Then there is always some $r > 0$ such that $B((x, y), r)$ does not intersect $B$ and hence $(x, y) \notin \overline{B}$. The proof is over.

(b) Alternatively and anticipating a result on continuity (to be seen in the next chapter) the given set $B$ is the graph of the function $x \mapsto x$ which is continuous on the usual $\mathbb{R}$ and the graph of a continuous function is closed.

(c) Also, $B$ is the diagonal of $\mathbb{R}$ and it is closed since usual $\mathbb{R}$ is Hausdorff (cf Exercise 4.3.38).

(3) The answer becomes obvious once we write $C = C_1 \times C_2$ where

$$C_1 = (-2, 2) \text{ and } C_2 = (-3, 3).$$

Hence

$$\overset{\circ}{C} = \overset{\overbrace{\phantom{C_1 \times C_2}}^{\circ}}{C_1 \times C_2} = \overset{\circ}{C_1} \times \overset{\circ}{C_2} = (-2, 2) \times (-3, 3) = C$$

(or simply because $C$ is open). Similarly

$$\overline{C} = \overline{C_1} \times \overline{C_2} = [-2, 2] \times [-3, 3].$$

(4) Since $\{(1, 1)\}$ is a singleton in usual $\mathbb{R}^2$, we immediately get

$$\overset{\circ}{D} = \overset{\overbrace{\phantom{\{(1,1)\} \times C}}^{\circ}}{\{(1, 1)\} \times C} = \overset{\overbrace{\phantom{\{(1,1)\}}}^{\circ}}{\{(1, 1)\}} \times \overset{\circ}{C} = \varnothing \times \overset{\circ}{C} = \varnothing.$$

Finally,

$$\overline{D} = \overline{\{(1,1)\}} \times \overline{C} = \{(1,1)\} \times [-2,2] \times [-3,3].$$

**SOLUTION 3.3.50.** Let $\varphi : X \to X/\mathcal{R}$ be the quotient map. Let

$$T = \{A \in X/\mathcal{R} : \varphi^{-1}(A) \text{ is open in } X\}.$$

Let us show that $T$ is a topology in $X/\mathcal{R}$.

(1) $\varnothing, X/\mathcal{R} \in T$ as: $\varphi^{-1}(\varnothing) = \varnothing$ and $\varphi^{-1}(X/\mathcal{R}) = X$.
(2) Let $A, B \in T$. Then

$$\varphi^{-1}(A \cap B) = \varphi^{-1}(A) \cap \varphi^{-1}(B)$$

is open in $X$ so that $A \cap B \in T$.
(3) Let $(A_i)_{i \in I}$ be a collection of elements in $T$ (i.e. $\varphi^{-1}(A_i)$ are all open in $X$). Then

$$\varphi^{-1}(\bigcup_{i \in I} A_i) = \bigcup_{i \in I} \varphi^{-1}(A_i)$$

is open in $X$ and so $\bigcup_{i \in I} A_i \in T$.

**SOLUTION 3.3.51.**

(1) Denote the quotient map by $p$, i.e. the map $p : \mathbb{R} \to \mathbb{R}/\mathbb{Q}$. Let $[s]$ and $[r]$ be two elements of $\mathbb{R}/\mathbb{Q}$. Let $U \in \mathcal{N}([s])$ and $V \in \mathcal{N}([r])$. By definition, $p^{-1}(U)$ and $p^{-1}(V)$ are two open sets in $\mathbb{R}$. Hence

$$\exists q, q' \in \mathbb{Q} : q \in p^{-1}(U), \ q' \in p^{-1}(V) \text{ (why?)}$$

Thus $[q] \in U$ and $[q'] \in V$. But $[q] = [q']$ since $q - q' \in \mathbb{Q}$. Therefore, $U \cap V$ is never empty and consequently $\mathbb{R}/\mathbb{Q}$ is not Hausdorff.

(2) Let $U$ be any non-empty open in $\mathbb{R}/\mathbb{Q}$. We must show that $U = \mathbb{R}/\mathbb{Q}$. Denote the quotient map by $p$. Then $p^{-1}(U)$ is open in $\mathbb{R}$. Now, the map $t \mapsto t + \alpha$ is continuous for each real $\alpha$. Whence the set $\{t \in \mathbb{R} : t + \alpha \in p^{-1}(U)\}$ is open in $\mathbb{R}$. Thus it must necessarily intersect $\mathbb{Q}$ and so

$$\exists q \in \mathbb{Q} : q + \alpha \in p^{-1}(U).$$

But $p(\alpha) = p(q + \alpha)$ (why?). This implies that $p(\alpha) \in U$, i.e. $\alpha \in p^{-1}(U)$ and hence $p^{-1}(U) = \mathbb{R}$. Therefore, $U = \mathbb{R}/\mathbb{Q}$. The proof is complete.

## 3.4. Hints/Answers to Tests

SOLUTION 7.

(1) We can write (can't we?) $A = \bigcup_{x \in A} U_x$...

(2) It reminds us of the definition of an open set in a metric space with $U_x$ playing the role of an open ball...

SOLUTION 8. Yes! why?...

SOLUTION 9. No! Consider e.g. $\{\pi\}$...

SOLUTION 10. No! (why?)...

SOLUTION 11. No, $T$ is not a topology on $X$. While $\varnothing$ and $X$ both belong to $T$ (why?), the union of two elements in $T$ need not remain in $T$ (consider $A_4$ and $A_3$ for example)...

SOLUTION 12. There are nine topologies having four open sets. Find them directly or just apply Exercise 3.5.1.

SOLUTION 13. Construct such a set using sets similar to $\{\frac{1}{n} : n \in \mathbb{N}\}$ which we know it has 0 as its *unique* limit point. What is left to do should be clear to the reader by now...

SOLUTION 14. In the co-finite topology, $A' = X$ if $A$ is infinite and $A' = \varnothing$ if $A$ is finite. The reason is that if $A$ is infinite, then $X - \{a\}$ is open and contains $x$ where $a \in X$...and if $A$ is a singleton, consisted of the element $a$ say, then $X - \{a\}$ is always open and does not intersect $A$ and similar arguments work for $A$ consisted of a finite number of elements...

SOLUTION 15. No. Why?...

SOLUTION 16. Let $T$ be the usual topology on $\mathbb{R}$ and let $T'$ be the co-countable topology on $\mathbb{R}$. These two topologies are not comparable: Indeed, $\mathbb{Q}$ is closed in $T'$ but not in $T$. Also, $[0, 1]$ is closed in $T$ but not in $T'$...

SOLUTION 17. The answer is no as every proper subset in this topology is closed...

SOLUTION 18.

(1) The empty set corresponds to the case $a = 0$. The rest is obvious too...

(2) Closed sets are of the form $(-\infty, -a] \cup [a, +\infty)$. For $[-1, 2]$, the smallest closed superset is $\mathbb{R}$ and the largest open subset is $(-1, 1)$.

(3) The closure and the interior of $\{0\}$ are given by $\mathbb{R}$ and $\varnothing$ respectively.

As for $\{1\}$ they are given by $\mathbb{R} \setminus (-1, 1)$ and $\varnothing$ respectively.

**SOLUTION** 19. Yes, if this set is clopen. Otherwise, this cannot occur as the frontier of a set is always *closed* (is it not?).

**SOLUTION** 20. Yes (why?)...

**SOLUTION** 21. Well, a similar example appeared in the "True or False" Section...

# CHAPTER 4

# Continuity and Convergence

## 4.2. True or False: Answers

ANSWERS.

(1) No, this is not always the case if the topologies endowing the domain and the co-domain are different. See Exercise 4.3.1. If, however, the identity mapping is between two identical spaces endowed with the same topologies, then it is continuous.

(2) It is asked whether each continuous function is open? Such is not the case. As a counterexample, let $f : \mathbb{R} \to \mathbb{R}$ defined by $f(x) = 0$ ($\mathbb{R}$ endowed with its standard topology). Then $f$ is continuous but for some (and here any!) open $U$ in $\mathbb{R}$, $f(U) = \{0\}$ is not open in $\mathbb{R}$.

(3) The left-to-right implication is correct and for a proof see Exercise 4.3.25. As for the backward implication, it is not true. For a counterexample, take the function

$$f(x) = \begin{cases} 1, & x \in \mathbb{Q}, \\ 0, & x \notin \mathbb{Q}. \end{cases}$$

Then $f$ is not continuous whilst $f_{\mathbb{Q}}$ is continuous (both in the usual topology).

(4) In general, only the right-to-left implication is verified. To see this, let $U$ be an open set containing $x$. Since $(x_n)$ converges to $x$, for any open set containing $x$, and in particular for $U$,

$$\exists N \in \mathbb{N}, \forall n \in \mathbb{N} \ (n \geq N \Longrightarrow x_n \in U)$$

and hence $A \cap U \neq \varnothing$ or $x \in \overline{A}$.

   The other implication may fail to hold. In $\mathbb{R}$ equipped with the co-countable topology, let $A = [0, 2]$. Then $\overline{A} = \mathbb{R}$ (cf. Exercise 3.3.33). Now, the only convergent sequences in this space are the eventually constant ones (see Exercise 4.3.20). Hence

$$\forall x_n \in [0, 2], \ x_n \nrightarrow -1 \text{ and yet } -1 \in \overline{A}.$$

If, however, we are dealing with metric spaces only (or metrizable!), then the equivalence always holds and it is a very useful result to use. A proof may be found in Exercise 4.3.26.

(5) This statement as it stands is something to avoid imperatively in Topology. One has to be more precise about the space in which the convergence is to be established. For instance, in the usual topology of $\mathbb{R}$, this sequence converges to zero while in other spaces (or/and topologies) it can have different limits (see Exercise 4.3.18).

(6) The result is true but the reasoning has a problem with the passage $\frac{1}{n} \notin U$ implying $\frac{1}{n} \in U^c$. This is wrong since $\frac{1}{n}$ not being in $U$ means that $\left(\frac{1}{n}\right) \not\subset U$ and this does not imply necessarily that $\left(\frac{1}{n}\right) \subset U^c$. For a correct proof see Exercise 4.3.18.

(7) First, $A$ is closed as its complement, being an arbitrary union of open sets, is open!

Second, the known result says that a set is closed in a metric space, if whenever a sequence in this set converges, then it must have a limit inside that set. In our case, that result cannot be used as $(x_n)$ does not even converge!

(8) The answer is no! For a counterexample, see Exercise 4.3.19.

(9) Only the left-to-right implication is true. In other words, the uniqueness of the limit of a sequence does not characterize the Hausdorffness property. We give a proof. Assume a given sequence $(x_n)$ in a separated space $X$ has two different limits, $x$ and $y$ say. Since $X$ is Hausdorff,

$$\exists U \in \mathcal{N}(x), \exists V \in \mathcal{N}(y) : U \cap V = \varnothing.$$

Since $(x_n)$ converges to $x$, for all open sets containing $x$ and in particular for $U$

$$\exists N_1 \in \mathbb{N}, \ \forall n \in \mathbb{N} : (n \geq N_1 \Longrightarrow x_n \in U).$$

Similarly,

$$\exists N_2 \in \mathbb{N}, \ \forall n \in \mathbb{N} : (n \geq N_2 \Longrightarrow x_n \in V).$$

So, for $n \geq \max(N_1, N_2)$, $x_n \in U \cap V$ which contradicts the fact that $U$ and $V$ are disjoint! Thus the limit is unique.

As for the other implication we present a counterexample. Consider $X = \mathbb{R}$ equipped with the co-countable topology. Then the only convergent sequences are the eventually constant ones (see Exercise 4.3.20). Let $(x_n)$ be a sequence in $X$ which converges to two different limits, $x$ and $y$ ($x \neq y$), say.

Then

$$\exists N_1 \in \mathbb{N}, \ \forall n \geq N_1 : \ x_n = x \text{ and } \exists N_2 \in \mathbb{N}, \ \forall n \geq N_2 : \ x_n = y.$$

Hence for $n \geq \max(N_1, N_2)$ we would have $x_n = x = y$ which contradicts the hypothesis $x \neq y$. Thus the limit is unique. However, we already know from Exercise 3.3.33 that $X$ is not Hausdorff.

(10) The answer is yes. Since $\{(a_i, b_i)\}_{i \in I}$ is a basis for $\mathbb{R}$, an open set $U$ in $\mathbb{R}$ is written as

$$U = \bigcup_{i \in I}(a_i, b_i).$$

Then

$$f^{-1}(U) = f^{-1}\left(\bigcup_{i \in I}(a_i, b_i)\right) = \bigcup_{i \in I} f^{-1}(a_i, b_i).$$

Since the arbitrary union of open sets is open, it suffices to have $f^{-1}(a_i, b_i)$ open which is the hypothesis.

(11) Two topological spaces ($X$ and $Y$ say) are said to be homeomorphic if there is a homeomorphism $f : X \to Y$ (or obviously $f : Y \to X$). Thus it becomes apparent that a priori it is fairly easy to show that two spaces are homeomorphic since it suffices for that purpose to exhibit *one* homeomorphism between the two spaces.

However, if one wants to show that two spaces $X$ and $Y$ are not homeomorphic, then one has to show that there is no homoeomorphism between the two spaces! And this is not always easy to establish. An advanced topology course, namely **algebraic topology**, is a powerful tool for proving that two spaces are not homeomorphic. This is not discussed in this book. However, in Chapters 5 and 6, some criteria will be introduced to prove that some spaces are not homeomorphic.

(12) Since $f$ is continuous, we have $f(\overline{A}) \subset \overline{f(A)}$. Since $A$ is dense in $X$, we have $\overline{A} = X$. Then

$$f(X) = f(\overline{A}) \subset \overline{f(A)}.$$

Hence, the closure of $f(A)$ in $f(X)$ is equal to $f(X)$. Thus $f(A)$ is dense in $f(X)$.

(13) True! To see this, let $f : X \to Y$ be a homeomorphism between two topological spaces where $X$ is separable. Let us show that $Y$ is separable. Since $X$ is separable, there exists a

countable subset $A$ such that $\overline{A} = X$. By the previous answer $f(A)$ is dense in $f(X) = Y$. But

$$f(A) = \{f(x) : x \in A\}$$

is countable (by Exercise 1.3.8). This completes the proof.

(14) The answer is again no. One has to distinguish between an algebraic notion and a topological one (even though in some cases, we can prove an algebraic property using a topological one and vice versa). The bijectivity is purely algebraic whilst the continuity is topological. There are many counterexamples which the reader will see below. Here is one more. Let

$$f(x) = \left\{ \begin{array}{ll} x, & 0 \le x < \frac{1}{2}, \\ \frac{3}{2} - x, & \frac{1}{2} \le x \le 1. \end{array} \right.$$

Then in the usual topology, $f : [0,1] \to [0,1]$ is not continuous at $1/2$ but $f$ is a bijection.

(15) No! Consider $f(x) = x$ from $X = \mathbb{R}$ (endowed with the usual topology) onto $Y = \mathbb{R}$ (endowed with the discrete topology). Then $f$ is bijective. It is also open and closed because every subset of $Y$ is open and closed. It is, however, not continuous since $\{2\}$ is open in $Y$ but its preimage (itself in this case) is not open in $X$.

(16) The answer is yes. This is a direct consequence of Proposition 4.1.14 and Theorems 4.1.9 & 4.1.15.

(17) The answer is no! We give a counterexample. Let $Y = \mathbb{R}$ endowed with the usual topology and let $X = \mathbb{R}$ be endowed with the discrete topology. Let $f : X \to Y$ defined for all $x \in \mathbb{R}$ by $f(x) = x$. Then $f$ is a bijection. Moreover, $f$ is continuous as for every open $U$ set in $Y$, $f^{-1}(U)$ is open in $X$. However, its inverse, i.e. $f^{-1} : Y \to X$ is not continuous (why?).

The reader *must not* think this is solely true with different topologies or that this cannot occur in the usual topology. For instance, let $f : X = [0,1) \cup \{3\} \to Y = [0,1]$ be defined for all $x \in X$ by

$$f(x) = \left\{ \begin{array}{ll} x, & 0 \le x < 1, \\ 1, & x = 3 \end{array} \right.$$

and both $X$ and $Y$ are endowed with the induced usual topology of $\mathbb{R}$. Details are left to the reader.

(18) The answer is yes!...

(19) The answer is yes. For a proof see Exercise 4.3.11.

(20) False! Let $f : (0, 1) \to \{1\}$ be the constant function (both sets with respect to the usual topology). Then, $f$ is continuous. Besides, 0 is a limit point for $(0, 1)$ while $f(0) = 1$ is not a limit point for $\{1\}$.

(21) False! Only the left-to-right implication holds. For a proof and for a counterexample to the other implication, see Exercise 4.3.40.

(22) False! Consider in the usual topology

$$f(x) = \begin{cases} 1, & x \in \mathbb{Q}, \\ 0, & x \notin \mathbb{Q} \end{cases} \quad \text{and} \quad g(x) = \begin{cases} 0, & x \in \mathbb{Q}, \\ 1, & x \notin \mathbb{Q} \end{cases}$$

Then $f$ and $g$ are both everywhere discontinuous on $\mathbb{R}$, and yet

$$(f + g)(x) = 1, \ \forall x \in \mathbb{R},$$

that is, $f + g$ is *everywhere continuous* on $\mathbb{R}$.

(23) The problem is that $p$ is not a homeomorphism. While $p$ is continuous, open and surjective, it is not injective as long as $\text{card} Y \geq 2$. Indeed, let $y, y' \in Y$ such that $y \neq y'$. Then (for some $x \in X$)

$$(x, y) \neq (x, y') \text{ and yet } p(x, y) = x = p(x, y').$$

## 4.3. Solutions to Exercises

**SOLUTION 4.3.1.**

(1) (a) If $\text{card} X \geq 2$, then the given function is not continuous. For if $U$ is an open (different from $\varnothing$ and $X$) set in $Y$, then $f^{-1}(U) = U$ is not open in $X$ as the only open sets in $X$ are $\varnothing$ and $X$.

   (b) Now if $\text{card} X = 1$, then $f$ is obviously continuous as the discrete and indiscrete topologies coincide in this case.

(2) No $f$ is not continuous. For example, $\{1\}$ is open in $Y$ but its preimage $\{0\}$ is not open in $X$.

(3) Let $U = (0, 1)$ be an open set in $Y$. Then

$$f^{-1}(U) = \{x \in \mathbb{R} : x^2 \in (0, 1)\} = (-1, 0) \cup (0, 1)$$

which is not open in $X$ (why?).

(4) The function $f$ in this case is continuous since for any open set $U$ in $Y$, $f^{-1}(U)$ is open since it is a subset of $X$. This means that if $X = [0, 3]$, say, is given the discrete topology, then a

function like

$$x \mapsto f(x) = \begin{cases} -1, & 0 \leq x < 1, \\ 0, & 1 \leq x < 2, \\ 2, & 2 \leq x < 3, \end{cases}$$

will be continuous. This type of functions (i.e., those defined on a discrete topology) will have little interest in practise, but it is quite enlightening as a source of counterexamples.

(5) In this case, $f$ is continuous. Indeed, let $f(x) = b$ for all $x \in X$ and let $U$ be open in $Y$. We have

$$f^{-1}(U) = \{x \in X : \ f(x) \in U\} = \{x \in X : \ b \in U\} = \varnothing \text{ or } X$$

depending on whether $b \notin U$ or $b \in U$. Anyway, in either case $f^{-1}(U)$ is open in $X$ and hence $f$ is continuous.

(6) Take $f : A \to X$, where $A \subset X$, such that $f(x) = x$ for all $x \in A$. For any open $U$ in $X$, $f^{-1}(U) = U \cap A$ is open in $A$ (in the subspace topology). Thus $f$ is continuous.

**SOLUTION** 4.3.2. First, recall that a function $f : X \to Y$ ($X$ and $Y$ being two topological spaces) is continuous at $x \in X$ if

$$\forall U \in \mathcal{N}(f(x)), \ f^{-1}(U) \in \mathcal{N}(x).$$

(1) $f$ is not continuous at $a$ for

$$\exists U = \{a, b\} \in \mathcal{N}(a = f(a)) \text{ and } f^{-1}(U) = \{a, c\} \notin \mathcal{N}(a).$$

(2) $f$ is continuous at $b$ because for any open set containing $f(b) = c$ (in this case there is only one, namely $X$), $f^{-1}(X) = X$ is an open set that contains $b$!

(3) $f$ is not continuous at $c$ since

$$\exists U = \{b\} \in \mathcal{N}(f(c)) = \mathcal{N}(b) \text{ and } f^{-1}(U) = \{c\} \notin \mathcal{N}(b)$$

as $b \notin \{c\}$ (and also $\{c\}$ is not open in $X$!).

**SOLUTION** 4.3.3.

(1) If $f$ is continuous, then the inverse image of all open sets in $Y$ is open in $X$. In particular, the inverse image of all elements of $\mathcal{S}$ is open in $X$, and this proves half of the equivalence.

(2) Conversely, assume that the inverse image of any element of $\mathcal{S}$ is open in $X$. To show that $f$ is continuous, let $U$ be an open set in $Y$. Since $\mathcal{S}$ is a subbasis for $Y$, we know that $U$ may be written as

$$U = \bigcup_{k \in I} (B_{i_1} \cap \cdots B_{i_k}), \text{ with } B_{i_k} \in \mathcal{S}.$$

Since the inverse image is stable under intersections and unions, we obtain

$$f^{-1}(U) = f^{-1}(\bigcup_{k \in I} (B_{i_1} \cap \cdots \cap B_{i_k}))$$

$$= \bigcup_{k \in I} f^{-1}(B_{i_1} \cap \cdots \cap B_{i_k})$$

$$= \bigcup_{k \in I} [f^{-1}(B_{i_1}) \cap \cdots \cap f^{-1}(B_{i_k})].$$

But by assumption, each $f^{-1}(B_{i_k})$ is open and hence so is their (finite!) intersection. Since the union of open sets is open, it follows that $f^{-1}(U)$ is open, establishing the continuity of $f$.

SOLUTION 4.3.4.

(1) "(1) $\Rightarrow$ (2)": Let $A \subset X$ and let $y \in f(\overline{A})$. Then $y = f(x)$ for some $x \in \overline{A}$. To show that $y \in \overline{f(A)}$, let $V$ be a neighborhood of $f(x)$. Since $f$ is continuous, $f^{-1}(V)$ is open in $X$. As $f^{-1}(V)$ contains $x$ (why?), then $f^{-1}(V) \cap A \neq \varnothing$. Hence

$$\varnothing \neq f(f^{-1}(V) \cap A) \subset V \cap f(A),$$

showing that $V \cap f(A) \neq \varnothing$. Therefore, $y = f(x) \in \overline{f(A)}$.

(2) "(2) $\Rightarrow$ (3)": Let $V$ be closed in $T'$. We have to show that $f^{-1}(V)$ is closed in $T$. It then suffices to show that $\overline{f^{-1}(V)} \subset f^{-1}(V)$. By assumption, we have

$$f(\overline{f^{-1}(V)}) \subset \overline{f(f^{-1}(V))} \subset \overline{V} = V$$

and so

$$\overline{f^{-1}(V)} \subset f^{-1}[f(\overline{f^{-1}(V)})] \subset f^{-1}(V),$$

as expected.

(3) "(3) $\Rightarrow$ (1)": Let $U$ be an open set in $T'$. Then $U^c$ is closed in $T'$. By hypothesis, $f^{-1}(U^c)$ too is closed (in $T$). But $f^{-1}(U^c) = [f^{-1}(U)]^c$. Hence $[f^{-1}(U)]^c$ is closed and so $f^{-1}(U)$ is open in $T$, showing that $f$ is continuous.

SOLUTION 4.3.5. Assume that $f^{-1}(\overset{\circ}{U}) \subset \overset{\circ}{\overline{f^{-1}(U)}}$ holds for all $U$ in $Y$. We must show that $f$ is continuous. Let $V$ be an open set in $Y$. Then $V = \overset{\circ}{V}$ and hence by hypothesis we obtain

$$\overset{\circ}{\overline{f^{-1}(V)}} \subset f^{-1}(V) = f^{-1}(\overset{\circ}{V}) \subset \overset{\circ}{\overline{f^{-1}(V)}}.$$

Therefore, $\widehat{f^{-1}(V)} = f^{-1}(V)$, proving that $f^{-1}(V)$ is open or that $f$ is continuous.

Conversely, suppose $f$ is continuous and let $U \subset Y$. Since $\overset{\circ}{U}$ is open, by continuity of $f$, $f^{-1}(\overset{\circ}{U})$ is open too. Besides, $\overset{\circ}{U} \subset U$ and so

$$f^{-1}(\overset{\circ}{U}) = \widehat{f^{-1}(\overset{\circ}{U})} \subset \widehat{f^{-1}(U)}.$$

The proof is complete.

**SOLUTION** 4.3.6.

(1) Assume that $f$ is closed and let $A \subset X$. Then

$$A \subset \overline{A} \Longrightarrow f(A) \subset f(\overline{A}) \Longrightarrow \overline{f(A)} \subset \overline{f(\overline{A})}.$$

Since $\overline{A}$ is closed and $f$ is closed, it follows that $\overline{f(\overline{A})} = f(\overline{A})$. Hence $\overline{f(A)} \subset f(\overline{A})$, as desired.

Conversely, let $V$ be a closed subset of $X$. Then

$$f(V) = f(\overline{V}) \supset \overline{f(V)}.$$

Since $f$ is continuous, we know that $f(V) \subset \overline{f(V)}$ (as we took a closed $V$). Therefore, $f(V) = \overline{f(V)}$, i.e. $f(V)$ is closed and this proves the closedness of $f$.

(2) A similar idea may be adopted by using Exercise 4.3.5. Details are left to the reader.

**SOLUTION** 4.3.7.

(1) Let $\overline{A} = \overline{B}$. Since $f$ is continuous, we have

$$f(\overline{A}) \subset \overline{f(A)} \text{ and } f(\overline{B}) \subset \overline{f(B)}.$$

Hence

$$A \subset \overline{A} = \overline{B} \Longrightarrow f(A) \subset f(\overline{B}) \subset \overline{f(B)}$$

and so

$$\overline{f(A)} \subset \overline{f(B)}.$$

Similarly,

$$B \subset \overline{B} = \overline{A} \Longrightarrow f(B) \subset f(\overline{A}) \subset \overline{f(A)}$$

and so

$$\overline{f(B)} \subset \overline{f(A)}.$$

Accordingly,

$$\overline{f(A)} = \overline{f(B)}.$$

(2) Since $A$ is dense in $X$, $\overline{A} = X$. Hence $\overline{A} = \overline{X}$. By the previous question and since $f(X)$ is dense in $Y$, we have:
$$\overline{f(A)} = \overline{f(X)} = Y,$$
that is, $f(A)$ is dense in $Y$.

**SOLUTION** 4.3.8. We use sequential continuity. Let $(x_n)$ and $(y_n)$ be two sequences converging to $x$ and $y$ respectively and with respect to the usual metric. Denote the function $(x, y) \mapsto x + y$ by $f$ and $(x, y) \mapsto xy$ by $g$. Then we may write
$$0 \le |x_n + y_n - (x + y)| \le |x_n - x| + |y_n - y|$$
and so
$$\lim_{n \to \infty} |f(x_n, y_n) - f(x, y)| = 0.$$
This shows that $f$ is continuous.

The proof of the other case is very similar once we observe that we can write:
$$|x_n y_n - xy| = |x_n(y_n - y) + (x_n - x)y| \le |x_n||y_n - y| + |x_n - x||y|.$$
The continuity of $g$ then follows by remembering that a convergent sequence is bounded (details are left to the reader).

**SOLUTION** 4.3.9. First, we note that $f$ is obviously a bijection. The function $f : T' \to T$ is continuous. To see this, take any nonvoid (the case of an empty set trivially holds) open set $U$ in $T$, then $U^c$ is finite and hence it is countable, i.e. $U \in T'$. But $f^{-1}(U) = U$. Hence $f$ is continuous. Now, since a countable set is not necessarily finite, we deduce that $f^{-1} : T \to T'$ is not continuous. Therefore, $f$ is not a homeomorphism.

**SOLUTION** 4.3.10.

(1) The bijectivity of $f$ is evident. Next, since $d$ and $d'$ are topologically equivalent (see Exercise 2.3.30), $f$ is a homeomorphism.
(2) If $X = \mathbb{R}$ and $d = |\cdot|$ the usual metric, then $(\mathbb{R}, d)$ is unbounded while $(\mathbb{R}, d')$ is bounded, yet these two spaces are homeomorphic.

**SOLUTION** 4.3.11. Let $f$ be a homeomorphism between two topological spaces $X$ and $Y$. Assume that $X$ is Hausdorff, and let us show that $Y$ is in its turn Hausdorff. Let $y, y' \in Y$ be such that $y \ne y'$. By the bijectivity of $f$, there exist *unique* $x, x' \in X$ such that $y = f(x)$ and $f(y') = x'$. By the bijectivity of $f^{-1}$, say, we see that $x$ and $x'$ must be *different*. By the Hausdorffness of $X$, we get
$$\exists (U, U') \in \mathcal{N}(x) \times \mathcal{N}(x') : U \cap U' = \varnothing.$$

Since $f$ is open (and since $U$ and $U'$ are open), $f(U)$ and $f(U')$ are also open. They obviously contain $y$ and $y'$ respectively. But $f$ is injective (cf. Exercise 1.3.3) and hence

$$f(U) \cap f(U') = f(U \cap U') = \varnothing,$$

proving that $f(U)$ and $f(U')$ are disjoint. On that account, $Y$ is Hausdorff.

SOLUTION 4.3.12.

(1) (a) Let $(a, b)$ and $(c, d)$ be any two open (finite) intervals in $\mathbb{R}$. Define a function $f : (a, b) \to (c, d)$ defined by

$$f(x) = c + (d - c)\frac{x - a}{b - a}$$

for $x \in (a, b)$. Then it is clear that $f$ is continuous and bijective. Its inverse, $f^{-1} : (c, d) \to (a, b)$ given by

$$f^{-1}(x) = a + (b - a)\frac{x - c}{d - c}$$

for each $x \in (c, d)$, is obviously continuous too. Thus $f$ is a homeomorphism and hence $(a, b)$ and $(c, d)$ are homeomorphic.

REMARK. Needless to repeat that this is in the usual topology and that in other topologies these two intervals may not be homeomorphic.

(b) We leave it to the reader to check that $\mathbb{R}$ is homeomorphic to $(-1, 1)$ via the *homeomorphism*

$$f(x) = \frac{x}{1 + |x|}, \ x \in \mathbb{R}.$$

Since the "homeomorphism relation" is transitive, then $\mathbb{R}$ is homeomorphic to any open interval by the previous question.

(2) The answer is yes! Remember that $\overline{\mathbb{R}} = \mathbb{R} \cup \{-\infty, +\infty\}$. For a possible homeomorphism consider

$$f(x) = \begin{cases} \frac{x}{1 + |x|}, & x \in \mathbb{R}, \\ -1, & x = -\infty, \\ 1, & x = \infty. \end{cases}$$

SOLUTION 4.3.13. We only give details in the first case.

(1) Let $f : X \times Y \to Y \times X$ be defined by $f(x, y) = (y, x)$. Then $f$ is clearly continuous (as the components are continuous). It is also clear that $f$ is bijective and that $g : Y \times X \to X \times Y$ given

by $g(y,x) = (x,y)$ is its inverse. Finally, $g$ is also continuous. Accordingly, $f$ is a homeomorphism.

(2) Consider $f : \{x\} \times Y \to Y$ defined by $f(x,y) = y$. Then show that $f$ is homeomorphism...

(3) Consider $f : X \times \{y\} \to X$ defined by $f(x,y) = x$ and show that $f$ is homeomorphism...

**SOLUTION** 4.3.14. This is easy. We have $A = f^{-1}(\{a\})$. Since $\{a\}$ is closed in $\mathbb{R}$, so is $A$ since it is the preimage of a closed set under a continuous function.

**SOLUTION** 4.3.15.

(1) The function $(x,y) \mapsto f(x,y) = xy$ defined on $\mathbb{R}^2$ since it is a polynomial. Now $A$ is closed as it is the inverse image of a closed set, which is $\{1\}$, by a continuous function (which is $f$).

(2) As before, the function $(x,y) \mapsto f(x,y) = x^2 + y^2$ defined on $\mathbb{R}^2$ is continuous since it is a polynomial. Hence

$$B = \{(x,y) \in \mathbb{R}^2 : x^2 + y^2 \leq 1\} = f^{-1}([0,1])$$

is closed for $[0,1]$ is closed in $\mathbb{R}$.

(3) First, this space has an algebraic dimension equal to $n^2$. The "function determinant" defined in $X = \mathcal{M}_n(\mathbb{R})$ and taking values in $\mathbb{R}$ is a polynomial of degree $n^2$ and hence it is continuous. The remaining part of the answer is a routine.

**SOLUTION** 4.3.16.

(1) The set $A$ is closed since $A = f^{-1}((-\infty, a])$, $f$ is continuous and $(-\infty, a]$ is closed in $\mathbb{R}$.

(2) The converse is not always true. Consider the *discontinuous* function (at $x = 0$)

$$f(x) = \begin{cases} 0, & x \leq 0, \\ 2, & x > 0. \end{cases}$$

Now we show that for any $a$, the resulting set $A$ will always be closed. We have
(a) $a < 0 \Rightarrow A = \varnothing$, i.e. $A$ is closed in $\mathbb{R}$.
(b) $0 \leq a < 2 \Rightarrow A = (-\infty, 0]$, i.e. $A$ is closed in $\mathbb{R}$.
(c) $a \geq 2 \Rightarrow A = \mathbb{R}$, i.e. $A$ is closed in $\mathbb{R}$.

**SOLUTION** 4.3.17. The left-to-right implication is evident. Let us prove the right-to-left implication. Since $\{(a,b) : a, b \in \mathbb{R}\}$ is a basis for usual $\mathbb{R}$, it suffices to prove that the inverse image of $(a,b)$ via $f$ is open. We have

$$(a,b) = (-\infty, b) \cap (a, \infty)$$

and hence

$$f^{-1}((a,b)) = f^{-1}((-\infty,b) \cap (a,\infty)) = f^{-1}((-\infty,b)) \cap f^{-1}((a,\infty))$$

which is open by our assumptions. Thus $f$ is continuous.

**SOLUTION** 4.3.18.

(1) Obviously, in the usual $\mathbb{R}$, the sequence $\left(\frac{1}{n}\right)_{n\geq 1}$ converges to 0.

(2) We have already proved that this topology is not Hausdorff (see Exercise 3.3.31) and hence if this sequence is convergent, then it need not have a unique limit.

The sequence $\left(\frac{1}{n}\right)_{n\geq 1}$ converges to every element of $\mathbb{R}$. To see this, let $U$ be an open set containing $x$, where $x \in \mathbb{R}$. Hence $U^c$ must be finite. Since

$$\frac{1}{n} \in \mathbb{R} = U \cup U^c,$$

we see that $\frac{1}{n}$ must be in $U$, for all, but finitely many, $n \in \mathbb{N}$. This proves the convergence of the sequence.

(3) On the contrary of the previous topology, this sequence does not converge to any point in $\mathbb{R}$. To illustrate this, let us show that $\frac{1}{n}$ does not converge to $a \in \mathbb{R}$ for any $a$. The question amounts to finding a $U \in \mathcal{N}(a)$ such that

$$\forall N \in \mathbb{N}, \exists n \ (n \geq N \wedge \frac{1}{n} \notin U).$$

It suffices to take $U = \{a\} \in \mathcal{N}(a)$ (which is of course open in this topology) and then

$$\forall N \in \mathbb{N}, \exists n = N \ (n \geq N \wedge \frac{1}{n} \notin \{a\}).$$

**REMARK.** In fact, the only convergent sequences in a discrete topological (or metric) space are the eventually constant ones.

(4) In the indiscrete topology, all sequences (and in particular ours) converge to every point in $\mathbb{R}$. For $\mathbb{R}$ is the *only* non-empty open set. Thus it will contain any sequence.

**SOLUTION** 4.3.19.

(1) Let $x \in \mathbb{R}$. Since $U = (x-1, x+1) - K$ is a neighborhood of $x$ and since $\frac{1}{n} \notin U$ for all $n$, we immediately deduce that $\frac{1}{n} \not\to x$.

On the contrary, $-\frac{1}{n}$ does converge to 0 in $\mathbb{R}_K$ since any neighborhood of zero will contain infinitely many points of $(x_n)$ by the Archimedean Property.

(2) $-K$ is not closed because $-\frac{1}{n} \in -K$ but $-\frac{1}{n} \to 0$, in $\mathbb{R}_K$, and $0 \notin K$.

(3) We leave it to you to show that $(x_n)$ converges to 0.

No, $(x_n)$ cannot have another limit as $\mathbb{R}_\ell$ is Hausdorff (see Exercise 3.3.38).

(4) No, $f$ is not continuous as $K$ is closed in $\mathbb{R}_K$ while its preimage $f^{-1}(K) = -K$ is not closed in $\mathbb{R}_K$.

(5) No, $f$ is not continuous since $[0, \infty)$ is open in $\mathbb{R}_\ell$ and it is not the case for its preimage $f^{-1}([0, \infty)) = (-\infty, 0]$ is not open in $\mathbb{R}_\ell$ (cf. Exercise 3.3.38).

**SOLUTION 4.3.20.**

(1) We already know that eventually constant sequences converge. Now, let $(x_n)$ be a convergent sequence to some $x$. Then

$$\forall U \in \mathcal{N}(x), \exists N \in \mathbb{N}, \forall n \, (n \geq N \Rightarrow x_n \in U).$$

In particular, for $U = X \setminus \{x_n : x_n \neq x\}$ (which is open and contains $x$). Hence there exists some $N_0$ such that for all $n$ :

$$n \geq N_0 \Longrightarrow x_n \in X \setminus \{x_n : x_n \neq x\}.$$

Therefore, $x_n = x$ for all $n \geq N_0$, i.e. $(x_n)$ is eventually convergent.

(2) (a) $1 \in [2, 3]'$ since every open set containing 1 intersects $[2, 3]$ (why?).

(b) Since $(x_n)$ takes its values in $[2, 3]$ and since $(x_n)$, if it converges, is eventually constant, we deduce directly that $x_n \nrightarrow 1$.

(c) The conclusion is: there are sets having a limit point to which no sequence in this set need to converge.

**SOLUTION 4.3.21.**

(1) A simple application of Exercise 2.5.2 gives, for all $n$

$$0 \leq |d(x_n, y_n) - d(x, y)| \leq d(x_n, x) + d(y_n, y).$$

Passing to the limit, as $n$ tends to infinity, finishes the proof.

(2) The previous result means that the function $d$, defined on $X \times X$, is a continuous function (something already known from the metric spaces chapter!).

**SOLUTION** 4.3.22.

(1) Assume $f$ is continuous and let $x_n \to x$. We are required to prove that $f(x_n) \to f(x)$ in $Y$. Let $U$ be an open neighborhood of $f(x)$. Then $x$ is in $f^{-1}(U)$ which is open by the continuity of $f$. Hence $f^{-1}(U)$ contains all but finitely many terms of $(x_n)$. Accordingly, $U$ contains all but finitely many terms of $f(x_n)$. This means that $f(x_n) \to f(x)$ in $Y$, establishing the sequential continuity of $f$.

(2) Let $f : X \to Y$, where $X$ is $\mathbb{R}$ equipped with the co-countable topology and $Y$ is the usual $\mathbb{R}$, be defined by $f(x) = x$. Then the only convergent sequences in $X$ are the eventually constant ones (see Exercise 4.3.20). These sequences also converge in usual $\mathbb{R}$ and hence

$$x_n \longrightarrow x \text{ in } X \Longrightarrow x_n \longrightarrow x \text{ in } Y \text{ or } f(x_n) \longrightarrow f(x).$$

This means that $f$ is sequentially continuous. However, it is not continuous for $U = (-1, 1)$ is open in $Y$ and it is not the case for its preimage in $X$.

(3) Thanks to Question 1, we only prove that $f$ is sequentially continuous implying that $f$ is continuous. Assume $f : X \to Y$ ($X$ and $Y$ being two metric spaces) is sequentially continuous and we show that $f$ is continuous. It is slightly better here to use closed sets (why?). Let $V$ be a closed set in $Y$. We need to establish the closedness of $f^{-1}(V)$ in $X$. Let $x_n \in f^{-1}(V)$ be converging to $x \in X$. Then $f(x_n)$ is in $V$ for all $n$ and the sequential continuity hypothesis implies that $f(x_n) \to f(x)$. But $V$ is closed and so $f(x) \in V$ or $x \in f^{-1}(V)$. The proof is complete.

**SOLUTION** 4.3.23. It is known that an open set $U$ in $\mathbb{R}$ is of the form

$$U = \bigcup_{i \in I}(a_i, b_i), \ a_i, b_i \in \mathbb{R}.$$

We have to show that $f(U)$ is open. But since

$$f(U) = f(\bigcup_{i \in I}(a_i, b_i)) = \bigcup_{i \in I} f((a_i, b_i)),$$

we need only show that $f((a_i, b_i))$ is open for every $i$. WLOG we may assume that $f$ is increasing. Now, since $f$ is *continuous* and *increasing*, we have

$$f((a_i, b_i)) = (f(a_i), f(b_i)), \ \forall i \in I$$

which are all open and hence so is their union. Thus $f(U)$ is open, i.e. $f$ is an open map.

**SOLUTION** 4.3.24. Let $C$ be a closed set in $\mathbb{R}$. We need to show that $P(C)$ is closed. Let $(y_n) \subset P(C)$ be a converging sequence to $y$. Hence, there is $(x_n) \subset C$ such that $P(x_n) = y_n$. This implies that $(P(x_n) = y_n)$ is bounded. But since a polynomial has an infinite limit only at $\pm\infty$, we get that $(x_n)$ is also bounded in $\mathbb{R}$. The Bolzano-Weierstrass Property tells us that we can extract from $(x_n)$ a convergent subsequence $(x_{n(k)})$. Call $x$ its limit. Whence, $x \in C$ as $C$ is closed. By the continuity of $P$ (it is a polynomial!), we obtain

$$y \longleftarrow y_{n(k)} = P(x_{n(k)}) \longrightarrow P(x).$$

Since we are in a Hausdorff space, the limit is unique and hence $y = P(x) \in P(C)$, i.e. $P(C)$ is closed which means that $P$ is a closed mapping, as desired.

**SOLUTION** 4.3.25. Denote the restriction of $f$ to $A$ by $f_A$. Let $U$ be an open set in $Y$. Then (by Exercise 1.2.12),

$$f_A^{-1}(U) = A \cap f^{-1}(U)$$

is open in $A$ (in the subspace topology) as $f^{-1}(U)$ is open in $X$ by the continuity of $f$. Thus, $f_A$ is continuous.

**SOLUTION** 4.3.26.

(1) We already showed one implication in the "True or False" Section (in a more general context). We recall it here for convenience.

(a) "$\Rightarrow$": Let $x \in \overline{A}$. Then

$$\forall \varepsilon > 0, \; B(x, \varepsilon) \cap A \neq \varnothing.$$

In particular,

$$\forall n \in \mathbb{N}, \; B(x, 1/n) \cap A \neq \varnothing.$$

This intersection has at least a point in common and it surely depends on $n$. Call it $x_n$. Hence for all $n$, $x_n \in A$ and $x_n \in B(x, \frac{1}{n})$. This latter implies that

$$\forall n \in \mathbb{N}: \; d(x, x_n) < \frac{1}{n}.$$

Passing to the limit yields $x_n \to x$, as desired.

(b) "$\Leftarrow$": Let $x \in X$ such that $x_n \to x$, where $x_n \in A$. Let $\varepsilon > 0$. We have to show that $A$ intersects all open balls $B(x, \varepsilon)$.

Since $\varepsilon > 0$, by the convergence of $(x_n)$, we know there is some $N \in \mathbb{N}$ such that for all $n \geq N$, we have $d(x, x_n) < \varepsilon$. Hence $x_n \in B(x, \varepsilon) \cap A$ (for all $n \geq N$). Therefore, $B(x, \varepsilon) \cap A \neq \varnothing$ or $x \in \overline{A}$, as required.

(2)  (a)  "$\Leftarrow$": Since $A \subset \overline{A}$ always holds, it only remains to show that $\overline{A} \subset A$. Let $x \in \overline{A}$. By the previous question,

$$\exists x_n \in A : \ x_n \longrightarrow x.$$

By assumption, all convergent sequences in $A$ have a limit inside of $A$. Thus, $x \in A$ or $\overline{A} \subset A$.

(b)  "$\Rightarrow$": Assume that $A$ is closed, i.e. $\overline{A} = A$. Let $x_n \in A$ be such that $x_n \to x$. By the previous question $x \in \overline{A}$. As we have supposed that $A$ is closed, we end up with $x \in A$. This completes the proof.

**SOLUTION** 4.3.27. None of the sets considered in this exercise is closed.

(1) To show that $A = (0, 1]$ is not closed in $\mathbb{R}$, it suffices to find a convergent sequence $(x_n)_n$ in $A$ having a limit outside of $A$. Take $x_n = \frac{1}{n}$ which obviously lies in $A$. Its limit in $\mathbb{R}$ is 0 and it is not in $A$. Hence $A$ is not closed.

(2) The same arguments (and even the same sequence) apply to show that $B$ is also not closed.

(3) $C$ is not closed. To see that we need a convergent sequence $(x_n, y_n)_n$ in $C$ having a limit outside of $C$. One choice is to take

$$(x_n, y_n) = \left( \sqrt{\frac{n}{1+n}}, 0 \right) \in C \text{ as } \frac{n}{1+n} + 0 < 1, \ \forall n \in \mathbb{N}.$$

Then, its limit is $(1, 0) \notin C$ since $1^2 + 0^2 \not< 1$.

(4) The same method again. The reader may easily show that $D$ is not closed (consider for instance the sequence $\left( \sqrt{1 + \frac{1}{n}}, 0 \right)_n$).

**SOLUTION** 4.3.28. First we show that $A$ is not closed. Since

$$(x_n, y_n) = \left( \frac{n+1}{n}, \frac{n}{n+1} \right) \in A$$

($n \geq 1$) with limit $(1, 1)$ not in $A$, we easily conclude that $A$ is not closed.

To show that $A$ is not open, we show instead (and equivalently) that $\mathbb{R}^2 \setminus A$ is not closed. Obviously

$$\left( \frac{2n+1}{n}, \frac{n-2}{2n} \right) \notin A, \text{ i.e. } \left( \frac{2n+1}{n}, \frac{n-2}{2n} \right) \in \mathbb{R}^2 \setminus A.$$

However,

$$\left(\frac{2n+1}{n}, \frac{n-2}{2n}\right) \longrightarrow (2, 1/2) \in A,$$

i.e. $(2, \frac{1}{2}) \notin \mathbb{R}^2 \setminus A$. Therefore $\mathbb{R}^2 \setminus A$ is not closed, i.e. $A$ is not open.

**SOLUTION** 4.3.29. A similar idea to the second part of the previous solution may be applied to prove that $A$ is not open. We show that $\mathbb{R} \setminus A$ is not closed. Consider the sequence $(1 + \frac{1}{p})_{p \geq 1}$. It certainly does not belong to $A$ and hence it belongs to $\mathbb{R} \setminus A$. Its limit in $\mathbb{R}$ is $1 \in A$, i.e. $1 \notin \mathbb{R} \setminus A$. Thus $\mathbb{R} \setminus A$ is not closed.

**SOLUTION** 4.3.30.

(1) There are different methods to answer this question. One is the following (the reader may try to give a different proof): Let $(x_n)_n \in A$ be such that $x_n \to x$ in $X$. We need to show that $x \in A$, i.e. $f(x) = g(x)$. Since $(x_n)_n \in A$, $f(x_n) = g(x_n)$. But $f$ and $g$ are both continuous. Hence

$$f(x) \longleftarrow f(x_n) = g(x_n) \longrightarrow g(x).$$

Since $Y$ is Hausdorff, the limit is unique and thus $f(x) = g(x)$, i.e. $x \in A$.

(2) The set $B$ is such that $\overline{B} = X$. Assume that $f$ and $g$ coincide on $B$ and let us show that this forces them to coincide on all of $X$. Let $x \in X$. Then there exists a sequence $(x_n)_n$ in $B$ such that $x_n \to x$ (in $X$). Hence one has $f(x_n) = g(x_n)$ and since $f$ and $g$ are continuous,

$$f(x) \longleftarrow f(x_n) = g(x_n) \longrightarrow g(x).$$

Thus $f$ and $g$ coincide everywhere.

**REMARK.** There is a tempting but completely false proof of the first question. We write $A$ as $(f - g)^{-1}(\{0\})$, then we say that $A$ is closed as it is equal to the inverse image of $\{0\}$ by a continuous function which is $f - g$. There are some false arguments here, mainly algebraic ones, e.g. who knows whether 0 is in $Y$? Is " $-$ " defined in $Y$? (in fact the function $f - g$ may make no sense at all). Of course if $Y = \mathbb{R}$, then this wrong proof becomes a true and nice one.

**SOLUTION** 4.3.31. If $f$ were continuous at $(0, 0)$, then we would have for any $(x_n, y_n)$ converging to $(0, 0)$, $f(x_n, y_n) \to f(0, 0)$. But

$$(1/n, 1/n) \to (0, 0) \text{ and } f(1/n, 1/n) = \frac{1}{2} \nrightarrow 0$$

which means that $f$ is discontinuous at $(0,0)$ (worse, taking $(\frac{1}{n}, 0) \to$ $(0,0)$ shows that in fact the limit at $(0,0)$ does not even exist!).

**SOLUTION 4.3.32.**

(1) Assume that $f : X \to Y$ is a continuous, one-to-one mapping and that $Y$ is Hausdorff. Let us show that $X$ is Hausdorff. Let $x, y \in X$ such that $x \neq y$. Since $f$ is one-to-one, $f(x) \neq f(y)$. But $Y$ is Hausdorff (and $f(x), f(y) \in Y$) and hence

$$\exists U \in \mathcal{N}(f(x)), \ \exists V \in \mathcal{N}(f(y)) \text{ such that } U \cap V = \varnothing.$$

Since $U$ and $V$ are open and $f$ is continuous, $f^{-1}(U)$ and $f^{-1}(V)$ are also open. Since $f(x) \in U$ and $f(y) \in V$, $x \in f^{-1}(U)$ and $y \in f^{-1}(V)$. We also have

$$f^{-1}(U) \cap f^{-1}(V) = f^{-1}(U \cap V) = f^{-1}(\varnothing) = \varnothing.$$

Hence $f^{-1}(U)$ and $f^{-1}(V)$ are two disjoint neighborhoods of the $x$ and $y$ respectively (remember that $x \neq y$). Thus $X$ is Hausdorff.

(2) On $\mathbb{R}$, consider the discrete topology (denoted by $Y$) and the indiscrete one (denoted $X$). Now let $f : X \to Y$ be the identity map. Then $f$ is one-to-one but it is *not* continuous (see Exercise 4.3.1). It is also known that $Y$ is Hausdorff whereas $X$ is not.

(3) Keeping the same topologies as in the previous answer but take $f(x) = a$ (the constant function). Then $f$ is continuous but not one-to-one and $Y$ is Hausdorff whilst $X$ is not.

**SOLUTION 4.3.33.**

(1) The set $Z(g)$ is closed since it is the inverse image of the closed set $\{0\}$ in $\mathbb{R}$ under the continuous function $g$.

(2) First, we must check that $f$ is well-defined, i.e. its denominator never vanishes. By Exercise 3.3.43 and since $A$ and $B$ are closed and disjoint, we immediately see that $d(x, A)$ and $d(x, B)$ cannot vanish simultaneously and hence

$$\forall x \in X : \ d(x, A) + d(x, B) > 0.$$

Now, since $x \mapsto d(x, A)$ and $x \mapsto d(x, B)$ are (uniformly) continuous by Exercise 2.3.24, the function $f$, being the quotient of two continuous functions, is therefore continuous.

(3) We have by Exercise 3.3.43

$$f(x) = 0 \Longleftrightarrow d(x, A) = 0$$
$$\Longleftrightarrow x \in \overline{A} = A$$

and hence $f^{-1}(\{0\}) = A$.

A quite similar reasoning applies to show that $f^{-1}(\{1\}) = B$. Indeed,

$$f(x) = 1 \iff d(x, B) = 0$$
$$\iff x \in \overline{B} = B,$$

as expected.

(4) We must show that any closed set is the zero set of some continuous function. Let $h : X \to \mathbb{R}$ be a given function. All possible cases are treated.

(a) If $Z(h) = \varnothing$, take $h(x) = 1$ for all $x \in X$ and this is a continuous function.

(b) If $Z(h) = X$, take $h(x) = 0$ for all $x \in X$ and this is a continuous function.

(c) If $Z(h) \neq X$ (and $Z(h) \neq \varnothing$) is closed, then there exists $a \in X$ such that $h(a) \neq 0$. This implies that the two sets $Z(h)$ and $\{a\}$ are disjoint. Moreover, they are both closed. Define $h$ by

$$h(x) = \frac{d(x, Z(h))}{d(x, \{a\}) + d(x, Z(h))}$$

for all $x \in X$. By Question 2, this is a continuous function. The proof is complete.

(5) Let

$$U = f^{-1}(-1/5, 1/4) \text{ and } V = f^{-1}(1/3, 3/2)$$

Since $f$ is continuous, both $U$ and $V$ are open. They are also disjoint for

$$U \cap V = f^{-1}(-1/5, 1/4) \cap f^{-1}(1/3, 3/2)$$
$$= f^{-1}(-1/5, 1/4) \cap (1/3, 3/2)$$
$$= f^{-1}(\varnothing)$$
$$= \varnothing.$$

In the end, $A$ is contained in $U$ for $0 \in (-\frac{1}{5}, \frac{1}{4})$ and so

$$A = f^{-1}(\{0\}) \subset f^{-1}(-1/5, 1/4) = U$$

Also, $V$ contains $B$ as $1 \in (\frac{1}{3}, \frac{3}{2})$ and so

$$B = f^{-1}(\{1\}) \subset f^{-1}(1/3, 3/2) = V.$$

**SOLUTION** 4.3.34. Let $n, m \in \mathbb{N}$ be such that $n \neq m$. The reader can check that
$$d_\infty(f_n, f_m) = 1...$$

**SOLUTION** 4.3.35. We recall that for $f, g \in X$
$$d(f, g) = \int_0^1 |f(x) - g(x)| dx \quad \text{and} \quad d'(f, g) = \sup_{x \in [0,1]} |f(x) - g(x)|.$$

First, we prove the closedness of $A$ with respect to $d'$. Let $f \in \overline{A}$. Then for some $f_n \in A$, $f_n$ converges uniformly to $f$. This implies two things. First, that $f$ must be continuous (a well-known result from the course of advanced calculus, or from Chapter 8).

Second,
$$\forall x \in [0, 1]: \lim_{n \to \infty} f_n(x) = f(x).$$
But, since $f_n(0) = 0$, we have $f(0) = 0$ too. Thus $f \in A$.

Now we show that $A$ is dense in $X$ with respect to $d$. We need only show that $X \subset \overline{A}$. Let $f \in X$, i.e. $f$ is continuous. Consider the sequence of functions $f_n$ defined by
$$f_n(x) = \begin{cases} xe^n f(e^{-n}), & 0 \leq x \leq e^{-n}, \\ f(x), & e^{-n} \leq x \leq 1. \end{cases}$$

The continuity of $f$ implies that of $(f_n)$. Also $f_n(0) = 0$. Hence $f_n \in A$ for all $n \in \mathbb{N}$. It only remains to show that $d(f_n, f) \to 0$ as $n$ tends to infinity. Let $n \in \mathbb{N}$. We have
$$d(f_n, f) = \int_0^{e^{-n}} |f_n(x) - f(x)| dx + \int_{e^{-n}}^1 |f_n(x) - f(x)| dx$$
$$= \int_0^{e^{-n}} |xe^n f(e^{-n}) - f(x)| dx.$$

But for all $(x, n) \in [0, e^{-n}] \times \mathbb{N}$ we have
$$|xe^n f(e^{-n}) - f(x)| \leq |xe^n f(e^{-n})| + |f(x)|$$
$$\leq e^{-n} e^n |f(e^{-n})| + |f(x)|$$
$$\leq 2 \sup_{x \in [0, e^{-n}]} |f(x)|$$
$$\leq 2 \sup_{x \in [0,1]} |f(x)|.$$

Thus
$$d(f_n, f) \leq 2 \sup_{x \in [0,1]} |f(x)| \int_0^{e^{-n}} dx = 2e^{-n} \sup_{x \in [0,1]} |f(x)| \longrightarrow 0$$
as $n \to \infty$. Therefore, $f \in \overline{A}$.

SOLUTION 4.3.36.

(1) We can prove that $A_a$ is closed as done in the foregoing exercise. Let $d$ be the supremum metric on $X$. Let $f_n \in A_a$ such that $d(f_n, f) \to 0$ as $n$ tends to infinity. Then $f$ is continuous on $[0, 1]$. We also have

$$|f(a)| \leq |f_n(a) - f(a)| + |f_n(a)| = |f_n(a) - f(a)| \leq d(f_n(a), f(a)) \to 0$$

as $n$ tends to infinity. This gives $f \in A_a$.

(2) Observe that for each $a \in I$, $B$ reduces to some $A_a$. Thus $B$ may be written as the arbitrary intersection of closed sets of the form $A_a$, i.e.

$$B = \bigcap_{a \in I} A_a \text{ is closed.}$$

SOLUTION 4.3.37.

(1) We show that $p$ is continuous. Let $U$ be an open set in $X$. We need to show that $p^{-1}(U)$ is open in $X \times Y$. We have

$$p^{-1}(U) = \{(x, y) \in X \times Y : p(x, y) = x \in U\} = U \times Y$$

which is open in $X \times Y$. Thus $p$ is continuous. The proof of the continuity of $q$ is very akin to that of $p$.

(2) None of the two projections is closed! We first note that the set $A$ (given in the hint) is closed (why?). Now we have

$$p(A) = \left\{ p\left(x, \frac{1}{x}\right) : x \neq 0 \right\} = \mathbb{R}^*.$$

Hence $p(A)$ is not closed and hence $p$ is not closed. Similar arguments apply to show that $q$ is not closed either.

(3) Let $U$ and $V$ be open sets in $X$ and $Y$ respectively. We have

$$p(U \times V) = \{p(x, y) = x : (x, y) \in U \times V\} = U.$$

Now, any open set $\Omega$ in $X \times Y$ is of the form $\cup_{i \in I}(U_i \times V_i)$ where $U_i$ and $V_i$ are open sets in $X$ and $Y$ respectively. Moreover,

$$p(\Omega) = p\left( \bigcup_{i \in I}(U_i \times V_i) \right) = \underbrace{\bigcup_{i \in I} p(U_i \times V_i)}_{\text{open in } X}$$

and hence $p$ is an open mapping. The proof of the openness of $q$ is very similar to that of $p$.

**SOLUTION** 4.3.38.

(1) We are required to prove that $X$ is Hausdorff if and only if $\triangle$ is closed, that is, if and only if $\triangle^c$ is open. Assume $X$ is separated. Let $(x, y) \in \triangle^c$. Then $x \neq y$. But $X$ is Hausdorff and hence

$$\exists (U, V) \in \mathcal{N}(x) \times \mathcal{N}(y) : \ U \cap V = \varnothing.$$

Now, $U \times V$ is open in $X \times X$, it contains $(x, y)$ and $(U \times V) \cap \triangle = \varnothing$ which implies that $U \times V \subset \triangle^c$. By Test 7, we get that $\triangle^c$ is open or that $\triangle$ is closed.

Conversely, suppose that $\triangle$ is closed. To show that $X$ is separated, let $x \neq y$. Then $(x, y) \in \triangle^c$. But $\triangle^c$ is a union of basis elements and hence

$$\exists U, V \in X : \ (x, y) \in U \times V \subset \triangle^c.$$

This implies that $U \cap V = \varnothing$ (if $a$ were in both $U$ and $V$, then $(a, a) \in U \times V$ and it would be in $\triangle^c$, a contradiction!). Since $U$ and $V$ are open sets that contain $x$ and $y$ respectively, we immediately get that $X$ Hausdorff, establishing the result.

(2) The function $(f, g) : X \to Y \times Y$ is continuous (Proposition 4.1.21). Now, let $\triangle$ be the diagonal of $Y$. Then

$$[(f, g)]^{-1}(\triangle) = \{x \in X : \ f(x) = g(x)\}$$

is closed, being the preimage of a closed set by a continuous function.

**SOLUTION** 4.3.39.

(1) (a) "$\Rightarrow$": Let $x \in A$. Adopting the notations of Exercise 4.3.37 we can write $g(x) = p(f(x))$ and $h(x) = q(f(x))$. Since $f$, $p$ and $q$ are all continuous, so are $g$ and $h$.

(b) "$\Leftarrow$": Assume that $h$ and $g$ are both continuous. Let $U$ be an open set in $X$ let $V$ be an open set in $Y$. It *suffices* to show that $f^{-1}(U \times V)$ is open in $A$ (why?). We have

$$x \in f^{-1}(U \times V) \iff f(x) \in U \times V$$
$$\iff g(x) \in U \wedge h(x) \in V$$
$$\iff x \in g^{-1}(U) \wedge x \in h^{-1}(V)$$
$$\iff x \in g^{-1}(U) \cap h^{-1}(V)$$

which shows that

$$f^{-1}(U \times V) = g^{-1}(U) \cap h^{-1}(V).$$

Since $g$ and $h$ are continuous, both $g^{-1}(U)$ and $h^{-1}(V)$ are open in $A$. Hence $f^{-1}(U \times V)$ is also open in $A$. Thus $f$ is continuous.

(2) The answer is no in this case, i.e. a function may well have partial continuous functions without being continuous itself. The following example elucidates that. Let $f : \mathbb{R}^2 \to \mathbb{R}$ be defined by

$$f(x,y) = \begin{cases} \frac{2xy}{x^2+y^2}, & (x,y) \neq (0,0), \\ 0, & (x,y) = (0,0). \end{cases}$$

The reader can easily check that $x \mapsto f(x,y)$ and $y \mapsto f(x,y)$ are both continuous on $\mathbb{R}$ while $f$ is not continuous at $(0,0)$.

**SOLUTION 4.3.40.**

(1) The proof is based on Exercise 4.3.38. We know that $p :$ $(x,y) \mapsto x$ and $q : (x,y) \mapsto y$ are both continuous. Then $f \circ p : (x,y) \mapsto f(x)$ is continuous too. Thus by Exercise 4.3.38, the set

$$\{(x,y) \in X \times Y : (f \circ p)(x,y) = q(x,y)\},$$

which is nothing but the graph of $f$, is closed. This finishes the proof.

(2) In usual $\mathbb{R}$, consider

$$f(x) = \begin{cases} 0, & x = 0 \\ \frac{1}{x}, & x \in \mathbb{R}^*. \end{cases}$$

Then $f$ is defined on $\mathbb{R}$ and it takes its values in $\mathbb{R}$ which is Hausdorff. Obviously $f$ is not continuous at $x = 0$. Its graph, given by

$$G_f = \{(x, f(x)) : x \in \mathbb{R}\} = \{(0,0)\} \cup \left\{ \left(x, \frac{1}{x}\right) : x \in \mathbb{R}^* \right\},$$

is closed. To show this, we observe that since $\{(0,0)\}$ is closed in $\mathbb{R}^2$, we need only show that $B = \left\{ \left(x, \frac{1}{x}\right) : x \in \mathbb{R}^* \right\}$ is closed. But since $xy = 1 \Rightarrow x \neq 0$, we can write

$$B = \{(x,y) \in \mathbb{R}^2 : xy = 1\}$$

and this is easily seen to be closed.

**SOLUTION 4.3.41.**

(1) (a) Assume that $f$ is continuous. By definition of the product topology, all projections $\pi_i$ (as defined in Definition 4.1.41) are continuous. Hence all $\pi_i \circ f$ are continuous, that is, each $f_i$ is continuous.

(b) Now, assume that each $f_i = \pi_i \circ f$ is continuous. If $U$ is an open subset of $X_i$, then by the continuity of $\pi_i \circ f$, we know that

$$(\pi_i \circ f)^{-1}(U) = f^{-1}(\pi_i^{-1}(U))$$

must be open in $X$. But, the subbasis defining the product topology of $\prod_{i \in I} X_i$ has elements of the type $\pi_i^{-1}(U)$ (where $U$ is open in $X_i$), which are also open. Proposition 4.1.5 then tells us that $f$ is continuous.

(2) As a counterexample (borrowed from [19]), take $f : \mathbb{R} \to \mathbb{R}^\omega$, where $\mathbb{R}^\omega = \mathbb{R} \times \mathbb{R} \times \cdots \times \mathbb{R} \times \cdots$ (a countably infinite product), defined by

$$f(x) = (x, x, \cdots, x, \cdots).$$

By the first question, $f$ is continuous with respect to the product topology since all its components are continuous. However, $f$ is not continuous with respect to the box topology since

$$\Omega = (-1, 1) \times \left(-\frac{1}{2}, \frac{1}{2}\right) \times \left(-\frac{1}{3}, \frac{1}{3}\right) \times \cdots \times \left(-\frac{1}{n}, \frac{1}{n}\right) \times \cdots$$

is open in $\mathbb{R}^\omega$ while its preimage $f^{-1}(\Omega)$ is not open in $\mathbb{R}$. For if it were and since $(0, 0, \cdots, 0, \cdots) \in \Omega$, we would have for some $r > 0$,

$$(-r, r) \subset f^{-1}(\Omega) \Longrightarrow f(-r, r) \subset \Omega.$$

Applying the $n^{\text{th}}$ projection to the previous inclusion would yield

$$(-r, r) \subset \left(-\frac{1}{n}, \frac{1}{n}\right), \ \forall n \in \mathbb{N},$$

which contradicts the Archimedian Property.

## 4.4. Hints/Answers to Tests

SOLUTION 22. No! $f$ is not bijective...

SOLUTION 23. It is closed since $A = \varnothing$ (the empty set) and this is the only reason here why it is closed!...

SOLUTION 24. The answer is no. Call the co-finite topology $X$ and the other one $Y$. Then $id : X \to Y$ is not continuous for $\{a\}$ is open in $Y$ but it is not open in $X$. Similar arguments show that $id : Y \to X$ is not continuous either...

**SOLUTION** 25. The given sequence does not converge to any point $x$ in this topology since

$$\exists U = \{x, a\} \in \mathcal{N}(x) : \ \forall N \in \mathbb{N}, \exists n \ (n \geq N \wedge \frac{1}{n} \notin U)...$$

**SOLUTION** 26. No! $[0, 1)$ is an open set that contains $0$...

**SOLUTION** 27. The non-empty open sets in $Y$ are $\{1\}$ and $Y$. So if $U$ is one of the latter open sets, then $f^{-1}(U) = A$ or $X$. What is left to do should be clear to the reader...

**SOLUTION** 28. $f$ is continuous iff it is constant...

**SOLUTION** 29.
   (1) Should be a routine by now...
   (2) No, it is not...

**SOLUTION** 30. Yes. Do it!...

**SOLUTION** 31. Well, do it...

**SOLUTION** 32. Very easy!...

**SOLUTION** 33. Use projections and Proposition 4.1.21...

**SOLUTION** 34. The idea of proof in the three cases is similar. For instance, to show that $f + g : X \to \mathbb{R}$ is continuous, consider the function $h : \mathbb{R} \times \mathbb{R} \to \mathbb{R}$ defined by $h(y, t) = y + t$. Also, consider $(f, g) : X \to \mathbb{R} \times \mathbb{R}$ defined by $(f, g)(x) = (f(x), g(x))$. Then $f + g$ is continuous as it is the composition of the two continuous functions $h$ and $(f, g)$...

# CHAPTER 5

# Compact Spaces

## 5.2. True or False: Answers

ANSWERS.

(1) True! If we want to show that $\{U_i\}_{i\in I}$ covers the whole space $X$, then we may write $X = \bigcup_{i\in I} U_i$ for we always have $X \supset \bigcup_{i\in I} U_i$.

(2) The answer is negative. In the definition of a compact set, it is asked to verify that *every* open cover has a finite subcover.

(3) The use of closed covers in the definition of compact spaces would be of little interest. For instance, in Hausdorff spaces, which is already a large class of interesting topological spaces, *the only compact spaces (using closed covers) would be the finite ones.* For if $X$ is a Hausdorff space, then $\{\{x\}\}_{x\in X}$ is a closed cover for $X$ and we immediately see that $X$ is compact if it is finite.

(4) True if $X$ is compact. In fact, we have an equivalence.

The answer is, however, not true in general. For instance, in $\mathbb{R}$ (which is not compact with respect to the usual topology), let $A_n = [n, \infty)$. Then, for each $n \in \mathbb{N}$, $A_n$ is non-empty, closed in $\mathbb{R}$, $A_{n+1} \subset A_n$ but

$$\bigcap_{n\in\mathbb{N}} A_n = [1, \infty) \bigcap \cdots \bigcap [n, \infty) \bigcap \cdots = \varnothing.$$

(5) The answer is no! In the induced usual metric of $X = [1, \infty)$, let $A_n = [n, \infty)$. Let $f : X \to X$ defined for all $x \geq 1$ by $f(x) = 1$. Then $f$ is continuous. Besides, the sequence $(A_n)$ is non-empty and decreasing for all $n$. But

$$\bigcap_{n\in\mathbb{N}} A_n = \varnothing \text{ and so } f(\varnothing) = \varnothing.$$

whereas $f(A_n) = \{1\}$ and so

$$\varnothing \neq \bigcap_{n\in\mathbb{N}} f(A_n) = \{1\}.$$

The result is, however, true if $X$ is assumed to be compact. See Exercise 5.3.32.

(6) The answer is yes. Take any open cover in any topological space. Then $\varnothing$ will always be contained in any finite union of elements of that cover.

(7) True in the usual topology of $\mathbb{R}$ (the Heine-Borel Theorem). But as soon as we leave the usual topology, this result may fail to hold. For instance, in the discrete topology, $[a, b]$ is not compact anymore (see Exercise 5.3.2).

(8) False! Consider the topology of Exercise 3.3.34 (with $X = \mathbb{N}$). Since $\{1\}$ is finite, it is compact. However, $\overline{\{1\}} = \mathbb{N}$ is not compact. For a proof see Test 39.

  The question has a positive answer in case the topological space is Hausdorff (by Theorem 5.1.9).

(9) The answer is no. In the usual topology, consider $f : \mathbb{R} \to \mathbb{R}$ defined by $f(x) = 1$ for all $x \in \mathbb{R}$. Then $A = [0, 1]$ is compact in $\mathbb{R}$ but $f^{-1}(A) = \mathbb{R}$ is not compact.

  In Exercise 5.5.9, some condition implying the compactness of $f^{-1}(A)$ is given.

(10) We try to give an exhaustive comment on this question. The well known result says that a set in $\mathbb{R}$ equipped with the usual topology (or $\mathbb{R}^n$ equipped with the euclidian metric) is compact if and only if it is bounded and closed. This result need not hold if we change $\mathbb{R}$ by $\mathbb{Q}$ even if we give $\mathbb{Q}$ the induced usual topology (see Exercise 5.3.26 where a closed and bounded set is not compact). Even in $\mathbb{R}$ equipped with a metric different from the usual metric, a closed and bounded set does not have to be compact (see also Exercise 5.3.27). See Question 15 below for further discussion.

(11) The answer is yes. The reason is simple. Every open cover for $X$ with respect to $T$ is one for $X$ with respect to $T'$.

  The converse is not true. The usual topology is finer than the co-finite topology (on $\mathbb{R}$). However, $\mathbb{R}$ is compact in the co-finite topology and it is not in the usual topology.

(12) The problem with the reasoning is purely algebraic. More precisely,

$$\mathbb{R} \not\subset \bigcup_{i=1}^{n} U_i \not\Longrightarrow \mathbb{R} \subset \left( \bigcup_{i=1}^{n} U_i \right)^c .$$

  A correct proof of the compactness of $\mathbb{R}$ in the co-finite topology, see Exercise 5.3.11.

(13) First, $\mathbb{R}$ has to be equipped with a metric to be able to talk about the possible boundedness of $f$. Second, what is the topology given to $[a, b]$? In general, the answer is no as shown by the following example: Let $f : [0, 1] \to \mathbb{R}$ defined by

$$f(x) = \begin{cases} \frac{1}{x}, & x \in (0, 1], \\ 1, & x = 0. \end{cases}$$

If we equip $[0, 1]$ with the discrete topology (and $\mathbb{R}$ with the usual metric), then $f$ is continuous and $f$ is clearly unbounded. What went wrong with our example is the fact that $[0, 1]$ is *not compact* in the discrete topology (see Exercise 5.3.2).

(14) If considered as a subspace of usual $\mathbb{R}$, then $(0, 2)$ is relatively compact. But, if considered, for instance, as a subspace of itself, then it is clearly not relatively compact.

(15) Not always! For a counterexample, see Exercise 5.3.29 or Test 59. We note that in *normed vector spaces*, the closed unit ball is compact iff the space in question has a finite (algebraic) dimension. It may be proved in an introductory functional analysis course that a normed vector space is locally compact iff the closed unit ball is compact.

(16) The answer is no! For instance, in $\mathbb{R}$ endowed with the co-finite topology, every subset is compact (see the remark below the solution of Exercise 5.3.11). Hence $\mathbb{R}^+$ is compact but it is not closed (observe that this topology is not Hausdorff).

(17) The answer is yes. A proof is given in Exercise 5.3.10.

(18) True. Let us show that. Let $f$ be a homeomorphism between two topological spaces $X$ and $Y$. Assume that $X$ is compact. Since $f$ is continuous, $f(X)$ is compact. Since $f$ is onto $Y = f(X)$. Hence $Y$ is compact. The same idea of proof can be applied to $f^{-1}$ (in lieu of $f$) so that if $Y$ is compact, so is $X$.

(19) True. The proof is based on a very similar case which may be found in Exercise 5.3.38.

(20) False! For a counterexample the reader is referred to Exercise 5.3.34.

(21) True! In fact, we can show the following: *Let $X$ be a locally compact space. If $f : X \to Y$ is continuous and open, then $f(X)$ is locally compact.* Here is the proof:

Let $y \in f(X)$, i.e. $y = f(x)$ for some $x \in X$. Since $X$ is locally compact, there is a compact set $A$ ($A \subset X$) and a neighborhood $U$ of $x$ such that $U \subset A$. Hence $f(U)$ is open because $U$ is open and $f$ is open. Moreover, $f(U) \subset f(A)$. Since $f$ is continuous, $f(A)$ is compact too and contains a neighborhood $f(U)$ of $y = f(x)$. This exactly means that $f(X)$ is locally compact.

(22) False! A counterexample may be found in Exercise 5.3.35.

(23) True. Let us show that. Let $A = \{x_1, x_2, \cdots, x_n\}$ be this finite part. Then obviously

$$A \subset \bigcup_{1 \leq i \leq n} \{x_i\} \text{ and } d(\{x_i\}) = 0, \ \forall i = 1, 2, \cdots, n.$$

Another result (in a restrained context) holds: *Every bounded set in $\mathbb{R}^n$ is totally bounded* (the reader is asked to give a proof in Exercise 5.5.17).

If the metric space is arbitrary, then this may not be true (see Exercise 7.5.12). Remember that the converse is always true, that is, *any totally bounded set is bounded*.

(24) False! Total boundedness is not a topological property. This is its main weakest point. As a counterexample, consider the function $f : (0, 1) \to (1, +\infty)$ defined by $f(x) = \frac{1}{x}$. Then $f$ is a homeomorphism, $(0, 1)$ is totally bounded whereas $(1, +\infty)$ is not for it is not bounded.

## 5.3. Solutions to Exercises

**SOLUTION** 5.3.1. Let $\mathcal{U} = \{(-n, n)\}_{n \in \mathbb{N}}$ be an open cover of $\mathbb{R}$ (cf. Exercise 1.2.26). Now no finite subcollection of $\mathcal{U}$ can cover all of $\mathbb{R}$. For if $\mathcal{U}_p = \{(-n_i, n_i)\}_{1 \leq i \leq p}$ is a finite subcover, then its union will be of the form $(-N, N)$ where $N = \max_{1 \leq i \leq p} n_i$.

For $[0, +\infty)$, we may consider the open cover $\mathcal{U} = \{(-1, n)\}_n$. It is in effect a cover since

$$[0, +\infty) \subset \bigcup_{n=1}^{\infty} (-1, n).$$

Now any finite subcover of $\mathcal{U}$ is of the form $\{(-1, n_1), (-1, n_2), \cdots, (-1, n_p)\}$ and its union is $(-1, N)$, $N = \max_{1 \leq i \leq p} n_i$. Lastly, it is plain that

$$[0, +\infty) \not\subset (-1, N), \ \forall N \in \mathbb{N}.$$

As for $(0, 1)$, the reader may take $\mathcal{U} = \{(0, 1 - \frac{1}{n})\}_{n \geq 2}$ (or just $\{(\frac{1}{n}, 1)\}_{n \geq 2}$) and then show that $(0, 1)$ is not compact.

**SOLUTION** 5.3.2.

(1) $\mathbb{Q}$ is not compact as it is not closed in $\mathbb{R}$ (or since it is not bounded).

(2) $A = \{\frac{1}{n} : n \in \mathbb{N}\}$ is not compact since it is not closed.

(3) $\mathbb{Q} \cap [0, 1]$ is not closed in $\mathbb{R}$ and hence it is not compact.

(4) In fact $[a, b]$ is not compact in the discrete topology and neither is $\mathbb{R}$ nor is an infinite set. We only do that in the case of $[a, b]$.

In $(\mathbb{R}, \mathcal{P}(\mathbb{R}))$, the singleton $\{x\}$ is open and hence $\mathcal{U} = \{x\}_{x \in [a,b]}$ is an open cover of $[a, b]$ and every finite subcollection of it will never cover $[a, b]$. Therefore, $[a, b]$ is not compact in $(\mathbb{R}, \mathcal{P}(\mathbb{R}))$.

(5) Since we are now in the standard topology of $\mathbb{R}^2$, then to show the compactness of the given sets it suffices to show that they are both closed and bounded in $\mathbb{R}^2$ and this in all the remaining questions of this exercise. The set $A$ is obviously closed and bounded and hence it is compact in $\mathbb{R}^2$. As for $B$ and $C$ none of them is compact since they are not closed (cf. Exercise 4.3.27) and $C$ is even unbounded .

(6) $A$ is not compact as it is not bounded in $\mathbb{R}^2$ (it cannot be contained in a ball in $\mathbb{R}^2$ of finite radius, show it!).

(7) Both $A$ and $B$ are not compact since $A$ is closed and not bounded while $B$ is bounded but not closed. Let us show that. The set $A$ can be modified to be written as

$$A = \{(x, y) \in \mathbb{R}^2 : x \geq 0, xy = 1\}$$

since $xy = 1$ implies $x \neq 0$. Now we can easily show that $A$ is closed. However it is not bounded as it cannot be contained in a ball of finite radius.

The set $B$ is obviously bounded since $B \subset (0, 1] \times [-1, 1]$. Let us show that it is not closed. The sequence $\left(\frac{1}{n\pi}, 0\right)$ does belong to $B$ while its limit (in $\mathbb{R}^2$) is $(0, 0)$ is not in $B$.

(8) The reader may show that $A$ is actually equal to $B_c(0_{\mathbb{R}^2}, 1)$ (the closed ball in $\mathbb{R}^2$ of center $(0, 0)$ and of radius 1). Thus $A$ is compact.

## SOLUTION 5.3.3.

(1) We use the Euclidean metric (we could have used the sup metric or the taxicab metric too). The set $A$ is closed (why?) but it is not bounded. Let us show that. The set $A$ cannot be bounded since for all $M > 0$,

$$(M, \sqrt[3]{1 - M}) \in A \text{ and } d_2((M, \sqrt[3]{1 - M}), (0, 0))^2 \geq M^2.$$

Therefore, $A$ is not compact in usual $\mathbb{R}^2$.

(2) Now we use the taxicab metric. The set $B$ is compact in $\mathbb{R}^2$. It is obviously closed (is it not?). Since $x^4 + y^2 = 1$, $|x| \leq 1$

and $|y| \leq 1$ (why?). Whence

$$\forall x, y \in B : \ d_1((x,y),(0,0)) = |x| + |y| \leq 2,$$

proving the boundedness of $B$ and hence its compactness.

**SOLUTION** 5.3.4. Set

$$U_n = \begin{cases} \left(\frac{2}{3}, \frac{3}{2}\right), & n = 1, \\ \left(\frac{1}{n+1}, \frac{1}{n-1}\right), & n \geq 2. \end{cases}$$

Observe that for any $n \geq 1$, $U_n$ is open. It is also clear that

$$A = \left\{ \frac{1}{n} : \ n \in \mathbb{N} \right\} \subset \bigcup_{n \geq 1} U_n.$$

Hence $\mathcal{U} = \{U_n\}_n$ is an open cover for $A$. Since each $U_n$ contains only one point of $A$ and since $A$ is *infinite*, we conclude that no finite subcover of $\mathcal{U}$ can contain $A$. Thus $A$ is not compact.

**SOLUTION** 5.3.5. We write what $x_n \to a$ means in a topological space

$$x_n \longrightarrow a \Longleftrightarrow \forall U \in \mathcal{N}(a), \exists N \in \mathbb{N}, \forall n \ (n \geq N \Longrightarrow x_n \in U).$$

Now let $\mathcal{U} = \{U_i\}_{i \in I}$ be an open cover of $A = \{x_n : \ n \in \mathbb{N}\} \cup \{a\}$, i.e. $A \subset \bigcup_{i \in I} U_i$. We can then write the following

$$A = \{x_n : \ n < N\} \cup \{x_n : \ n \geq N\} \cup \{a\}.$$

With the $x_n$ for $n < N$ we can associate $N - 1$ elements of the cover $\mathcal{U}$. Also we observe that $\{a\}$ is contained in some $U_j$ and since the latter is open (and contains $a$), it is a neighborhood of $a$. By definition of the limit (see above), we also have $x_n \in U_j$ for $n \geq N$. Therefore, we deduce that

$$A \subset \bigcup_{i=1}^{N-1} U_i \cup U_j,$$

establishing the compactness of $A$.

**SOLUTION** 5.3.6. Two methods are given (another one will follow in Exercise 7.3.30).

(1) *A method based on the completeness axiom*: Let $\mathcal{U} = \{U_i\}_{i \in I}$ be an open cover of $[a, b]$. Set

$$A = \{x \in [a, b] : \ [a, x] \text{ can be covered by a } \textit{finite} \text{ subcollection of } U_i\}.$$

It is clear that we have to show that $b \in A$. We first note that $A$ is non void for $a \in A$ because $[a, a] = \{a\}$ is covered by one $U_i$ for some $i \in I$. Besides, $A$ is bounded above by $b$

as $A \subset [a,b]$. Thus, $A$ has the least upper bound property. Hence its sup, denoted by $M$, satisfies $M \leq b$. If we come to show that $M \in A$ and $M = b$, then we are done, i.e. $[a,b]$ will be compact.

Let us first show that $M \in A$. Since $M \in [a,b]$, there exists some $j \in I$ such that $M \in U_j$. But $U_j$ is open in the usual topology of $\mathbb{R}$ and so

$$\exists r > 0 : \ (M - r, M + r) \subset U_j.$$

Moreover, by definition of the least upper bound we have

$$\exists x \in A : \ M - r < x \leq M.$$

Since $x \in A$, $[a,x]$ is covered by a finite number of $U_i$ ($i = 1, 2, \cdots, n$, say), $\{U_1, U_2, \cdots, U_n, U_j\}$ covers the interval $[a, M]$ for

$$[a, M] = [a, x] \cup [x, M] \subset [a, x] \cup (M - r, M + r) \subset \underbrace{\bigcup_{i=1}^{n} U_i \cup U_j}_{\text{a finite union!}}.$$

Therefore, $M \in A$. To finish the proof, we need to verify that $M = b$. Assume $M \neq b$, i.e. $M < b$. Select a $y$ such that $y < b$ and $M < y < M + r$. Then we immediately see that $\{U_1, U_2, \cdots, U_n, U_j\}$ covers $[a, y]$ too (why?), i.e. $y \in A$. We have then reached a contradiction as we have $y \in A$ and $M < y$!

Thus $b = M \in A$, i.e. $[a,b]$ is covered by a *finite* subcollection of $U_i$, that is, $[a,b]$ is compact. The proof is complete.

(2) *A method based on the Nested Interval Property*: Assume $[a,b]$ is not compact, i.e. there is an open cover $U = \{U_i\}_{i \in I}$ from which no finite subcover can be extracted. We bisect $[a,b]$ at its middle point, i.e. at $\frac{1}{2}(a + b)$. We then obtain the two closed intervals

$$\left[a, \frac{a+b}{2}\right] \text{ and } \left[\frac{a+b}{2}, b\right]$$

(observe that their length is $(b - a)/2$).

One (at least!) of the two segments cannot be covered by a finite number of $U_i$ otherwise the two intervals and hence their union will be covered by a finite number of $U_i$! Call this interval $I_1$ and write it as $[a_1, b_1]$. Now, bisect it into two intervals as done previously and obtain the two intervals

$$\left[a_1, \frac{a_1 + b_1}{2}\right] \text{ and } \left[\frac{a_1 + b_1}{2}, b_1\right]$$

(with this time intervals having for length $(b_1 - a_1)/4$ regardless of what $a_1$ and $b_1$ can be). As before, at least one of them cannot be covered by a finite number of $U_i$. Continuing this process, we obtain a sequence of closed and bounded intervals $(I_n)$ (with $I_0 = [a, b]$) verifying
  (a) $(I_n)$ is decreasing;
  (b) the length of $I_n$ is $(b - a)/2^n$ for each $n$;
  (c) each $I_n$ cannot be covered by a finite number of $U_i$.
Now, by construction, the sequences $(a_n)$ and $(b_n)$ satisfy

$$a \leq a_n \leq a_{n+1} \leq \cdots \leq b_{n+1} \leq b_n \leq b$$

for all $n$. Thus $(a_n)$ is increasing and bounded above by $b$ and $(b_n)$ is decreasing and bounded below by $a$. Hence they both converge (to different limits), but the condition

$$b_n - a_n = (b - a)/2^n$$

ensures that their limits must be identical, call it $l$. It is certainly (by the sandwich rule!) inside $[a, b]$ and so it lies in some $U_j$, $j \in I$, of the initial open cover. By the openness of $U_j$ we have

$$\exists r > 0 : (l - r, l + r) \subset U_j.$$

Since $(b-a)/2^n$ tends to zero as $n$ goes to infinity, we have for large $n$, $(b-a)/2^n < r$ which gives

$$[a_n, b_n] \subset (l - r, l + r)$$

(remember that $l \in [a_n, b_n]$ for all $n$). Therefore, we have reached the desired contradiction for $[a_n, b_n]$ cannot be covered by a *finite* number of $\{U_i\}$. The proof is complete.

**SOLUTION** 5.3.7. We answer both questions together. We claim that any set containing $K$ is not compact. The proof is simple. Let $A$ be a set which contains $K$. Since $K' = \varnothing$ (see Exercise 3.3.39), we immediately deduce that $A$ is not limit point compact and hence it is not compact (cf. Proposition 5.1.13).

**SOLUTION** 5.3.8.

(1) Let $A$ be a finite set having $n$ elements, say. That is,

$$A = \{x_1, x_2, \cdots, x_n\}$$

where all $x_i$ $(1 \leq i \leq n)$ belong to $X$. Let $\mathcal{U} = \{U_i\}_{i \in I}$ be an open cover of $A$, i.e.

$$A = \{x_1, x_2, \cdots, x_n\} \subset \bigcup_{i \in I} U_i.$$

WLOG, each $x_i$ then belongs to some $U_i$ (it could happen that two or more of the $x_i$ belong to one $X_i$). Hence

$$A = \{x_1, x_2, \cdots, x_n\} \subset U_1 \cup U_2 \cup \cdots \cup U_n,$$

establishing the compactness of $A$.

(2) Question 4 of Exercise 5.3.2 combined with the preceding question does the job...

**SOLUTION** 5.3.9.

(1) Let $A$ and $B$ be compact. Let $\mathcal{U} = \{U_i\}_{i \in I}$ be an open cover of $A \cup B$, i.e.

$$A \cup B \subset \bigcup_{i \in I} U_i.$$

Hence

$$A \subset \bigcup_{i \in I} U_i \text{ and } B \subset \bigcup_{i \in I} U_i.$$

This tells us that $\mathcal{U} = \{U_i\}_{i \in I}$ is an open cover for both $A$ and $B$. Since they are compact, we get

$$A \subset \bigcup_{i=1}^{n} U_i \text{ and } B \subset \bigcup_{j=1}^{m} U_j.$$

Hence

$$A \cup B \subset \underbrace{\bigcup_{i=1}^{n} U_i \cup \bigcup_{j=1}^{m} U_j}_{\text{a finite union}}.$$

Therefore, $A \cup B$ is compact.

(2) The previous result is not true for an arbitrary union. For instance, in standard $\mathbb{R}$, the intervals $[-n, n]$ are all compact for any $n \in \mathbb{N}$. However, $\bigcup_{n=1}^{\infty} [-n, n] = \mathbb{R}$ is not compact.

(3) Let $(A_i)_{i \in I}$ be an arbitrary family of compact sets in a Hausdorff space. Since they are all compact, they are all closed.

Hence $\bigcap_{i \in I} A_i$ is closed. But for some (and all) $j \in I$ one has

$$\underbrace{\bigcap_{i \in I} A_i}_{\text{closed}} \subset \underbrace{A_j}_{\text{compact}} .$$

Thus $\bigcap_{i \in I} A_i$ is compact since a closed subset of a compact set is compact.

**SOLUTION 5.3.10.**

(1) Let $t \in X$. Recall that $x \mapsto d(x, t)$ is continuous on $X$ (see Exercise 2.3.24). Since $A$ is compact, $\inf_{x \in A} d(t, x) = d(t, A)$ is attained. So

$$\exists x \in A : \ d(t, A) = d(t, x).$$

Therefore,

$$d(t, A) \leq r \Longleftrightarrow \exists x \in A : \ d(t, x) \leq r,$$

which, in its turn, implies that

$$\bigcup_{x \in A} B_c(x, r) = \{t \in X : \ d(t, A) \leq r\}.$$

(2) Again by Exercise 2.3.24, this time $t \mapsto d(t, A)$ is continuous. Hence $\{t \in X : \ d(t, A) \leq r\}$ is closed as it is the inverse image of the closed $(-\infty, r]$ by a continuous function. Accordingly, $\bigcup_{x \in A} B_c(x, r)$ is closed, as desired.

**SOLUTION 5.3.11.** Yes, $\mathbb{R}$ is compact in the co-finite topology. Let $\mathcal{U} = \{U_i\}_{i \in I}$ be an open cover of $\mathbb{R}$ in $X$, i.e. $\mathbb{R} = \bigcup_{i \in I} U_i$. We need to show that $\mathbb{R}$ can be covered by a finite subcollection of $\mathcal{U}$. We have $\mathbb{R} = U_j \cup U_j^c$ ($j \in I$) and since $U_j^c$ is finite, it has the form $\{x_1, x_2, \cdots, x_p\}$, say, where all $x_i$ are real. Hence by assuming that each $U_i$ contains *one* $x_i$ (if not, then the cardinal of the subcover will be even smaller!)

$$\{x_1, x_2, \cdots, x_p\} \subset U_1 \cup U_2 \cup \cdots \cup U_p$$

and thus

$$\mathbb{R} = U_j \cup U_j^c \subset U_j \cup U_1 \cup U_2 \cup \cdots \cup U_p.$$

This shows that $\mathbb{R}$ is compact in the co-finite topology.

**REMARK.** The same method applies to show that any infinite set (finite sets are already compact!) is compact in the co-finite topology.

**SOLUTION** 5.3.12.

(1) $\mathbb{R}$ is not compact. Let $U = \{x_n : n \in \mathbb{N}\}$ be a countable set. Set
$$U_n = U^c \cup \{x_1, x_2, \cdots, x_n\}.$$
Since
$$\bigcup_{n=1}^{\infty} U_n = U^c \cup U = \mathbb{R},$$
i.e. $\{U_n\}_n$ constitutes an *open* cover for $\mathbb{R}$ (doesn't it?). However, no finite subcollection of $\{U_n\}_n$ can cover all of $\mathbb{R}$. To see this explicitly,
$$\bigcup_{p=1}^{N} U_{n(p)} \neq \mathbb{R} \text{ as, for instance, } x_{n(N)+1} \notin \bigcup_{p=1}^{N} U_{n(p)}$$
where $N = \max_{1 \leq p \leq N} n(p)$.

(2) No, for the same reason as before. In fact, the previous method can be applied to show that any infinite set is not compact in the co-countable topology.

(3) No and also as before. An alternative way of seeing this is the following: If $A$ is countable, then the induced topology is the discrete one and in a discrete topology a set is compact iff it is finite!

**SOLUTION** 5.3.13.

(1) Let us show that $T$ is a topological space on $\mathbb{R}$. $\varnothing$ belongs to $T$ by definition of $T$ and $\mathbb{R} \in T$ since $\mathbb{R}^c = \varnothing$ is compact. Let $U$ and $V$ be both in $T$.

(a) If $U$ or $V$ is empty, then $U \cap V$ is empty and hence $U \cap V \in T$.

(b) If $U$ and $V$ are both non-empty, then $U^c$ and $V^c$ are both compact in usual $\mathbb{R}$. Since $(U \cap V)^c = U^c \cup V^c$, it follows that $(U \cap V)^c$ is compact in usual $\mathbb{R}$ as it is a finite union of compact sets. Hence $U \cap V \in T$.

Finally, let $(U_i)_{i \in I}$ be an arbitrary collection of elements of $T$. We need to show that $\bigcup_{i \in I} U_i \in T$. As before, we need to discuss two cases.

(a) If all $U_i$ are empty, then so is their union $\bigcup_{i \in I} U_i$ yielding
$$\bigcup_{i \in I} U_i \in T.$$

(b) If at least one $U_j$ $(j \in I)$ is non-empty, then $U_j^c$ is compact. Moreover,

$$\left(\bigcup_{i \in I} U_i\right)^c = \bigcap_{i \in I} U_i^c \subset U_j^c.$$

Since $\bigcap_{i \in I} U_i^c$ is compact in usual $\mathbb{R}$, it is closed in usual $\mathbb{R}$. Since $U_j^c$ is compact, it follows that (in usual $\mathbb{R}$) $\left(\bigcup_{i \in I} U_i\right)^c$ is compact, i.e. $\bigcup_{i \in I} U_i \in T$. Thus we have proved that $T$ is a topological space.

(2) No, $T$ cannot be separated as no two open sets of $T$ can be disjoint. To see this, take any two (non-empty) open sets $U$ and $V$ of $T$. If $U \cap V = \varnothing$, then $U^c \cup V^c = \mathbb{R}$ where $U^c$ and $V^c$ are compact in standard $\mathbb{R}$ and hence they must be bounded. Hence their (finite!) union remains bounded and so it cannot be equal to $\mathbb{R}$. Thus, $T$ is not Hausdorff.

(3) We need to find a dense subset of $\mathbb{R}$ (density with respect to $T$) which is countable. The desired set is $\mathbb{Q}$. It is of course countable. It is also dense in $\mathbb{R}$. To see this, take any $x \in \mathbb{R}$. We need to show that

$$\forall U \in \mathcal{N}(x): \ U \cap \mathbb{Q} \neq \varnothing.$$

If $U$ is an open set (containing $x$), then $U^c$ is compact with respect to standard $\mathbb{R}$ and hence it is closed with respect to standard $\mathbb{R}$. This tells us that $U$ is open in usual $\mathbb{R}$ and so $U$ satisfies (why?)

$$U \cap \mathbb{Q} \neq \varnothing.$$

This means that $\mathbb{R} \subset \overline{\mathbb{Q}}$. The other inclusion being obvious, we conclude that $\mathbb{Q}$ is dense in $\mathbb{R}$. Consequently, $(\mathbb{R}, T)$ is separable.

(4) Let us show that $\mathbb{R}$ is compact in this topology. Let $\mathcal{U} = \{U_i\}_{i \in I}$ be an open cover of $\mathbb{R}$ in $T$, i.e. $\mathbb{R} \subset \bigcup_{i \in I} U_i$ where $U_i^c$ are compact in $\mathbb{R}$. But we observe that for some $j \in I$

$$U_j^c \subset \mathbb{R} \subset \bigcup_{i \in I} U_i.$$

Since all $U_i$ are open in usual $\mathbb{R}$ (why?), the compact $U_j^c$ is covered by $\mathcal{U} = \{U_i\}_{i \in I}$ and hence $U_j^c \subset \bigcup_{i=1}^{p} U_i$. Therefore,

$$\mathbb{R} = U_j \cup U_j^c \subset U_j \cup \bigcup_{i=1}^{p} U_i \text{ (a finite subcollection of } \mathcal{U}).$$

This establishes the compactness of $\mathbb{R}$ in $T$.

**SOLUTION 5.3.14.** We recall that the topology $T$ was defined on $[-1, 1]$ as

$$U \text{ open in } T \iff \{0\} \not\subset U \text{ or } (-1, 1) \subset U.$$

Let us show that $[-1, 1]$ is compact in $T$. Let $\mathcal{U} = \{U_i\}_{i \in I}$ be an open cover of $[-1, 1]$, i.e. $[-1, 1] \subset \bigcup_{i \in I} U_i$. Since $0 \in [-1, 1]$, this implies the existence of a $j_0$ in $I$ such that $\{0\} \subset U_{j_0}$. By definition of this topology, this forces us to have $(-1, 1) \subset U_{j_0}$. Similarly,

$$\exists j_1, j_2 \in I : \{-1\} \subset U_{j_1} \text{ and } \{1\} \subset U_{j_2}.$$

Thus,

$$[-1, 1] = \{-1\} \cup (-1, 1) \cup \{1\} \subset U_{j_1} \cup U_{j_0} \cup U_{j_2},$$

i.e. proving the compactness of $[-1, 1]$, as required.

**SOLUTION 5.3.15.** Let $x_n = \frac{1}{n}$ $(n \geq 2)$ be a sequence in $(0, 1)$. It obviously converges to $0$ and hence so do all its subsequences! But $0 \not\in (0, 1)$ and hence $(0, 1)$ is not sequentially compact. Therefore $(0, 1)$ is not compact.

**SOLUTION 5.3.16.** Let $\{U_i\}_{i \in I}$ be an open cover for $f(X)$, i.e. $f(X) \subset \bigcup_{i \in I} U_i$. Hence

$$X \subset f^{-1}(f(X)) \subset f^{-1}\left(\bigcup_{i \in I} U_i\right) = \bigcup_{i \in I} f^{-1}(U_i).$$

As each $U_i$ is open (in $Y$) and $f$ is continuous, each $f^{-1}(U_i)$ is open (in $X$). Hence the compact $X$ has been covered by a family of open sets, namely $\{f^{-1}(U_i)\}_{i \in I}$. Compactness then implies that $X$ is covered only by a finite number of them. This leads to

$$X \subset \bigcup_{i=1}^{n} f^{-1}(U_i) = f^{-1}\left(\bigcup_{i=1}^{n} U_i\right),$$

so that

$$f(X) \subset f\left(f^{-1}\left(\bigcup_{i=1}^{n} U_i\right)\right) \subset \bigcup_{i=1}^{n} U_i,$$

proving that $f(X)$ is compact.

**SOLUTION 5.3.17.** Let $x \in X$. Then (by the Archimedian Property, say!),

$$\forall a \in A, \exists n \in \mathbb{N} : d(x, a) < n.$$

Hence $A \subset \bigcup_{n \in \mathbb{N}} B(x, n)$, that is, $\{B(x, n)\}_{n \in \mathbb{N}}$ constitutes an open cover for the compact $A$. Hence

$$A \subset \bigcup_{i=1}^{k} B(x, n_i).$$

But, all $B(x, n_i)$ have a common center. Therefore,

$$\bigcup_{i=1}^{k} B(x, n_i) = B(x, N) \text{ where } N = \max(n_1, \cdots, n_k) < \infty.$$

Accordingly, $A \subset B(x, N)$, showing that $A$ is indeed bounded, as desired.

**SOLUTION** 5.3.18. We know by Theorem 5.1.20, that $f(X)$ is compact in $\mathbb{R}$. Hence $f(X)$ is bounded and so $\sup f(X)$ and $\inf f(X)$ exist. Besides, by Proposition 3.1.24, we have

$$\sup f(X), \inf f(X) \in \overline{f(X)}.$$

Since $f(X)$ is compact and $\mathbb{R}$ is separated, we also know that $f(X)$ must be closed (cf. Theorem 5.1.9). Accordingly,

$$\sup f(X), \inf f(X) \in f(X),$$

as required.

**SOLUTION** 5.3.19. Let $X$ be compact and let $A \subset X$ be closed. We want to show that $A$ is compact, so let $\{U_i\}_{i \in I}$ be an open cover for $A$, that is, $A \subset \bigcup_{i \in I} U_i$. We now want to exploit the compactness of $X$. Since $A$ is closed, $A^c$ is open. We can then write

$$X = A \cup A^c \subset \left( \bigcup_{i \in I} U_i \right) \cup A^c = \bigcup_{i \in J} V_i,$$

where $V_i$ are $U_i$ and one of them is $A^c$. This means that $\{V_i\}_{i \in J}$ constitutes an open cover for the compact $X$. Hence

$$X \subset \bigcup_{i=1}^{n} V_i \text{ so that } A \subset X \subset \bigcup_{i=1}^{n} V_i.$$

If none of the $V_i$ with $i = 1, \cdots, n$ is $A^c$, then this means that we have covered $A$ by a finite subcover. Otherwise, if $A^c$ is one of the elements of the finite subcover, call it $V_n$ for convenience, then $A$ may be covered by $V_1, \cdots, V_{n-1}$. Consequently, in either case $A$ has been covered by a finite subcover, meaning that $A$ is compact.

**SOLUTION** 5.3.20.

(1) Fix $a \in A$. Since $x \in A^c$, $a \neq x$ and so as $X$ is separated, then
$$\exists U_a, V_a \text{ (both open)}, \ x \in U_a, \ a \in V_a : U_a \cap V_a = \varnothing.$$
Hence $A \subset \bigcup_{a \in A} V_a$, which by compactness of $A$, permits us to have
$$A \subset V_{a_1} \cup \cdots \cup V_{a_n},$$
say. Now, set
$$V = V_{a_1} \cup \cdots \cup V_{a_n} \text{ and } U = U_{a_1} \cap \cdots \cap U_{a_n}.$$
Then clearly $U$ and $V$ are open (why?). Also, $x \in U$ ($x$ is in each $U_a$) and $A \subset V$. It only remains to see why $U \cap V = \varnothing$, but this immediately follows for
$$U \cap V = U \cap (V_{a_1} \cup \cdots \cup V_{a_n}) = (U \cap V_{a_1}) \cup \cdots \cup (U \cap V_{a_n}) = \varnothing \cup \cdots \cup \varnothing = \varnothing,$$
as required.

(2) The required $W$ is $U$: Indeed, by the previous question, there are two open sets $U$ and $V$ s.t.
$$x \in U, \ A \subset V : U \cap V = \varnothing.$$
Hence
$$U \cap A = \varnothing$$
and so
$$x \in U \subset A^c.$$

(3) To show that $A$ is closed, we show that $A^c$ is open. Let $x \in A^c$. By the previous question, there is an open set $W_x$ such that $x \in W_x \subset A^c$. Proposition 1.1.24 then tells us that
$$A^c = \bigcup_{x \in A^c} W_x,$$
that is, $A^c$ is a union of open sets and this establishes the openness of $A^c$ or the closedness of $A$.

**SOLUTION** 5.3.21. Let $V$ be a closed set in $X$ (we ought to show that $f(V)$ is closed in $Y$). Since $X$ is compact, $V$ too is compact (Theorem 5.1.9). Hence $f(V)$ is compact for $f$ is continuous (Theorem 5.1.20). Finally, $f(V)$, being compact in a Hausdorff space, is itself closed (Theorem 5.1.9).

**SOLUTION** 5.3.22. Let $x \in A$. Since $W$ is open (remember that it is then a union of elementary open sets of $X \times Y$) and contains $A \times \{y\}$, it follows that there are open sets $U_x \subset X$ and $V_x \subset Y$ such that
$$(x, y) \in U_x \times V_x \text{ and } U_x \times V_x \subset W.$$

The collection $\{U_x\}_{x \in A}$ is then easily seen to be an open cover for $A$. Since $A$ is compact, we deduce that $A \subset \bigcup_{x \in I} U_x$ where $I$ is a *finite* subset of $A$. Now, set

$$U = \bigcup_{x \in I} U_x \text{ and } V = \bigcap_{x \in I} V_x.$$

Then $U$ is open as it is a union of open sets. The set $V$ too is open but because it is a finite (thanks to the finiteness of $I$!) intersection of open sets. Do $U$ and $V$ satisfy the required conditions? The answer is yes. Indeed, $A \subset U$ is evident. Now, by construction, for all $x \in A$, $y \in V$ (as for all $x \in I$, $y \in V_x$, that is, $y \in \bigcap_{x \in I} V_x = V$).

So, it only remains to show that $U \times V \subset W$. This follows from (for all $x \in I$):

$$U_x \times V \subset U_x \times V_x \subset W \Longrightarrow U \times V = \bigcup_{x \in I}(U_x \times V) \subset W.$$

**SOLUTION 5.3.23.** Let

$$q : X \times Y \to Y \text{ be such that } q(x, y) = y.$$

Let $V$ be a closed set in $X \times Y$ and set

$$W = V^c = (X \times Y) \setminus V.$$

We are required to show that $q(V)$ is closed, i.e. $\overline{q(V)} \subset q(V)$ or $[q(V)]^c \subset [\overline{q(V)}]^c$. Let $y \notin q(V)$. Then $(x, y) \notin V$ for all $x \in X$, i.e. $X \times \{y\} \subset W$. The assumptions of the "Tube Lemma" are realized. Hence there is some open set $U$ in $Y$, with $y \in U$ and such that

$$X \times U \subset W = (X \times Y) \setminus V.$$

So if $z \in U$, then $(x, z) \notin V$ so that $z \notin q(V)$, that is, $z \in [q(V)]^c$. Therefore

$$U \cap q(V) = \varnothing$$

for some neighborhood $U$ of $y$ which precisely means that $y \notin \overline{q(V)}$. This completes the proof.

**SOLUTION 5.3.24.**

(1) Suppose that $X \times Y$ is compact. We know that the projections

$$p : X \times Y \to X$$
$$(x, y) \mapsto p(x, y) = x$$

and

$$q : X \times Y \to Y$$
$$(x, y) \mapsto q(x, y) = y$$

are continuous functions (see Theorem 4.1.7). Since $X \times Y$ is compact and $p$ is *continuous* and *onto*, $p(X \times Y) = X$ is compact. An analogous reasoning using the map $q$ shows that $Y$ is also compact.

(2) (partly inspired by [24]) Conversely, assume that $X$ and $Y$ are two compact spaces. To show that $X \times Y$, let $\{U_i\}_{i \in I}$ be an open cover for $X \times Y$, i.e. $X \times Y = \bigcup_{i \in I} U_i$. By taking the complement, we get

$$\bigcap_{i \in I} V_i = \varnothing \text{ where } V_i = U_i^c$$

(that is, $V_i$ are all closed). Let $q$ be the projection defined in the first part of the proof. Let $t \in Y$. Then $X \times \{t\}$ is homeomorphic to $X$ (for each $t$). Since $X$ is compact, $X \times \{t\}$ too is compact. But

$$X \times \{t\} \subset X \times Y = \bigcup_{i \in I} U_i$$

and so by compactness of $X \times \{t\}$, we know that

$$X \times \{t\} \subset \bigcup_{i \in J_t} U_i$$

where $J_t$ is a finite part of $I$. Hence

$$(X \times \{t\}) \cap \bigcap_{i \in J_t} V_i = \varnothing.$$

Setting $W_t = \bigcap_{i \in J_t} V_i$ (which is closed) gives us $(X \times \{t\}) \cap W_t = \varnothing$. Thus, $t \notin q(W_t)$ (why?) and so

$$\bigcap_{t \in Y} q(W_t) = \varnothing.$$

But each $q(W_t)$ is closed in $Y$ by Exercise 5.3.23. Since $Y$ is compact, it follows that there a finite subset $Z$ of $Y$ such that

$$\bigcap_{t \in Z} q(W_t) = \varnothing.$$

Now, put $J = \bigcup_{t \in Z} J_t$ (*which is a finite part of* $I$). Let $(x, y) \in X \times Y$. As $y \notin \bigcap_{t \in Z} q(W_t)$, there is some $t \in Z$ such that $y \notin q(W_t)$. But this gives $(x, y) \notin W_t$. Whence

$$\exists i \in J_t \subset J : (x, y) \notin V_i.$$

Accordingly, $(x, y) \notin \bigcap_{i \in J} V_i$. Since this holds for all $(x, y)$, it finally follows that $\bigcap_{i \in J} V_i = \varnothing$ or merely

$$X \times Y = \bigcup_{i \in J} U_i,$$

which proves the compactness of $X \times Y$, as desired.

**SOLUTION 5.3.25.**

(1) Let $f : \mathbb{R}^n \times \mathbb{R}^n \to \mathbb{R}^n$ defined by $f(x, y) = x + y$. It is known that $f$ is continuous (Corollary 4.1.25). Since $A$ and $B$ are compact, so is $A \times B$. Therefore,

$$f(A \times B) = A + B$$

too is compact.

(2) We show that $\overline{A + B} \subset A + B$. Let $x \in \overline{A + B}$. Then there exists $x_n \in A + B$ such that $x_n \to x$, that is, there exist $a_n$ in $A$ and $b_n$ in $B$ such that $a_n + b_n \to x$.

But, $A$ is compact and hence there exists a subsequence $(a_{nk})$ of $(a_n)$ such that $a_{nk} \to a \in A$. Now we have a subsequence $(a_{nk} + b_{nk})$ of $(a_n + b_n)$. Since $(a_n + b_n)$ converges, so does $(a_{nk} + b_{nk})$ (it converges to $x$!). Thus

$$(b_{nk}) = (a_{nk} + b_{nk} - a_{nk})$$

converges too. Since $B$ is closed, $(b_{nk})$ has a limit $b$ belonging to $B$. Therefore,

$$x = a + b \in A + B,$$

establishing the closedness of $A + B$.

**SOLUTION 5.3.26.**

(1) That $A$ is bounded is clear. Let us show that it is closed. We can write

$$A = \{x \in \mathbb{Q} : \sqrt{2} < x < \sqrt{3}\} \cup \{x \in \mathbb{Q} : -\sqrt{3} < x < -\sqrt{2}\}.$$

Call the first set in the union $A_1$ and the second one $A_2$. Now $A_1$ is closed in $\mathbb{Q}$ since one can write

$$A_1 = [\sqrt{2}, \sqrt{3}] \cap \mathbb{Q}.$$

Also $A_2$ is closed in $\mathbb{Q}$ and hence $A$ is closed in $\mathbb{Q}$. To show that $A$ is not compact, we can consider the open covering $\mathcal{U} = \{U_n\}_{n \geq 1}$ where

$$U_n = \left\{ x \in \mathbb{Q} : 2 + \frac{1}{n} < x^2 < 3 - \frac{1}{n} \right\}$$

and we can verify that it has no finite subcover...

(2) Finally, $A$ is open in $\mathbb{Q}$ as one can write

$$A = [(\sqrt{2}, \sqrt{3}) \cup (-\sqrt{3}, -\sqrt{2})] \cap \mathbb{Q}.$$

**SOLUTION 5.3.27.**

(1) $(\mathbb{R}, d)$ is bounded if there exists an $r > 0$ such that $\mathbb{R} \subset B_c(0, r)$. This "$r$" is 1 since

$$\forall x \in \mathbb{R} : \; d(x, 0) \leq 1.$$

Thus $(\mathbb{R}, d)$ is bounded.

(2) Let us show that $(\mathbb{R}, d)$ is not sequentially compact. We show that $(x_n)$ does not have any convergent subsequence. Before that we obviously have

$$d(x_n, x_m) = d(n, m) = \inf(|n - m|, 1) = \begin{cases} 0, & n = m, \\ 1, & n \neq m \end{cases}$$

($d(n, m)$ is equal to 1 in the case $n \neq m$ since $n$ and $m$ are *integers*). In order to reach a contradiction, assume that $(x_{n(k)})$ converges to a real number $x$, say. Hence

$$\forall \varepsilon > 0, \exists K \in \mathbb{N}, \forall k \; (k \geq K \Longrightarrow d(x_{n(k)}, x) < \varepsilon).$$

In particular for $\varepsilon = \frac{1}{4}$, there is $K \in \mathbb{N}$ such that

$$d(x_{n(k)}, x) < \frac{1}{4}.$$

whenever $k \geq K$. Hence

$$d(x_{n(K)}, x_{n(K+1)}) \leq d(x_{n(K)}, x) + d(x, x_{n(K+1)}) < \frac{1}{4} + \frac{1}{4} = \frac{1}{2}.$$

Now, remember that $k \mapsto n(k)$ is strictly increasing and so $n(K + 1) > n(K)$ (as $K + 1 > K$). Hence $n(K + 1) \neq n(K)$. Therefore,

$$d(n(K + 1), n(K)) = 1 \not< \frac{1}{2},$$

which is a contradiction and this proves the non-compactness of $\mathbb{R}$.

(3) We have just seen that $\mathbb{R}$ was bounded with respect to $d$. Since it is the whole set, it is obviously closed. This is another example of a closed and bounded set which is not compact.

**SOLUTION** 5.3.28.

(1) Set $x_n = n$ for each $n \in \mathbb{N}$. Then $(x_n)$ is a sequence in $\mathbb{R}$ from which no convergent subsequence can be extracted for

$$\lim_{n \to \infty} \delta(n, x) = \lim_{n \to \infty} |\arctan n - \arctan x| = \left| \frac{\pi}{2} - \arctan x \right| > 0$$

for all $x \in \mathbb{R}$. Thus $\mathbb{R}$ is not sequentially compact.

(2) This is yet another example of a closed and bounded set which is not compact (that $\mathbb{R}$ is bounded was done in Exercise 2.3.29 and that $\mathbb{R}$ is closed is plain).

**SOLUTION** 5.3.29.

(1) We know from a basic linear algebra course that $\dim X = \infty$.
(2) The unit closed ball is not sequentially compact in $C([0, 1], \mathbb{R})$. To see this, let $(f_n)$ be the moving bump (see Exercise 4.3.34). Then for all different $n, m$

$$d_\infty(f_n, f_m) = 1.$$

Since $f_n \in B_c(0, 1)$ for all $n$, we deduce from the equality just above that $(f_n)$ is a sequence in the closed unit ball which cannot have a convergent subsequence. Thus the closed unit ball in $C([0, 1], \mathbb{R})$ is not sequentially compact.

(3) Since we are in a metric space, the closed unit ball is not compact either.

(4) No, it is not locally compact. The proof relies on results from normed vector spaces, so we shall not include it here.

**SOLUTION** 5.3.30.

(1) No, $f(\mathbb{R}) = \mathbb{R}^+$ is not compact (see Exercise 5.3.2).
(2) Since $\mathbb{R}$ is compact in $X$ and $f(\mathbb{R})$ is not compact, we deduce that $f$ cannot be continuous.

**SOLUTION** 5.3.31. The answer is no in both cases, that is, $[0, 1]$ is neither homeomorphic to $[0, \infty)$ nor to $(0, 1]$. The reason is that $[0, 1]$ is compact and the other two sets are not. Hence no homeomorphism between the two sets exists.

**SOLUTION** 5.3.32. Since the intersection of all $A_n$ is contained in each $A_n$, we have

$$f(\bigcap_{n \in \mathbb{N}} A_n) \subset \bigcap_{n \in \mathbb{N}} f(A_n).$$

To prove the reverse inclusion, we first observe that since $X$ is compact, the hypotheses on $A_n$ guarantee that $\bigcap_{n \in \mathbb{N}} A_n \neq \varnothing$ (by Corollary 5.1.6) and hence $f(\bigcap_{n\in\mathbb{N}} A_n) \neq \varnothing$. This yields

$$\bigcap_{n\in\mathbb{N}} f(A_n) \neq \varnothing \text{ since } f\left(\bigcap_{n\in\mathbb{N}} A_n\right) \subset \bigcap_{n\in\mathbb{N}} f(A_n).$$

Now, let $y \in \bigcap_{n\in\mathbb{N}} f(A_n)$. Then $y \in f(A_n)$ for all $n$. Hence for all $n$, $y = f(x_n)$ with $x_n \in A_n \subset X$. Therefore, $(x_n)$ being a sequence in $X$ which is compact or equivalently sequentially compact, it possesses a convergent subsequence, denoted by $(x_{n(k)})$. Let $x \in X$ be its limit.

Since $f$ is continuous or equivalently sequentially continuous, we have

$$\lim_{k\to\infty} f(x_{n(k)}) = f(x) = y.$$

The proof will be complete as soon as we verify that $x$ is in each $A_n$ for all $n$. To see this, fix an $m \in \mathbb{N}$. Then

$$x_n \in A_n \subset A_m, \ \forall n \geq m$$

for $(A_n)$ is decreasing. A fortiori, the previous holds true for each $k$. Passing to the limit (as $k \to \infty$) we immediately observe that $x$ must be in $\overline{A_m} = A_m$ (why?) and this is for each $m$. Thus

$$x \in \bigcap_{m\in\mathbb{N}} A_m = \bigcap_{n\in\mathbb{N}} A_n \text{ and so } y \in f(x) \in f\left(\bigcap_{n\in\mathbb{N}} A_n\right).$$

This completes the proof.

**SOLUTION** 5.3.33.

(1) (a) It is clear that

$$\forall x \in X, \ \forall U \in \mathcal{N}(x): \ U \subset X.$$

Since $X$ is compact, we immediately see that $X$ is locally compact.

(b) Let $x \in \mathbb{R}$. In usual $\mathbb{R}$, $[x-1, x-1]$ is a compact set. It contains $(x-1, x-1)$ which is a neighborhood of $x$. Therefore, $\mathbb{R}$ is locally compact (remember that it is not compact!).

(c) If $\mathbb{Q}$ were locally compact, there would exist a compact set $V$ such that a neighborhood of $0$ (for example!) $U$, say, would satisfy $U \subset V$. But $U = (-r, r) \cap \mathbb{Q}$ for some $r > 0$. For irrational $r$, $[-r, r] \cap \mathbb{Q}$ would be a closed set in $V$ and hence it would be compact. This is evidently not true for we can take a sequence of *rationals*, $(r_n)$ say, which converges to the (irrational!) $r$. Thus $(r_n)$ cannot have a converging subsequence as all subsequences would converge to $r$ too (and $r \notin [-r, r] \cap \mathbb{Q}$!).

If $r$ is not irrational, then replace $[-r, r]$ by $[-s, s]$ where $s < r$ is irrational and follow a similar reasoning!

(d) A similar proof as before applies to show that $\mathbb{R} \setminus \mathbb{Q}$ is not locally compact.

(e) Let $X$ be a discrete topological space and let $x \in X$. Then $\{x\}$ is a compact set (why?) and it is also a neighborhood of $x$. Consequently, $X$ is locally compact.

(f) Let $x \in X$. It is clear that $\{a, x\}$ is compact and a neighborhood of $x$ simultaneously. Thus $X$ is locally compact.

(2) From the previous answers, $\mathbb{Q}$ is Hausdorff but it is not locally compact. Also, $\mathbb{R}$ is Hausdorff and locally compact.

**SOLUTION** 5.3.34. Let $f$ be the identity mapping from $X$ onto $Y$ where $X$ is $\mathbb{Q}$ endowed with the discrete topology and $Y$ is $\mathbb{Q}$ equipped with the usual topology.

Then $f$ is continuous and we have $f(X) = Y$. Now from Exercise 5.3.33, $X$ is locally compact while $Y$ is not.

**SOLUTION** 5.3.35.

(1) Let $A$ be a closed set in $X$ and let $x \in A$. Let $K$ be a compact set which contains a neighborhood $U$ of $x$. Then $K \cap A$ is closed in $K$ (in the subspace topology) and thus it is compact. Besides

$$x \in U \cap A \subset K \cap A$$

and so $U \cap A$ is a neighborhood of $x$ in $A$, proving the local compactness of $A$.

(2) Let $U$ be an open set in $X$ and let $x \in U$. From Proposition 5.1.16, there exists $V \in \mathcal{N}(x)$ such that $\overline{V}$ is compact and $\overline{V} \subset U$. Since $V \subset \overline{V}$, $U$ becomes locally compact.

(3) Well, $\{(0, 0)\}$ is compact and hence locally compact. The set $\{(x, y) \in \mathbb{R}^2 : x > 0\}$ is locally compact by the previous question for it is open in usual $\mathbb{R}^2$ (which is Hausdorff!).

(4) The given set is not locally compact. It is not locally compact at 0 (which is the only point causing the problem of non-local compactness) as the reader may check.

(5) We deduce from the previous two questions that the union of two locally compact spaces need not remain locally compact.

**SOLUTION** 5.3.36.

(1) No! For example, take $X = [0, 1/2]$ and $f(x) = x^2$...

(2) Define a function

$$\varphi : (X, d) \to (\mathbb{R}, |\cdot|) \text{ by } \varphi(x) = d(x, f(x)).$$

We claim that $\varphi$ is continuous on $X$. To see this, let $x, y \in X$. Then we have by Exercise 2.5.2

$$|\varphi(y) - \varphi(x)| \leq d(x, y) + d(f(x), f(y)) < 2d(x, y)$$

(for all $x, y \in X$). Therefore, $\varphi$ is continuous on $X$ which is compact and so the function $\varphi$ attains its minimum (and its maximum too!) at some $a \in X$.

Let us show that $a$ is the expected fixed point. Assume $f(a) \neq a$. Then

$$\varphi(f(a)) = d(f(a), (f \circ f)(a)) < d(a, f(a)) = \varphi(a)$$

and we realize that $\varphi$ does not attain its minimum at $a$ anymore! Therefore, $f(a) = a$, showing that the $f$ admits a fixed point. Let us now show its uniqueness. Assume there were two *different* fixed points $x$ and $y$ say, i.e. $f(x) = x$ and $f(y) = y$ with $x \neq y$. Then

$$d(x, y) = d(f(x), f(y)) < d(x, y)$$

which is impossible and hence $x = y$ necessarily. The proof is complete.

**REMARK.** An example showing the importance of the compactness hypothesis may be found in Exercise 7.3.28.

**SOLUTION** 5.3.37. Let $(X, d)$ be a compact metric space. Then it is not hard to see that $X = \bigcup_{x \in X} B(x, 1/n)$, and so $\{B(x, 1/n)\}_{x \in X}$ constitutes an open cover for $X$. By compactness, we can cover $X$ by $\bigcup_{k=1}^{m} B(x_k, 1/n)$ which we denote by $S_n$. Now set $S = \bigcup_{n \in \mathbb{N}} S_n$. Then it is plain that $S$ is countable. It remains to verify that $S$ is dense in $X$. Observe that for all $(x \in X)$

$$\forall n \in \mathbb{N}, \exists y \in S_n \subset S : d(x, y) < \frac{1}{n}$$

which implies that

$$d(x, S) = \inf_{s \in S} d(x, s) < \frac{1}{n}$$

for all $n \in \mathbb{N}$. Thus $d(x, S) = 0$ and hence Exercise 3.3.43 implies that $x \in \overline{S}$ which completes the proof.

**SOLUTION** 5.3.38. Let $(y_n)$ be a sequence in $f(X)$ (we must show that $(y_n)$ has a convergent subsequence). Hence

$$\exists x_n \in X : y_n = f(x_n).$$

Since $X$ is sequentially compact, the sequence $(x_n)$ contains a convergent subsequence $(x_{n_k})$ to an $x \in X$. By (sequential) continuity,

$$y_{n_k} = f(x_{n_k}) \longrightarrow f(x) \in f(X),$$

showing that $f(X)$ is sequentially compact.

**SOLUTION** 5.3.39. The answer depends on the cardinality of $X$.

(1) If card$X$ is finite, then $(X, d)$ is totally bounded (see the "True/False" section).

(2) If card$X$ is infinite, then let $0 < \varepsilon < \frac{1}{3}$ (for instance). Then the only subsets of $X$ whose diameter is smaller than $\varepsilon$ are the singletons $\{x\}$, where $x \in X$ (and evidently $\varnothing$). It thus becomes clear that it is impossible to cover the *infinite* $X$ by a *finite* number of singletons! This tells us that $(X, d)$ is not totally bounded and that only finite discrete metric spaces are totally bounded.

**SOLUTION** 5.3.40. Assume that such an $f$ is not uniformly continuous. Then for some $\varepsilon > 0$ and all $\alpha > 0$, there exist $x, y \in X$ such that

$$d(x, y) < \alpha \text{ and } d'(f(x), f(y)) \geq \varepsilon.$$

In particular for $\alpha = \frac{1}{n}$, there are sequences $(x_n)$ and $(y_n)$ such that

$$d(x_n, y_n) < \frac{1}{n} \text{ and } d'(f(x_n), f(y_n)) \geq \varepsilon.$$

Since $X$ is compact, it is sequentially compact and so the sequence $(x_n)$ has a convergent subsequence $(x_{n_k})_k$. Call $x$ its limit (in $(X, d)$). It turns out that $(y_{n_k})_k$ too converges to $x$ because for all $k$

$$0 \leq d(x_{n_k}, y_{n_k}) < \frac{1}{n_k}$$

and passing to the limit as $k$ tends to $\infty$ yields the desired property (write $d(x, y_{n_k}) \leq d(x, x_{n_k}) + d(x_{n_k}, y_{n_k})$).

As $f$ is assumed to be continuous, it is also sequentially continuous and so

$$\lim_{k \to \infty} f(x_{n_k}) = \lim_{k \to \infty} f(y_{n_k}) = f(x).$$

All this would lead in the end to

$$\varepsilon \leq d'(f(x_{n_k}), f(y_{n_k})) \leq d'(f(x_{n_k}), f(x)) + d'(f(x), f(y_{n_k})) \longrightarrow 0$$

which is absurd! Accordingly, $f$ under the circumstances of this exercise must be uniformly continuous.

## 5.4. Hints/Answers to Tests

SOLUTION 35. Yes, $A \cap K$ is compact. Here is an outline of the proof

(1) $K$ is compact and hence closed (why?)...
(2) $A \cap K$ is closed because...
(3) Finally, use
$$A \cap K \subset K...$$

SOLUTION 36. Well, do it!...

SOLUTION 37. Yes $\{a\}$ is compact since it is finite. $X$ is not compact (if it is infinite!). To see this we observe that $\{a, x\}$ are open sets for each $x \in X$ (why?) whose union covers $X$ from which no finite subcollection can cover the whole of $X$.

We deduce from that $\{a\}$ is not relatively compact since from Exercise 3.3.34, $\overline{\{a\}} = X$.

SOLUTION 38. It suffices to show $S \subset T$. If $V$ is a closed set in $S$, then $V$ is compact in $S$. By Question 11 of the "True or False" Section, $V$ is compact in $T$ (as $T \subset S$). But compact sets in Hausdorff spaces are closed...Alternatively, use the identity mapping between $S$ and $T$...

SOLUTION 39. Consider the open cover consisted of the sets $\{1\}, \{1, 2\} \cdots, \{1, 2, \cdots, n\}, \cdots$ which is not reducible to a finite cover...

SOLUTION 40. No, $\mathbb{R}$ is not compact in $\mathbb{R}_\ell$. One reason is that $\{[-n, n)\}_{n \in \mathbb{N}}$ is an open cover of $\mathbb{R}$ etc...

SOLUTION 41. Yes...

SOLUTION 42. No! (Why?)...

SOLUTION 43. Fairly simple...

SOLUTION 44. Yes (why?)...

SOLUTION 45. Yes by the Heine Theorem!...

SOLUTION 46.

(1) Yes! In fact, the question is true for any topology endowing $A_i$. Why?...
(2) No! Each $A_i$ is compact... Since each $A_i$ is equipped with the discrete topology, the topology of $\prod_{i \in I} A_i$ too is discrete (by Exercise 4.5.19). Therefore, $\prod_{i \in I} A_i$ is not compact as it is infinite...

# CHAPTER 6

# Connected Spaces

## 6.2. True or False: Answers

ANSWERS.

(1) If $(X, T)$ is not connected, then $(X, T')$ is not connected. The reason is simple, that is, if $U$ and $V$ are open in $T$ such that $U \cup V = X$ and $U \cap V = \varnothing$ (meaning that $(X, T)$ is not connected), then $U$ and $V$ are open in $T'$ with $U \cup V = X$ and $U \cap V = \varnothing$ as well, leading to the non-connectedness of $(X, T')$. The other implication does not always hold. As a counterexample, take usual $\mathbb{R}$ and the discrete $\mathbb{R}$, denoted by $\mathbb{R}_{dis}$. Then $\mathbb{R}$ is connected while $\mathbb{R}_{dis}$ is not and yet $\mathbb{R} \subset \mathbb{R}_{dis}$.

(2) The answer is no! In usual $\mathbb{R}$, $\{1, 2\}$ is closed and not connected (it is not an interval) yet $\mathbb{R}$ is connected.

(3) The answer is no! For instance, in the usual topology, $A = [-1, 1]$ is connected while its boundary $\{-1, 1\}$ is not.

(4) The answer is no! A counterexample is $\mathbb{Q}$ which is not connected (as will be seen below) while $\overline{\mathbb{Q}} = \mathbb{R}$ is connected. The converse is always true, i.e. if $A$ is connected, so is its closure $\overline{A}$. See Exercise 6.3.10.

(5) The answer is no! For example, in the usual topology, both $A = [0, 1]$ and $B = [2, 3]$ are connected while their union is not.

(6) The answer is no! Let $A$ be the unit circle (in $\mathbb{R}^2$ endowed with its euclidian metric) and let $B$ be the real axis. They are both connected whilst their intersection, equal to $\{-1, 1\}$, is not connected.

   If you are a little perplexed about this (mainly because they are not subspaces of the same topological space!), then take again the unit circle and intersect it with the circle whose center is $(0, 1)$ and its radius 1, say. The intersection reduces to two points and hence it cannot be connected.

(7) The answer is no in general. Take

$$B = B_c((0, 1), 1) \cup B_c((0, -1), 1).$$

Then $B$ is connected (see Exercise 6.3.1). It can be shown that $\overset{\circ}{B} = B((0,1),1) \cup B((0,-1),1)$ which is not connected (see again Exercise 6.3.1).

It is worth stating that the result is true in the usual topology of $\mathbb{R}$. It is known that the only connected parts of standard $\mathbb{R}$ are the intervals and the interior of an interval in $\mathbb{R}$ remains an interval and hence it is connected.

(8) False! Let $A = \{1,2\}$. Then

$$A + A = \{2,3,4\} \neq \{2,4\} = 2A.$$

But we observe that $2A \subset A + A$. In fact, this is always true. Indeed, if $x \in 2A$, then for some $a \in A$, $x = 2a = a + a$ and hence $x \in A + A$.

For the reverse inclusion $A + A \subset 2A$ to be true, we need to assume e.g. that $A$ is convex. For the proof, let $x \in A + A$. Then for some $a$ in $A$ and some $b$ in $A$, $x = a + b$. But $A$ is convex and hence for all $t \in [0,1]$, $(1-t)a + tb \in A$. Taking $t = \frac{1}{2}$, yields

$$\frac{1}{2}a + \frac{1}{2}b \in A \text{ or } x = 2\left(\frac{1}{2}a + \frac{1}{2}b\right) \in 2A.$$

In fact, a more general result holds, namely:

$$A \text{ is convex} \iff (s+t)A = sA + tA, \ \forall s,t \geq 0.$$

(9) If both $[-1,1]$ and $\mathbb{R}$ are endowed with the usual topology, the answer is yes. But if we just endow $[-1,1]$ with the discrete topology (and leave $\mathbb{R}$ equipped with the usual topology), then the given statement becomes false. Consider the function $f$ defined by

$$f(x) = \begin{cases} -1, & -1 \leq x \leq 0, \\ 1, & 0 < x \leq 1. \end{cases}$$

The discreteness of the topology of $[-1,1]$ guarantees the continuity of $f$. Besides, $f$ verifies

$$f(-1)f(1) = -1 < 0.$$

Observe in the end that $\forall x \in [-1,1] : \ f(x) \neq 0$.

(10) True (in usual $\mathbb{R}$)! Polynomials on $\mathbb{R}$ are continuous. Since the degree is odd, then the limits at $\pm\infty$ are $\pm\infty$ or $\mp\infty$. Thus the polynomial in question will have to pass through the real axis at least once.

(11) No! In usual $\mathbb{R}$, it will be shown in Exercise 6.3.17 that $\mathbb{R}^*$ is not path-connected. However, $\overline{\mathbb{R}^*} = \mathbb{R}$ is path-connected.

(12) The answer is no! See e.g. Exercise 6.3.21.

(13) The answer is no! Consider the identity map from $X$ into $Y$ (denoted by $f$) where $X$ is the usual topology of $\mathbb{R}$ and $Y$ is $\mathbb{R}$ equipped with the co-finite topology. Then $f$ is continuous. But, $[0,1] \cup [2,3]$ is connected in $Y$ as it is infinite (cf. Exercise 6.3.1 below) whereas

$$f^{-1}([0,1] \cup [2,3]) = [0,1] \cup [2,3]$$

is not a connected set in $X$ as it is not an interval. The answer is also no in usual $\mathbb{R}$. Take $f : \mathbb{R} \to \mathbb{R}$ defined by $f(x) = x^2$. Then $\{2\}$ is connected but $f^{-1}(\{2\}) = \{\sqrt{2}, -\sqrt{2}\}$ is not connected.

(14) True! Let $f : X \to Y$ be such a homeomorphism. Let $x \in X$ and let $C_x$ be its component. We ought to show that

$$C_{f(x)} = f(C_x).$$

(a) Since $C_x$ is connected, $f(C_x)$ is connected for $f$ is continuous. But $f(x) \in f(C_x)$. Thus $f(C_x) \subset C_{f(x)}$.

(b) Conversely, assume that $C_{f(x)} \subset A$ for some connected $A \neq f(C_x)$. Then $f^{-1}(C_{f(x)}) \subset f^{-1}(A)$. By the proof of the other part,

$$C_x = f^{-1}[f(C_x)] \subset f^{-1}(C_{f(x)}) \subset f^{-1}(A).$$

But $f^{-1}$ is continuous and hence $f^{-1}(A)$ is connected. Since $x \in f^{-1}(A)$, we see that we have arrived at a contradiction for $C_x$ is the largest connected set which contains $x$! The proof is over.

REMARK. Thanks to this result, we can prove interesting results on the non-homeomorphy of some spaces.

(15) True! In fact, this a simple corollary of the fact that the continuous image of a connected set (path-connected respectively) is connected (path-connected respectively).

(16) True! Both implications are trivial. The right-to-left implication is the most trivial. For the other one, if $X$ is connected, then it is obviously the largest connected set in $X$ containing $x$!

(17) No, there is no contradiction between this question and the closedness of the components. First $\mathbb{R}^*$ is not connected since it has two components (this is one proof among others). Of

course both $(0, \infty)$ and $(-\infty, 0)$ are not closed in $\mathbb{R}$ but they are closed in $\mathbb{R}^*$ with respect to the induced topology (which is the meant topology in the definition).

## 6.3. Solutions to Exercises

**SOLUTION** 6.3.1.

(1) The only open and closed parts simultaneously in $T$ are $A$ and $\varnothing$. Hence $A$ is connected.

(2) In the discrete topology (on a set $X$) not only $X$ and $\varnothing$ are clopen, but *every* subset is clopen! Therefore, unless $\mathrm{card} X = 1$, the discrete topology is always disconcerted.

(3) Let $T = \{\varnothing\} \cup \{U \subset \mathbb{R} : U^c \text{ finite}\}$. Then $\mathbb{R}$ is connected in $T$. If $\mathbb{R}$ were not connected in this topology, then there would exist two non-empty and open sets $U$ and $V$ in $T$ (i.e. $U^c$ and $V^c$ are both finite) such that

$$\begin{cases} U \cup V = \mathbb{R}, \\ U \cap V = \varnothing. \end{cases}$$

Since $U \cap V = \varnothing$, $U^c \cup V^c = \mathbb{R}$ and this clearly contradicts the infiniteness of $\mathbb{R}$! Similarly, we can show that an infinite set $A$ is connected.

Finally, a finite set (with two elements at least) here is not connected. Indeed, the induced topology in such case is the discrete one.

(4) Let $A = B((0, 1), 1) \cup B((0, -1), 1)$. The intersection of the two balls is empty as we will show. Let $(x, y) \in B((0, 1), 1) \cup B((0, -1), 1)$. Then

$$\begin{cases} x^2 + (y-1)^2 < 1, \\ x^2 + (y+1)^2 < 1 \end{cases} \text{ which gives } \begin{cases} x^2 + y^2 - 2y < 0, \\ x^2 + y^2 + 2y < 0 \end{cases}$$

and hence $x^2 + y^2 < 0$ which clearly cannot occur for any real couple $(x, y)$.

Thus $A$ is not connected since $\{B((0, 1), 1), B((0, -1), 1)\}$ is an open partition for $A$.

(5) The given set $A$ is connected since it is the union of two connected sets whose intersection is not empty (the intersection being equal to $\{(0, 0)\}$).

(6) $A$ is also connected. From Proposition 6.1.13, we know that if $X$ and $Y$ are two non-empty connected sets such that $\overline{X} \cap Y \neq \varnothing$, then $X \cup Y$ is connected. Now, we have (as we are in

Euclidian $\mathbb{R}^2$)

$$\overline{B((0,1),1)} \cap B_c((0,-1),1) = B_c((0,1),1) \cap B_c((0,-1),1) = \{(0,0)\} \neq \varnothing.$$

Thus $A = B((0,1),1) \cup B_c((0,-1),1)$ is in effect connected.

**SOLUTION** 6.3.2. First, we can write

$$A = \{M \in \mathcal{M}_n(\mathbb{R}) : \det M \neq 0\}$$

which can also be written as

$$A = \underbrace{\{M \in \mathcal{M}_n(\mathbb{R}) : \det M < 0\}}_{A_1} \cup \underbrace{\{M \in \mathcal{M}_n(\mathbb{R}) : \det M > 0\}}_{A_2}.$$

But, the function "det" is real-valued and continuous (see Exercise 4.3.15). Hence, $A_1$ and $A_2$ are open (why?). Lastly, it is plain that $A_1 \cap A_2 = \varnothing$. Thus, $\{A_1, A_2\}$ is an "open partition" of $A$ which means that $A$ cannot be connected.

**SOLUTION** 6.3.3. We give three proofs.

- First proof: $\mathbb{Q}$ is not connected since it is not an interval (Exercise 1.2.8)...
- Second proof: $\mathbb{Q}$ is not connected since there is another open and closed set in the subspace topology of $\mathbb{Q}$ (apart from $\mathbb{Q}$ itself and $\varnothing$). Take e.g. $A = \mathbb{Q} \cap (-\infty, \sqrt{2})$. Then $A$ is clopen in the subspace topology of $\mathbb{Q}$ for

$$A = \mathbb{Q} \cap (-\infty, \sqrt{2}) = \mathbb{Q} \cap (-\infty, \sqrt{2}]$$

  and obviously $(-\infty, \sqrt{2})$ is open in $\mathbb{R}$ and $(-\infty, \sqrt{2}]$ is closed in $\mathbb{R}$.
- Third proof: The following two sets $A = \mathbb{Q} \cap (-\infty, \sqrt{2})$ and $B = \mathbb{Q} \cap (\sqrt{2}, \infty)$ are open in $\mathbb{Q}$ in the subspace topology and constitute an open partition for $\mathbb{Q}$.

**SOLUTION** 6.3.4.

(1) As for $\mathbb{R} \backslash \mathbb{Q}$ one can proceed exactly as in the foregoing exercise to show that this set is not connected.
(2) The set $\{\frac{1}{n} : n \geq 1\}$ is not connected since it is not an interval, say.
(3) No, since $[0,1) \cup (1,2)$ is not an interval.
(4) $\mathbb{N}$ is not connected since

$$U = (0, 3/2) \cap \mathbb{N}, \ V = (3/2, +\infty) \cap \mathbb{N}$$

are both open in $\mathbb{N}$ and they constitute a partition for $\mathbb{N}$.

**SOLUTION** 6.3.5. Assume $[0, 2)$ is not connected, then it could be written as $[0, 2) = U \cup V$ with $U \cap V = \varnothing$ where $U$ and $V$ are open. But $U \cap V = \varnothing$ does not hold since every non-empty open set in this topology contains at least 0. Thus, $[0, 2)$ is connected with respect to the given topology.

**SOLUTION** 6.3.6.

(1) We have $z \in A$ iff $\operatorname{Im} z \neq 0$. Hence there are two components corresponding to the cases $\operatorname{Im} z > 0$ and $\operatorname{Im} z < 0$.

(2) The components of $A$ are

$$B = \{(x, y) \in \mathbb{R}^2 : x < y\} \text{ and } C = \{(x, y) \in \mathbb{R}^2 : x > y\}.$$

(3) $A$ has two components, to wit $B((0, 1), 1)$ and $B((0, -1), 1)$. $B$ is connected and hence it has one component, viz itself.

**SOLUTION** 6.3.7. There are different proofs of the first part of the exercise. One of them is the following: We know that every interval is convex and that every convex set is path-connected and that every path-connected set is connected. We have therefore finished with one implication.

Conversely, let $A$ be a connected part of $\mathbb{R}$. We have to show that $A$ is an interval. Let $a, c \in A$ and $b \in \mathbb{R}$ such that $a < b < c$. We must show that $b \in A$. Assume for the sake of contradiction that $b \notin A$. Set

$$U = (-\infty, b) \cap A, \ V = (b, \infty) \cap A.$$

Then $U$ and $V$ are both open in the subspace topology of $A$. Besides, $U$ is not empty for $a \in U$ and $V$ is not empty as $c \in V$. Next,

$$U \cup V = [(-\infty, b) \cap A] \cup [(b, \infty) \cap A] = A \cap (\mathbb{R} \setminus \{b\}) = A$$

as we have supposed that $b \notin A$. Since $A$ is connected, we must have $U \cap V \neq \varnothing$ (why?). But

$$U \cap V = (-\infty, b) \cap A \cap (b, \infty) = \varnothing.$$

So we have seen that assuming that $b \notin A$ has led to a contradiction. In other words, $A$ must be an interval.

**SOLUTION** 6.3.8. Let $f : X \to Y$ be a continuous function. Assume that $X$ is connected and let's show that $f(X)$ is connected. Let $U$ be a clopen subset in $f(X)$ w.r.t. the induced topology of $Y$ (we are required to prove that $U = \varnothing$ or $U = f(X)$).

Since $f$ is continuous:

$$U \text{ is open} \implies f^{-1}(U) \text{ is open in } X$$

and

$$U \text{ is closed} \implies f^{-1}(U) \text{ is closed in } X.$$

But, $X$ is connected and so there cannot be other connected subsets apart from $\varnothing$ or $X$ itself. Therefore, either $f^{-1}(U) = \varnothing$ or $f^{-1}(U) = X$. The former clearly *implies* that $U = \varnothing$ (cf. Proposition 1.1.32), and the latter yields

$$f(X) = f[f^{-1}(U)] \subset U$$

or simply $U = f(X)$ as we already know that $U \subset f(X)$. This completes the proof.

**SOLUTION** 6.3.9. Set $A = \bigcup_{i \in I} A_i$. To show that $A$ is connected, let $f : A \to \{0, 1\}$ be continuous (we must show that $f$ is constant). Let $a \in \bigcap_{i \in I} A_i$. Then $a$ is in each $A_i$. Since $f$ is continuous, every restriction of $f$ to each $A_i$ is also continuous. But each $A_i$ is connected and so each restriction is constant, with a constant equal to $f(a)$. Therefore, $f$ is also constant, that is, $A = \bigcup_{i \in I} A_i$ is connected.

**SOLUTION** 6.3.10.

(1) First, we recall that $\overline{A}^B = B \cap \overline{A}^X$. Since $B \subset \overline{A}^X$, we get that $\overline{A}^B = B$. Now, let $f : B \to \{0, 1\}$ be a continuous function. We ought to show that $f$ is constant. Since $A \subset B$, we deduce that $f$ is continuous on $A$ too. By the connectedness of $A$, we know that $f$ must be constant, i.e. $f(A) = \{0\}$ or $f(A) = \{1\}$. Assume $f(A) = \{1\}$. The continuity of $f$ and the closedness of $\{1\}$ lead to

$$f(B) = f(\overline{A}^B) \subset \overline{f(A)} = \overline{\{1\}} = \{1\}.$$

Thus $f(B) = \{1\}$, i.e. $f$ is constant and hence $B$ is connected. The case $f(A) = \{0\}$ is very analogous to the foregoing one and hence we leave it to the reader. Therefore, in either case $B$ is connected.

(2) Since $A \subset \overline{A} \subset \overline{A}$, the previous result applies to $\overline{A}$, i.e. $\overline{A}$ is connected whenever $A$ is connected.

**SOLUTION** 6.3.11. Set $Z = \overline{X} \cap Y$ (which is non empty by assumption). We clearly have

$$X \subset X \cup Z = X \cup (\overline{X} \cap Y) \subset \overline{X}.$$

Since $X$ is connected, by Exercise 6.3.10, we infer that $X \cup Z$ is connected.

Next, as

$$\overline{X} \cap Y \subset (X \cup Z) \cap Y,$$

then $(X \cup Z) \cap Y \neq \varnothing$. Finally (since $Z \subset Y$),

$$(X \cup Z) \cup Y = X \cup (Z \cup Y) = X \cup Y.$$

Accordingly, since $(X \cup Z) \cap Y \neq \varnothing$ and both $Y$ and $X \cup Z$ are connected, Proposition 6.1.12 tells us that $(X \cup Z) \cap Y$ or $X \cup Y$ is connected, establishing the result.

**SOLUTION** 6.3.12.

(1) Assume that $X \times Y$ is connected. If $p : X \times Y \to X$ is the projection defined by $p(x, y) = x$, then

$$p(X \times Y) = X$$

is connected as $p$ is continuous. For the connectedness of $Y$, a similar argument applies by using the other projection $q : X \times Y \to Y$...

(2) Suppose that both $X$ and $Y$ are connected. Fix $y \in Y$ and let $x \in X$. Now, set

$$A_x = (\{x\} \times Y) \cup (X \times \{y\}).$$

By Proposition 4.1.26, both $\{x\} \times Y$ and $X \times \{y\}$ are connected for they are homeomorphic to $Y$ and $X$ respectively. Since the intersection $(\{x\} \times Y) \cap (X \times \{y\})$ has a point in common, namely $(x, y)$, Proposition 6.1.12 gives the connectedness of $A_x$. Now, we can easily show that

$$X \times Y = \bigcup_{x \in X} A_x.$$

Finally, as (check it!)

$$\bigcap_{x \in X} A_x = X \times \{y\} \neq \varnothing,$$

then Proposition 6.1.12 tells us that $X \times Y$ is connected, and this completes the proof.

**SOLUTION** 6.3.13.

(1) It is clear that $(-\infty, 0)$ inherits the usual topology as a subspace of $\mathbb{R}_K$ and so does $(0, \infty)$. Thus, both spaces are connected.

(2) Since $(-\infty, 0)$ and $(0, \infty)$ are connected, so are their closures, i.e. $(-\infty, 0]$ and $[0, \infty)$. But

$$(-\infty, 0] \cap [0, \infty) = \{0\} \neq \varnothing$$

and so $(-\infty, 0] \cup [0, \infty) = \mathbb{R}_K$ is connected.

**SOLUTION** 6.3.14. Assume that such function exists. Since $\mathbb{Q}$ is countable, so is $f(\mathbb{Q})$ by Exercise 1.3.8. The set $f(\mathbb{R} \setminus \mathbb{Q})$ is also countable, but this time, since it is by assumption a subset of $\mathbb{Q}$. We hence end up with $f(\mathbb{R})$ being countable.

Now, $\mathbb{R}$ is connected and so is its direct image $f(\mathbb{R})$ by the continuity of $f$. Thus $f(\mathbb{R})$ must be an interval. But, it is clear that the only countable intervals are the very particular intervals $[a, a]$, which are singletons $\{a\}$ $(a \in \mathbb{R})$. This forces $f$ to be constant and hence $f(x) = a$ for all $x \in \mathbb{R}$.

Going back to the hypotheses again we see that

$$f(\mathbb{Q}) \subset \mathbb{R} \setminus \mathbb{Q} \Longrightarrow a \in \mathbb{R} \setminus \mathbb{Q}$$

and

$$f(\mathbb{R} \setminus \mathbb{Q}) \subset \mathbb{Q} \Longrightarrow a \in \mathbb{Q},$$

which is a clear contradiction. Thus no such function $f$ verifying the hypotheses exists.

**SOLUTION** 6.3.15. Assume that $X$ is not connected, i.e. the function $f : X \to \{0, 1\}$ is continuous and surjective. Let $x, y \in X$ be such that $f(x) = 0$ and $f(y) = 1$. Let $g : [0, 1] \to X$ be a path from $x$ to $y$ (with $g(0) = x$ and $g(1) = y$). Then the composite function $f \circ g$ is continuous from $[0, 1]$ into $\{0, 1\}$.

Moreover, $f \circ g$ is surjective. Indeed, if $\alpha = 0$, then take $\beta = 0$ so that $g(\beta) = g(0) = x$ and so $f(g(\beta)) = f(x) = 0$. If $\alpha = 1$, take $\beta = 1$. Hence $g(1) = y$ and so $f(g(1)) = f(y) = 1$. Thus

$$\forall \alpha \in \{0, 1\}, \ \exists \beta \in [0, 1] : \ (f \circ g)(\beta) = \alpha.$$

Finally, since $[0, 1]$ is connected, we know that there cannot be a continuous function from $[0, 1]$ *onto* $\{0, 1\}$, and yet, this is what we have been dragged into! Thus, $X$ is connected.

**SOLUTION** 6.3.16. If $\mathbb{N}$ were path-connected, then there would exist a continuous mapping $f : [0, 1] \to \mathbb{N}$. Now we have

$$\mathbb{N} = \bigcup_{n \in \mathbb{N}} \{n\} \Longrightarrow [0, 1] = f^{-1}(\mathbb{N}) = f^{-1}(\bigcup_{n \in \mathbb{N}} \{n\}) = \bigcup_{n \in \mathbb{N}} f^{-1}(\{n\}).$$

Since $f$ is continuous and the singletons $\{n\}$ are closed for each $n$ (they are finite), the sets $f^{-1}(\{n\})$ are closed. Hence, we would end up with a countable union of disjoint and closed sets equal to $[0, 1]$ which is *continuum*. But this clearly contradicts the Sierpinski Theorem (Theorem 6.1.17). Therefore, $\mathbb{N}$ is not path-connected with respect to the co-finite topology.

**SOLUTION** 6.3.17. Assume that $\mathbb{R}^*$ is path-connected. Since $-1, 1 \in \mathbb{R}^*$, there exists a continuous function $f$ defined on $[0, 1]$, taking values in $\mathbb{R}^*$ such that $f(0) = -1$ and $f(1) = 1$. By the Intermediate Value Theorem there exists at least an $\alpha$ in $[0, 1]$ such that $f(\alpha) = 0$ which clearly leads to a contradiction as $f(x) \neq 0$, for all $x \in [0, 1]$.

SOLUTION 6.3.18.

(1) Let us show that if $C$ is convex, then it is path-connected. Let $x, y \in C$ and consider the function $f$ defined on $[0, 1]$ and taking its values in $C$ such that $f(t) = (1 - t)x + ty$.

Then $f$ is continuous and $f(0) = x$ and $f(1) = y$. Thus $C$ is path-connected (and hence connected).

(2) The set $\mathbb{R}^n$ is obviously convex (why?) and hence it is path-connected and so it is connected too.

Now, we say why the closed ball in $\mathbb{R}^n$ is convex (the case of the open ball is very similar). WLOG, we may do the proof for $n = 2$ only. Let $B_c((a, b), r)$ be the closed ball of center $(a, b) \in \mathbb{R}^2$ and radius $r > 0$. Let $(x, y), (x', y') \in B_c((a, b), r)$. We need to show that $(1 - t)(x, y) + t(x', y') \in B_c((a, b), r)$, i.e.

$$[(1 - t)x + tx' - a]^2 + [(1 - t)y + ty' - b]^2 \leq r^2,$$

and details are left to the reader...

SOLUTION 6.3.19.

(1) If $n = 1$, then we have already shown above that $\mathbb{R}^*$ is not path connected. If $n \geq 2$, then any two elements of $\mathbb{R}^n$ can always be joined by a segment if this segment does not pass through the origin. If it does pass through it, then we may avoid the origin by considering another point $z$ and draw a segment from $x$ to $z$, then one from $z$ to $y$.

(2) If $n = 1$, then $S^{n-1}$ is reduced to $S^0$, which in its turn, is reduced to the set $\{-1, 1\}$. Hence it is not path-connected. If $n \geq 2$, then $S^{n-1}$ becomes connected via the following arguments. Consider the function $f : \mathbb{R}^n \setminus \{0_{\mathbb{R}^n}\} \to S^{n-1}$ defined by

$$f(x_1, x_2, \cdots, x_n) = \frac{(x_1, x_2, \cdots, x_n)}{\sqrt{x_1^2 + x_2^2 + \cdots + x_n^2}}.$$

Then $f$ is continuous and onto. But $\mathbb{R}^n \setminus \{0_{\mathbb{R}^n}\}$ is path-connected (for $n \geq 2$) and hence $f(\mathbb{R}^n \setminus \{0_{\mathbb{R}^n}\}) = S^{n-1}$ is path-connected too.

SOLUTION 6.3.20.

(1) The functions $f$ and $g$ are obviously continuous on their domains since their components are so. Now $f \circ g : A \to A$ and $g \circ f : [a, b] \times \mathbb{T} \to [a, b] \times \mathbb{T}$ are given respectively by

$$(f \circ g)(x, y) = (x, y) \text{ and } (g \circ f)(t, (x, y)) = (t, (x, y)).$$

Hence $f$ is a continuous bijection. Since, its inverse $g$ is also continuous, $f$ becomes a homeomorphism.

(2) Since $[a, b]$ and $\mathbb{T}$ are both path-connected, so is $[a, b] \times \mathbb{T}$. Hence $A$ is path-connected too (Theorem 6.1.26).

**SOLUTION** 6.3.21.

(1) First, $\overline{f}$ is continuous since it is a restriction of $f$. So is $(\overline{f})^{-1}$ too as it is a restriction of $f^{-1}$. Now, it is clear that $\overline{f}$ is also a bijection. Therefore, $\overline{f}$ is a homeomorphism.

(2) Assume that $f : \mathbb{R} \to \mathbb{R}^2$ is a homeomorphism. According to the previous question, $\overline{f} : \mathbb{R} \setminus \{0\} \to \mathbb{R}^2 \setminus \{f(0)\}$ should remain a homeomorphism. But $(\overline{f})^{-1}$ is continuous, $\mathbb{R}^2 \setminus \{f(0)\}$ is path-connected (cf. Exercise 6.3.19) and $\mathbb{R}^*$ is not path-connected (see Exercise 6.3.17). This a clearly a contradiction since the image of a path-connected set under a continuous function has to stay path-connected. Thus $\mathbb{R}$ and $\mathbb{R}^2$ are not homeomorphic.

**SOLUTION** 6.3.22.

(1) Consider $f : \mathbb{R} \to C$ with $f(x) = e^{2i\pi x}$. Then clearly $f$ is continuous. Besides, it is onto. Since $\mathbb{R}$ is connected, so is $f(\mathbb{R}) = C$.

(2) Assume that $f : C \to \mathbb{R}$ is continuous and consider $g : C \to \mathbb{R}$ defined by $g(z) = f(z) - f(-z)$. Since $C$ is connected, $g(C)$ too is connected (in $\mathbb{R}$). In fact, $g(C)$ is an interval. So if $z \in C$, then $g(C)$ contains both $g(z)$ et $g(-z)$ so that $g(C)$ contains $\frac{g(z)+g(-z)}{2}$. But

$$\frac{g(z) + g(-z)}{2} = 0 \text{ and hence } 0 \in g(C).$$

Thus for some $\lambda \in C$ ($\lambda \neq 0$), $g(\lambda) = 0$, i.e. $f(\lambda) = f(-\lambda)$, that is, $f$ is not injective.

(3) Suppose there were a homeomorphism $f : \mathbb{R}^2 \to \mathbb{R}$. Then $f$ restricted to $C$ would be injective! (Wouldn't it?) This is the required contradiction. Accordingly, $\mathbb{R}^2$ is not homeomorphic to $\mathbb{R}$.

**SOLUTION** 6.3.23. X and S are not homeomorphic. If there were a homeomorphism between X and S, then

$$\overline{f} : \mathsf{X} \setminus \{a\} \to \mathsf{S} \setminus \{f(a)\} \ (a \in \mathsf{X})$$

would remain one. But if $a$ is the junction point of X, then $\mathsf{X} - \{a\}$ has *four* components while S deprived of one point has either one or two components (depending on the point's position on S).

A similar reasoning shows that E and W cannot be homeomorphic.

**SOLUTION 6.3.24.**

(1) Let $x \mapsto (x, \sin 1/x)$ be a map from $(0, 1]$ to $\mathbb{R}^2$. Then this map is continuous. Since $A$ is the image of $(0, 1]$ under this continuous map and $(0, 1]$ is connected, it follows that $A$ too is connected.

(2) We claim that

$$\overline{A} = A \cup (\{0\} \times [-1, 1]).$$

To see this, notice first that $A$, being the graph of a continuous function, is closed. Also, $\{0\} \times [-1, 1]$ is clearly closed. Hence

$$\overline{A} \subset A \cup (\{0\} \times [-1, 1]).$$

Conversely, we can write

$$A \subset \overline{A} \Longrightarrow A \cup (\{0\} \times [-1, 1]) \subset \overline{A} \cup (\{0\} \times [-1, 1]).$$

But $(\{0\} \times [-1, 1]) \subset \overline{A}$. Indeed, for any $y \in [-1, 1]$, there exists $t \in (0, 2\pi]$ such that $y = \sin t \ (= \sin(t + 2n\pi)$ for all $n \in \mathbb{N})$. Hence

$$\lim_{n \to \infty} (1/(t + 2n\pi), y) = (0, y) \in \overline{A}$$

because clearly $(1/(t + 2n\pi), y) \in A$ (for all $n$). Consequently,

$$A \cup (\{0\} \times [-1, 1]) \subset \overline{A} \cup (\{0\} \times [-1, 1]) = \overline{A},$$

proving the second inclusion.

(3) By drawing the graph of $\overline{A}$ we can see that $\overline{A}$ is not path connected. Let's prove this in a more rigorous way.

To show that $\overline{A}$ is not path-connected, assume it is. Let $f : [0, 1] \to \overline{A}$ defined by $f(t) = (\alpha(t), \beta(t))$ be continuous and such that $f(0) = (0, 0)$ and $f(1) = (1/\pi, 0)$ (notice that both $(0, 0)$ and $(1/\pi, 0)$ lie on $\overline{A}$). A repeated use of the Intermediate Value Theorem allows us to obtain a sequence $(t_n)$ such that $(\alpha(t_n))_n$ tends to 0 and $\beta(t_n) = (-1)^n$. More explicitly,
  (a) $\exists t_1 \in (0, 1) : \ \alpha(t_1) = \frac{2}{3\pi}$, then
  (b) $\exists t_2 \in (0, t_1) : \ \alpha(t_2) = \frac{2}{5\pi}$, and so on until we get
  (c) $\exists t_n \in (0, t_{n-1}) : \ \alpha(t_n) = \frac{2}{(2n+1)\pi}$

With these details, $(t_n)$ is decreasing and as it is bounded below by 0. Therefore, $(t_n)$ must converge (to some $\gamma$). But $f$ is assumed to be continuous and hence the limit of $f(t_n)$ has to exist and to be finite as well (in fact, equal to $f(\gamma)$). This is clearly not the case because $\lim_{n\to\infty} \beta(t_n)$ does not exist. Accordingly, $\overline{A}$ is not path connected.

## 6.4. Hints/Answers to Tests

SOLUTION 47. Well, do it!...For a two-point set find a counterexample...

SOLUTION 48. No! One reason is that $[0,1)$ is clopen in $\mathbb{R}_\ell$ (cf. Exercise 3.3.38)...

SOLUTION 49. Any subset $A$ $(a, b \in A)$ is not connected as $A \cap [a, b]$ is a *proper* clopen set of $A$...

SOLUTION 50. Well, $\mathbb{R}$ is connected in the co-countable topology... As for $\mathbb{Q}$, it is not connected (Is the discrete topology connected?)...

SOLUTION 51. The result follows from Exercise 6.3.10...

SOLUTION 52. We already know that $\mathbb{R}$ is connected in this topology...

SOLUTION 53. No! Deprive $[-1, 1]$ of one point (not 1 and not $-1$)...

SOLUTION 54. Consider the function $f : \mathbb{R}^n \times \mathbb{R}^n \to \mathbb{R}^n$ defined by $f(x,y) = x + y$. Is $f$ continuous? Is $A \times B$ connected? What is $f(A \times B)$?...

# CHAPTER 7

# Complete Metric Spaces

## 7.2. True or False: Answers

ANSWERS.

(1) Only the implication "$\Rightarrow$" is true (and trivial!). As for the implication "$\Leftarrow$", it is not always true. One way of seeing this, is to take $B = \overline{A}$. Then an appeal to Proposition 7.1.9 gives

$$d(A) = d(\overline{A}) = d(B).$$

Therefore, for a counterexample, it suffices to take any non-closed $A$.

(2) We **cannot** take $\varepsilon \geq 0$! But we can take $d(x_n, x_m) \leq \varepsilon$ with no problem at all.

(3) Let $X$ be a a metric space. To establish its completeness, we take an *arbitrary* Cauchy sequence $(x_n)$ in $X$ and we show that this sequence converges to a limit $x$ which *belongs* to $X$.

REMARK. **Do not** take a particular Cauchy sequence and show it converges! The sequence must be arbitrary.

(4) On the contrary to the previous answer, if we want to show that a given metric space $X$ is not complete, it suffices to find *one* Cauchy sequence $(x_n)$ which *converges*. But, its limit *does not* belong to $X$.

Alternatively, we can show that $X$ is not closed.

(5) If we want to use the completeness of a space $X$, we try to define a sequence in $X$ and then show it is Cauchy. Now, since the space is complete, this sequence converges to a limit in $X$ (do not say in the beginning, let $(x_n)$ be an arbitrary Cauchy sequence. Then it converges. Then what?)

(6) The importance is about convergence. For example, since usual $\mathbb{R}$ is complete, to prove a sequence is convergent it suffices to show that it is Cauchy. For instance, to prove the series $\sum_{n \geq 1} \dfrac{1}{n^a}$, $a > 1$ is convergent it suffices to show that it is Cauchy instead of trying to find its limit. But what is its limit anyway?

We recall here, that the exact value of this limit is known for example for $a = 2$ and for other even numbers in general (via the so-called **Bernoulli Numbers**). But what is the limit in case $a$ is an odd number? In fact, how would we proceed just with the case of $a = 3$? Fairly recently mathematically speaking (in 1978!), R. Apéry, a French mathematician, proved that the value of this sum (in the case $a = 3$) is irrational (see [2] for a shorter proof).

Another major point, still about convergence, is the Fixed Point Theorem. Solutions to many problems can be obtained via the Fixed Point Theorem, as we already discussed in the "Essential Background" Section.

(7) False! As a "lazy" counterexample, we can just endow $(0, \infty)$ with the discrete metric (see Exercise 7.3.3). But, there are more interesting counterexamples. For instance, if we equip $(0, \infty)$ with the following *metric*

$$d(x, y) = |\ln x - \ln y|, \ \forall x, y > 0,$$

then $(0, \infty)$ will be complete (see Exercise 7.3.19).

Evidently, $(0, \infty)$ is not complete in the usual metric.

(8) The answer is no! For instance, in usual $\mathbb{R}$, consider $\mathbb{Q}$ and $\mathbb{R} \setminus \mathbb{Q}$. They are both dense in $\mathbb{R}$ and yet

$$\mathbb{Q} \cap (\mathbb{R} \setminus \mathbb{Q}) = \varnothing.$$

(9) The answer is yes. For example, in $\mathbb{R}$ equipped with its usual metric, none of $\mathbb{Q}$ and $\mathbb{R} \setminus \mathbb{Q}$ is complete. However, their union, i.e. $\mathbb{R}$, is complete.

(10) Only the left-to-right implication holds. Let us show that. Since $(x_n)$ is Cauchy, for all $\varepsilon > 0$ and in particular for $\varepsilon = 1$,

$$\exists N \in \mathbb{N}, \ \forall n, m \ (n, m \geq N \implies d(x_n, x_m) \leq 1).$$

If $a = x_N$, then for all $m \geq N$: $d(a, x_m) \leq 1$. This means that $x_m \in B_c(a, 1)$. Thus it becomes clear that

$$\{x_n : \ n \in \mathbb{N}\} \subset \underbrace{\{x_1, x_2, \cdots, x_{N-1}\}}_{\text{finite hence bounded}} \cup B_c(a, 1)$$

proving the boundedness of the sequence as its range will be included in the union of *two* bounded sets.

The other implication need not be true in general. From basic real analysis, it is known that $((-1)^n)_n$ is bounded but not convergent hence not Cauchy (usual $\mathbb{R}$ is complete!).

(11) The answer is yes. Since the two metrics are equivalent, we have

$$\exists \alpha, \beta > 0, \ \forall n, m \in \mathbb{N} : \ \alpha d(x_n, x_m) \leq d'(x_n, x_m) \leq \beta d(x_n, x_m)$$

and the desired result then easily follows.

(12) The answer is yes if one of the inequalities involved in the definition of equivalent metrics is satisfied. For example, assume that $d$ and $d'$ are two metrics on a set $X$. If *only* (for some $\alpha > 0$) the inequality

$$d(x, y) \leq \alpha d'(x, y), \ \forall x, y \in X$$

holds, then $d$ and $d'$ are not equivalent metrics. Nevertheless, the "Cauchyness" of a sequence $(x_n)$ in $(X, d')$ implies that of $(x_n)$ in $(X, d)$.

   The answer is, however, *false* in general if none of the inequalities intervening in the definition of equivalent metrics is satisfied.

(13) It is false. While the fact that $(f_n)$ is a Cauchy sequence of continuous functions can be seen in Exercise 7.3.14, the remaining part is false for a simple reason. That is, the convergence used to get $f$ is the pointwise one, and so we are not using the right metric which should be $d$.

(14) False again! In this case, the metric is the right one but the argument used is not quite right. What tells us that there is not another *continuous* function which is the $d$-limit of $f_n$? One possible answer is that we are in a metric space (hence Hausdorff) and the limit of a convergent sequence is always unique in a metric space. Digging more into this, the function $f$ is Riemann integrable and $d$ is not a metric on the set of Riemann integrable functions (see Exercise 2.3.14) and more particularly $d(f, g) = 0 \Rightarrow f = g$ (which is essential for proving uniqueness of limits!) is the axiom which *does not* hold. The reader may find a correct proof is given in Exercise 7.3.14.

(15) The answer is no! Let $f : \mathbb{R}_+^* \to \mathbb{R}_+^*$ be defined by $f(x) = \frac{1}{x}$ for all $x > 0$. Then $f$ is evidently a homeomorphism. Now, let $x_n = \frac{1}{n}$. Then $(x_n)$ is Cauchy. Nevertheless, $f\left(\frac{1}{n}\right) = n$ is not Cauchy (for example, since it is not bounded).

(16) True! The reader is asked to prove this in Exercise 7.5.8.

(17) False! There are many counterexamples. Consider the function $f : [1, +\infty) \to (0, 1]$ defined for all $x \geq 1$ by $f(x) = \frac{1}{x}$.

Then $f$ is uniformly continuous and obviously $[1, +\infty)$ is complete (why?). But its image which is $(0, 1]$ is not complete as it is not closed.

(18) False! For a counterexample, see Exercise 7.3.19.

(19) False! This is just Theorem 7.1.26 but with two important conditions which have gone missing, namely: $Y$ complete and $f$ uniformly continuous. Let us therefore give two counterexamples which show that these assumptions cannot be dispensed with.

  (a) Let $f$ be the identity function from $\mathbb{Q}$ onto $\mathbb{Q}$ in the usual topology. Then $f$ is uniformly continuous (note that the codomain $\mathbb{Q}$ is not complete). Assume that $f$ has an extension $\tilde{f}$, then $\tilde{f}_{|\mathbb{Q}}(x) = x$. Since $\mathbb{Q}$ is dense in $\mathbb{R}$, an irrational $y$ is a limit of a sequence $(y_n) \subset \mathbb{Q}$. We would then have

  $$\tilde{f}(y) = \lim_{n\to\infty} \tilde{f}(y_n) = \lim_{n\to\infty} f(y_n) = \lim_{n\to\infty} y_n = y \notin \mathbb{Q}.$$

  This illustrates the importance of $Y$ being complete.

  (b) Let $X = [0, 1]$, $A = (0, 1]$ and $Y = \mathbb{R}$. Define a function $f$ on $A$ as $f(x) = \ln x$. Then $A$ is dense in $X$, but $f$ is "only" continuous (the continuity is not uniform). If $\tilde{f}$ were such an extension of $f$, then we would have

  $$\tilde{f}(0) = \lim_{x\to 0^+} \tilde{f}(x) = \lim_{x\to 0^+} f(x) = \lim_{x\to 0^+} \ln x = -\infty.$$

  This tells us that such an extension does not exist showing the importance of the uniform continuity.

(20) There are many properties shared by compact and complete metric spaces. An important result is that *every compact metric space is complete* (the reader is asked to give a proof in Exercise 7.3.7). Some of the properties illustrating the analogy are listed below:

  (a) Any closed subspace of a compact space is compact and any closed set in a complete metric space is complete.

  (b) Any compact subspace of a metric space is closed and so is any complete subspace.

  (c) The *finite* union of complete metric spaces is complete and a *finite* union of compact spaces is compact too.

  (d) The arbitrary intersection of complete metric spaces is complete. The same result holds with "compact" instead of "complete".

  (e) etc...

(21) The answer is no to both questions. So, we need only give one counterexample (to both). Consider the Euclidean metric $d_2$ on $\mathbb{R}^2$, i.e.

$$d_2((x,y),(x',y')) = \sqrt{(x-x')^2 + (y-y')^2},$$

which is equivalent to the metric $d_1$ given by

$$d_1((x,y),(x',y')) = |x-x'| + |y-y'|.$$

Then $f : \mathbb{R}^2 \to \mathbb{R}^2$ defined by

$$f(x,y) = \frac{1}{10}(8x+8y, x+y)$$

is a contraction in $(\mathbb{R}^2, d_1)$ as (for all $(x,y),(x',y') \in \mathbb{R}^2$)

$$d_1(f(x,y),f(x',y')) \leq \frac{9}{10}d_1((x,y),(x',y')).$$

Nonetheless, $f$ is not a contraction anymore in $(\mathbb{R}^2, d_2)$ as no $k \in [0,1)$ would satisfy

$$d_2(f(x,y),f(x',y')) \leq kd_2((x,y),(x',y'))$$

for all $(x,y),(x',y') \in \mathbb{R}^2$ (think of large $|x-y|$ and $|x'-y'|$).

## 7.3. Solutions to Exercises

**SOLUTION 7.3.1.**

(1) The implication "$\Leftarrow$" is evident. Conversely, if there *exist* $a, b \in A$ with $a \neq b$, then

$$d(A) \geq d(a,b) > 0.$$

(2) It is very easy to see that since $A \subset B$, we have

$$d(A) = \sup_{x,y \in A} d(x,y) \leq \sup_{x,y \in B} d(x,y) = d(B).$$

(3) One inequality is straightforward. Indeed, by the previous question:

$$A \subset \overline{A} \implies d(A) \leq d(\overline{A}).$$

Conversely, let $(x,y) \in \overline{A} \times \overline{A}$. Then there sequences $(x_n)$ and $(y_n)$ (both in $A$) converging to $x$ and $y$ respectively. But $d$ is continuous, and so

$$\lim_{n \to \infty} d(x_n, y_n) = d(x,y).$$

But

$$\forall n \in \mathbb{N} : d(x_n, y_n) \leq d(A).$$

Whence (for all $x, y \in \overline{A}$)

$$d(x, y) \leq d(A).$$

Therefore

$$d(\overline{A}) = \sup_{x, y \in \overline{A}} d(x, y) \leq d(A).$$

**SOLUTION 7.3.2.**

(1) $\mathbb{Q}$ is not complete as it is not closed (or, for instance, the sequence $\left(1 + \frac{1}{n}\right)^n$ belongs to $\mathbb{Q}$ and it is Cauchy in $\mathbb{R}$ and hence Cauchy in $\mathbb{Q}$ but its limit is $e \notin \mathbb{Q}$).

(2) It is not complete as it is not closed. Another way of seeing this is the following: Consider the sequence $(x_n)$ defined by

$$\begin{cases} x_{n+1} = \sqrt{2 + x_n}, & n \in \mathbb{N}, \\ x_1 = \sqrt{2}. \end{cases}$$

Then, it can easily be established that $(x_n)$ is an increasing and bounded above. Hence it converges (and thus Cauchy!). It can also be shown that $(x_n)$ takes its values in $\mathbb{R} \setminus \mathbb{Q}$ and that $x_n \to 2 \notin \mathbb{R} \setminus \mathbb{Q}$.

(3) The set $\mathbb{Q} \cap [3, 4]$ is not complete since it is not closed as we can show that its closure in $\mathbb{R}$ is $[3, 4]$ (cf. Exercise 3.3.28).

(4) The set $(0, \infty)$ is not complete since it is not closed.

(5) The set $\{n : n \geq 1\}$ (which is nothing but $\mathbb{N}$!) is closed in $\mathbb{R}$ since its complement is an infinite union of open sets and thus $\{n : n \geq 1\}$ is complete.

(6) Again, $\{(-1)^n : n \geq 1\}$ is closed (why?) in $\mathbb{R}$ and hence complete.

(7) $\{\frac{1}{n} : n \in \mathbb{N}\} \cup \{0\}$ is complete (why?).

(8) The given set can be written as

$$\{(x, y) \in \mathbb{R}^2 : x \geq 1, y(x - 1) \geq 1\}.$$

Adopting the method of Exercise 4.3.40, we can show that it is closed in $\mathbb{R}^2$. The completeness of $\mathbb{R}^2$ therefore yields the completeness of the given set.

**SOLUTION 7.3.3.** Let $(x_n)_n$ be a Cauchy sequence in $X$. Hence for any $\varepsilon > 0$, there exists $N \in \mathbb{N}$ such that for all $n, m \in \mathbb{N}$

$$n, m \geq N \implies d(x_n, x_m) < \varepsilon.$$

Take for instance $\varepsilon = \frac{1}{4}$, then this gives

$$\forall n, m \geq N : x_n = x_m$$

which is an eventually constant sequence and in a discrete metric space the only convergent sequences are the eventually constant ones (cf. Exercise 4.3.18). Thus $X$ is complete.

**SOLUTION** 7.3.4. Let $(x_n)$ be a Cauchy sequence in $X$ and let $(x_{n(k)})$ be a convergent subsequence to $x$. Let $\varepsilon > 0$. Since $(x_n)$ is Cauchy, there is some integer $N$ such that

$$\forall n, m \geq N : \ d(x_n, x_m) < \frac{\varepsilon}{2}.$$

Since $(x_{n(k)})$ converges to $x$, there is some integer $K$ such that

$$\forall k \geq K : \ d(x_{n(k)}, x) < \frac{\varepsilon}{2}.$$

For any $n \geq N$, choose $k \geq K$ verifying $n(k) \geq N$ so that we have

$$d(x_n, x) \leq d(x_n, x_{n(k)}) + d(x_{n(k)}, x) < \frac{\varepsilon}{2} + \frac{\varepsilon}{2} = \varepsilon.$$

The proof is complete.

**SOLUTION** 7.3.5. Let $(x_n)$ be a Cauchy sequence in $\mathbb{R}$. This implies that it is bounded. The Bolzano-Weierstrass Property then implies that $(x_n)$ has a convergent subsequence. But, a Cauchy sequence having a convergent subsequence is itself convergent to the same limit (see Proposition 7.1.16). Therefore, $(\mathbb{R}, |\cdot|)$ is complete.

**SOLUTION** 7.3.6. Let $(x_n)_n$ be a Cauchy sequence in $(\mathbb{C}, |\cdot|_{\mathbb{C}})$. Since $x_n$ is complex-valued, one can write $x_n = a_n + ib_n$ (where of course $(a_n)_n$ and $(b_n)_n$ are two real-valued sequences). Since $(x_n)_n$ is Cauchy,

$$\forall \varepsilon > 0, \exists N \in \mathbb{N}, \forall m, n \ (m, n \geq N \Longrightarrow |x_n - x_m|_{\mathbb{C}} < \varepsilon).$$

Hence both $(a_n)_n$ and $(b_n)_n$ are Cauchy as

$$|a_n - a_m|_{\mathbb{R}} = |\operatorname{Re}(x_n - x_m)|_{\mathbb{R}} \leq |x_n - x_m|_{\mathbb{C}}$$

and

$$|b_n - b_m|_{\mathbb{R}} = |\operatorname{Im}(x_n - x_m)|_{\mathbb{R}} \leq |x_n - x_m|_{\mathbb{C}}.$$

Since $(\mathbb{R}, |\cdot|)$ is complete, both $(a_n)_n$ and $(b_n)_n$ are convergent to reals $a$ and $b$ respectively. We set $x = a + ib$. Then obviously $x \in \mathbb{C}$ and we need only show that $x_n \to x$ in $(\mathbb{C}, |\cdot|_{\mathbb{C}})$. This follows easily from

$$0 \leq |x_n - x|_{\mathbb{C}} = |(a_n - a) + i(b_n - b)|_{\mathbb{C}} \leq |a_n - a|_{\mathbb{R}} + |b_n - b|_{\mathbb{R}} \longrightarrow 0.$$

**SOLUTION** 7.3.7.

(1) Call this space $X$. Let $(x_n)$ be a Cauchy sequence in $X$. Since $X$ is compact, it is sequentially compact and so $(x_n)$ has a subsequence $(x_{n(k)})$ which converges to a point $x \in X$, say. Thus, by Proposition 7.1.16, $(x_n)$ too converges to $x \in X$, proving the completeness of $X$.

(2) The answer is no in general. For instance, usual $\mathbb{R}$ is complete but not compact.

(3) By the first answer, we know that for instance $[-2, 2]$ is complete. To prove $\mathbb{R}$ is complete, let $(x_n)$ be a Cauchy sequence in $\mathbb{R}$. Then for all $\varepsilon > 0$ and in particular for $\varepsilon = 2$,

$$\exists N \in \mathbb{N}: \ \forall n, m \in \mathbb{N} \ (n, m \geq N \Rightarrow |x_n - x_m| \leq 2).$$

Set $y_n = x_{N+n} - x_N$ for all $n$. Then $(y_n)$ is Cauchy in $[-2, 2]$ and hence it converges, call $y$ its limit. Hence

$$y = \lim_{n \to \infty} y_n = \lim_{n \to \infty} x_n - x_N \implies \lim_{n \to \infty} x_n = y + x_N \in \mathbb{R},$$

proving the completeness of $\mathbb{R}$.

(4) None of the implication is true. For example, in the usual metric $(-1, 1)$ is locally compact but it is not complete. Conversely, the space $C([0, 1], \mathbb{R})$ (see Exercise 7.3.15) is complete w.r.t. the supremum metric, but not locally compact (see Exercise 5.3.29).

**SOLUTION** 7.3.8. Let us show that $(x_n = n)_n$ is a Cauchy sequence in $(\mathbb{R}, d)$ not converging to any point in $\mathbb{R}$. We have

$$d(n, m) = \left| \frac{n}{1 + |n|} - \frac{m}{1 + |m|} \right| = \left| \frac{n}{1 + n} - \frac{m}{1 + m} \right| = \left| \frac{n - m}{(1 + n)(1 + m)} \right|.$$

WLOG, we may assume that $n \geq m$. Hence

$$0 \leq d(n, m) = \left| \frac{n - m}{(1 + n)(1 + m)} \right| \leq \frac{n - m}{nm} \leq \frac{n}{nm} = \frac{1}{m} \longrightarrow 0$$

as $n, m \to \infty$.

Now if $(n)_n$ converges to some $a \in \mathbb{R}$ with respect to $d$, then we have

$$0 = \lim_{n \to \infty} \left| \frac{n}{1 + |n|} - \frac{a}{1 + |a|} \right| = \left| 1 - \frac{a}{1 + |a|} \right|.$$

By considering the cases $a \geq 0$ and $a < 0$, we can easily show that the equation

$$\left| 1 - \frac{a}{1 + |a|} \right| = 0$$

has no solution ($a \in \mathbb{R}$). Thus $(\mathbb{R}, d)$ is not complete.

**SOLUTION 7.3.9.** Recall that the $d$ is defined for every $x, y \in \mathbb{N}$, by

$$d(x, y) = \begin{cases} 0, & x = y, \\ 3 + \frac{1}{x} + \frac{1}{y}, & x \neq y. \end{cases}$$

Now, we proceed to show the completeness of $(\mathbb{N}, d)$. Let $(x_n)$ be a Cauchy sequence in $(\mathbb{N}, d)$. Then

$$\forall \varepsilon > 0, \exists N \in \mathbb{N}, \forall n, m \in \mathbb{N} : (n, m \geq N \Longrightarrow d(x_n, x_m) < \varepsilon)$$

or

$$\forall \varepsilon > 0, \exists N \in \mathbb{N}, \forall n, m \in \mathbb{N} : (n, m \geq N \Longrightarrow 3 + \frac{1}{x_n} + \frac{1}{x_m} < \varepsilon).$$

Since this is true for all $\varepsilon > 0$, it is true for $\varepsilon = 3$, say. This gives us a contradiction unless we take $d(x_n, x_m) = 0$. Hence $x_n = x_m$ for all $n, m \geq N$. This is an eventually constant sequence and hence it converges.

**SOLUTION 7.3.10.** Let $(x^p)_p$ be a Cauchy sequence in $\ell^2$ where $x^p = (x_{1p}, x_{2p}, \cdots, x_{np}, \cdots)$. Let $\varepsilon > 0$, then there exists an $N \in \mathbb{N}$ such that for all $p, q \in \mathbb{N}$

$$p, q \geq N \Longrightarrow d_2(x^p, x^q) < \varepsilon.$$

Thus for any $n$ we have

$$\sum_{k=1}^{n} |x_{kp} - x_{kq}|^2 \leq d_2(x^p, x^q) < \varepsilon^2.$$

Taking $q \to \infty$ (and keeping $n$ and $p$ fixed) gives us

$$\sum_{k=1}^{n} |x_{kp} - x_k|^2 \leq \varepsilon^2$$

and hence $\displaystyle\sum_{k=1}^{n} |x_{kp} - x_k|^2$ converges to a limit which is less than or equal to $\varepsilon^2$. Thus the sequence $y_k = x_{kp} - x_k$ belongs to $\ell^2$. But $(x^p)_p \in \ell^2$ and hence $x = (x_k)_k \in \ell^2$. In the end one has

$$d_2(x^p, x) \leq \varepsilon \text{ for all } p \geq N.$$

**SOLUTION 7.3.11.**

(1) Since $(\ell^\infty, d_\infty)$ is complete (Theorem 7.1.15), to show that $(c_0, d_\infty)$ is complete, it suffices to show that it is closed, i.e. $\overline{c_0} \subset c_0$. So, let $x \in \overline{c_0}$ (remember that $x = (x_n) \in \ell^\infty$ for the moment). Let $\varepsilon > 0$. Hence there is a sequence $y$ in $c_0$ such that

$$d_\infty(x, y) < \varepsilon.$$

Since $y \in c_0$, $y = (y_n)$ with $y_n \to 0$. Hence for some $N \in \mathbb{N}$ we have $|y_n| < \varepsilon$ when $N \geq n$. Accordingly, for $n \geq N$ we have

$$|x_n| = |x_n - y_n + y_n| \leq |x_n - y_n| + |y_n| \leq d_\infty(x, y) + |y_n| < 2\varepsilon.$$

Thus, $x_n \to 0$, i.e. $x \in c_0$, i.e. $\overline{c_0} \subset c_0$.

(2) Anticipating a little bit, by the next question, we must necessarily show that $c_0 \not\subset \ell^p$. Let

$$x = \left( \frac{1}{\ln 2}, \frac{1}{\ln 3}, \cdots, \frac{1}{\ln n}, \cdots \right).$$

Then $x \in c_0$ because $\frac{1}{\ln n} \to 0$. Assume now for the sake of contradiction that $x$ is in some $\ell^p$, where $1 \leq p < \infty$. Then this means that $\sum_{n \geq 2} \frac{1}{(\ln n)^p}$ converges. But, this is a particular case of a Bertrand's series, i.e. $\sum_{n \geq 2} \frac{1}{n^q (\ln n)^p}$, which diverges if $q < 1$ regardless the value of $p$ (which is our case!). Thus, $x$ is not in any $\ell^p$, where $p \geq 1$.

(3) Let $x \in \ell^p$. Then $\sum_{n=1}^{\infty} |x_n|^p$ converges. This implies that $|x_n|^p \to 0$ when $n \to \infty$ or simply $|x_n| \to 0$ or $x_n \to 0$, that is, $x \in c_0$.

(4) We already know that $(c_0, d_\infty)$ is closed. By the foregoing question, we may then write

$$\ell^p \subset c_0 \implies \overline{\ell^p} \subset \overline{c_0} = c_0$$

where the closures are with respect to the topology of $d_\infty$.

So, it only remains to show that $c_0 \subset \overline{\ell^p}$. To achieve this, let $x = (x_n) \in c_0$, i.e. $x_n$ tends to zero. Take now $x^k = (x_1, x_2, \cdots, x_k, 0, 0, \cdots)$ (the superscript is a label and not a power!) which is clearly in $\ell^p$. Hence

$$d_\infty(x, x^k) = \sup_{n \geq k+1} |x_n| \longrightarrow 0,$$

proving that $x \in \overline{\ell^p}$, i.e. $c_0 \subset \overline{\ell^p}$ so that $c_0 = \overline{\ell^p}$ (w.r.t. the topology of $d_\infty$).

(5) It is clear that $c_{00} \subset \ell^p$ so that $\overline{c_{00}} \subset \overline{\ell^p} = \ell^p$ (w.r.t. the topology of $\ell^p$). So we need only show that $\ell^p \subset \overline{c_{00}}$. Let $x \in \ell^p$. Consider $(x^k)$ where $x^k = (x_1, x_2, \cdots, x_k, 0, 0, \cdots)$, i.e. $(x^k)$ is in $c_{00}$. Now, we have

$$[d_p(x, x^k)]^p = \sum_{n=k+1}^{\infty} |x_n|^p \longrightarrow 0$$

and this completes the proof.

(6) No, $c_{00}$ is not dense in $\ell^\infty$, that is, $\ell^\infty \not\subset \overline{c_{00}}$. To see this, let $x = (1, 1, 1, \cdots) \in \ell^\infty$. Then this $x$ cannot be in $\overline{c_{00}}$ as for any $(x^k)$ in $c_{00}$, then $d_\infty(x, x^k)$ cannot be arbitrarily small as it will at least be bigger than one (why?).

Finally, we claim that $\overline{c_{00}} = c_0$ (w.r.t. the topology of $\ell^\infty$). We already know that $c_0$ is closed so that

$$c_{00} \subset c_0 \implies \overline{c_{00}} \subset c_0.$$

So let $x = (x_n)$ be in $c_0$. Consider now $(x^k)$ where $x^k = (x_1, x_2, \cdots, x_k, 0, 0, \cdots)$ (i.e. $(x^k)$ is in $c_{00}$). Remembering that $x_n \to 0$, we get

$$d_\infty(x, x^k) = \sup_{n \geq k+1} |x_n| \longrightarrow 0,$$

i.e. $x \in \overline{c_{00}}$, and the proof is over.

**SOLUTION 7.3.12.** The given set $A$ is not complete as it is not closed. Let us show that. Consider the sequence

$$(x_n)_n = \left(1, \frac{1}{2}, \frac{1}{3}, \cdots, \frac{1}{n}, 0, 0, \cdots\right).$$

Then $(x_n)_n$ belongs to $A$. Its limit with respect to $d$ is

$$x = \left(1, \frac{1}{2}, \frac{1}{3}, \cdots, \frac{1}{n}, \frac{1}{n+1}, \frac{1}{n+2}, \cdots\right)$$

as

$$d(x_n, x)^2 = \sum_{k=n+1}^\infty \frac{1}{k^2} \to 0 \text{ as } n \to \infty,$$

since this is a remainder of a convergent series.

Evidently, $x \notin A$ and hence $A$ is not complete.

**REMARK.** We ask the reader the following question: We could (couldn't we?) have taken as "another limit"

$$y = \left(1, \frac{1}{2}, \frac{1}{3}, \cdots, \frac{1}{n}, \frac{1}{2(n+1)}, \frac{1}{2(n+2)}, \cdots\right) \notin A$$

since $d(x_n, y) \to 0$, as $n \to \infty$. We can then have two limits for a convergent sequence in a metric space!! What is then wrong with this point?

**SOLUTION 7.3.13.** We first show that $(x_n)$ is Cauchy. Let $\varepsilon > 0$. We proceed as follows:

$$d(x_n, x_m) = |e^{-n} - e^{-m}| \leq |e^{-n}| + |e^{-m}| = e^{-n} + e^{-m}.$$

One can choose $N$ such that $e^{-N} < \frac{\varepsilon}{2}$ (why?). Then for all $n, m \geq N$, one has

$$e^{-n} + e^{-m} \leq e^{-N} + e^{-N} < \frac{\varepsilon}{2} + \frac{\varepsilon}{2} = \varepsilon,$$

i.e. $(x_n)$ is Cauchy.

Now, let us show that $(x_n)$ does not converge in $(\mathbb{R}, d)$. Assume it were and let $x$ be its limit. Then we would have on the one hand

$$\lim_{n \to \infty} |e^{-n} - e^x| = 0.$$

On the other hand, we know that

$$\lim_{n \to \infty} |e^{-n} - e^x| = e^x.$$

Hence we ended up with $e^x = 0$ which of course does not hold for any $x \in \mathbb{R}$. Thus the limit of the sequence does not exist and $(X, d)$ is not complete.

**SOLUTION 7.3.14.**

(1) We show that $(f_n)$ is a Cauchy sequence. We can show the "Cauchyness" of $(f_n)$ either by doing some arithmetic or geometrically by observing that $d(f_n, f_{n+p})$ is actually the area of the triangle of apexes $\left(\frac{1}{2} - \frac{1}{n}, 0\right)$, $\left(\frac{1}{2} - \frac{1}{n+p}\right)$ and $\left(\frac{1}{2}, 0\right)$ where $p, n \in \mathbb{N}$. Hence

$$d(f_n, f_{n+p}) = \frac{1}{2} \times 1 \times \left[\frac{1}{2} - \frac{1}{n+p} - \left(\frac{1}{2} - \frac{1}{n}\right)\right]$$

$$= \frac{1}{2}\left(\frac{1}{n} - \frac{1}{n+p}\right) \leq \frac{1}{2n}.$$

Hence

$$\forall \varepsilon > 0, \exists N = \left[\frac{\varepsilon}{2}\right] + 1 \in \mathbb{N}, \forall n, p \in \mathbb{N} \ (n \geq N \Rightarrow d(f_n, f_{n+p}) < \varepsilon),$$

establishing the "Cauchyness" of $(f_n)$.

(2) The pointwise limit of $(f_n)$ is

$$f(x) = \begin{cases} 0, & 0 \leq x < \frac{1}{2}, \\ 1, & \frac{1}{2} \leq x \leq 1. \end{cases}$$

Hence

$$d(f_n, f) = \int_0^1 |f_n(x) - f(x)| dx$$

$$= \int_0^{\frac{1}{2} - \frac{1}{n}} 0 dx + \int_{\frac{1}{2} - \frac{1}{n}}^{\frac{1}{2}} \left[ nx + \left( 1 - \frac{1}{2}n \right) \right] dx + \int_{\frac{1}{2}}^1 0 dx$$

$$= \frac{1}{2n} \longrightarrow 0$$

as $n \to \infty$. Now, to show $(X, d)$ is not complete, assume that there is a continuous function $g$ on $[0, 1]$ such that $d(f_n, g) \to 0$ and we will reach a contradiction. Since

$$|f(x) - g(x)| \le |f(x) - f_n(x)| + |f_n(x) - g(x)|, \ \forall x \in [0, 1],$$

we have

$$0 \le \int_0^1 |f(x) - g(x)| dx \le \int_0^1 |f(x) - f_n(x)| dx + \int_0^1 |f_n(x) - g(x)| dx.$$

Passing to the limit and using our hypotheses yield

$$\int_0^1 |f(x) - g(x)| dx = 0$$

and hence $f(x) = g(x)$ at each $x$ for which $f - g$ is continuous, i.e.

$$f(x) = g(x), \ \forall x \in [0, 1] \setminus \left\{ \frac{1}{2} \right\}.$$

But, $g$ is assumed to be continuous on $[0, 1]$ and hence it must be continuous at $\frac{1}{2}$ and we would have

$$\lim_{x \to \frac{1}{2}^-} g(x) = \lim_{x \to \frac{1}{2}^+} g(x) = g(1/2).$$

However,

$$\lim_{x \to \frac{1}{2}^-} g(x) = \lim_{x \to \frac{1}{2}^-} f(x) = 0 \ne 1 = \lim_{x \to \frac{1}{2}^+} g(x) = \lim_{x \to \frac{1}{2}^+} f(x).$$

Therefore, such a continuous function $g$ cannot exist and hence $(X, d)$ is not a complete metric space. The solution is over.

**SOLUTION 7.3.15.** Let $(f_n)$ be a Cauchy sequence in $(X, d_\infty)$. Then

$$\forall \varepsilon > 0, \exists N \in \mathbb{N}, \forall n, m \in \mathbb{N} \ (n, m \ge N \implies d_\infty(f_n, f_m) < \varepsilon)$$

or

$$\forall \varepsilon > 0, \exists N \in \mathbb{N}, \forall n, m \in \mathbb{N} \ (n, m \ge N \implies \sup_{x \in [0,1]} |f_n(x) - f_m(x)| < \varepsilon).$$

This implies that for each particular $x$ in $[0, 1]$ one has

$$|f_n(x) - f_m(x)| < \varepsilon \text{ for all } n, m \geq N.$$

Thus for each $x \in [0, 1]$, $(f_n(x))_n$ is a Cauchy sequence in usual $\mathbb{R}$. Since the latter is complete, we deduce that $(f_n(x))_n$ must converge to some $f(x) \in \mathbb{R}$ (for each $x \in [0, 1]$). Thus we have a function $f$ defined for all $x \in [0, 1]$. Fixing $n$ in the last displayed equation, and letting $m$ tend to infinity yield

$$|f_n(x) - f(x)| < \varepsilon \text{ whenever } n \geq N$$

and for each $x$ (where $N$ is independent of $x$). Thus

$$d_\infty(f_n, f) = \sup_{x \in [0,1]} |f_n(x) - f(x)| < \varepsilon \text{ for all } n \geq N.$$

Therefore, we have shown that $(f_n)$ converges in the supremum metric to $f$, i.e. that $(f_n)$ converges uniformly to $f$. From the hint, $f$ must therefore be continuous.

In the end, the *arbitrary* Cauchy sequence $(f_n)$ converges in $(X, d_\infty)$ to a *continuous* function $f$. In other words, $(X, d_\infty)$ is complete.

**SOLUTION 7.3.16.** Let $(f_n)$ be a Cauchy sequence in $X$. As in the previous exercise, we can easily obtain that $(f_n(x))_n$ converges to a function $f$ defined on $[0, 1]$. To finish the proof, two claims are to be shown.

(1) **Claim:** *f is bounded*: Since $(f_n)$ is Cauchy, for $\varepsilon = 1 > 0$, there is some $N \in \mathbb{N}$ such that for all $n, m \geq N$ we have $d(f_n, f_m) < 1$. Hence for $m = N$, we have $d(f_n, f_N) < 1$ whenever $n \geq N$. Remember that $f_N$ is bounded and hence for some $M \geq 0$, $|f_N(x)| \leq M$ for all $x \in [0, 1]$. Now we have

$$\forall x \in [0, 1] : |f_n(x)| \leq |f_n(x) - f_N(x)| + |f_N(x)| < 1 + M$$

whenever $n \geq N$. Passing to the limit as $n$ tends to $\infty$ gives

$$\forall x \in [0, 1] : \lim_{n \to \infty} |f_n(x)| = |\lim_{n \to \infty} f_n(x)| = |f(x)| \leq 1 + M,$$

proving the boundedness of $f$.

(2) **Claim:** $\lim_{n \to \infty} d(f_n, f) \to 0$: Since $(f_n)$ is Cauchy, we know that

$$\forall \varepsilon > 0, \exists N \in \mathbb{N}, \forall n, m \in \mathbb{N} \ (n, m \geq N \implies d(f_n, f_m) < \varepsilon).$$

But $d$ is the supremum metric, and hence

$$|f_n(x) - f_m(x)| < \varepsilon, \ \forall x \in [0, 1]$$

and whenever $m, n \geq N$. Keeping $n$ fixed and taking the limit as $m$ goes to infinity yield

$$|f_n(x) - f(x)| \leq \varepsilon.$$

whenever $n \geq N$ and for all $x \in [0, 1]$. Thus we are led to

$$d(f_n, f) \leq \varepsilon$$

for $n \geq N$. That is $\lim\limits_{n \to \infty} d(f_n, f) \to 0$.

The proof is therefore complete.

SOLUTION 7.3.17.

(1) To show that $x \mapsto e^x$ is not a polynomial we use a contradiction argument. Assume that $x \mapsto e^x$ is a polynomial of degree $m$, say, i.e. for some real $a_0, a_1, \cdots, a_m$ we have

$$e^x = a_0 + a_1 x + \cdots + a_m x^m.$$

Since $x \mapsto e^x$ is $C^{m+1}$ (it is in fact $C^\infty$), differentiating $(m+1)$-times both sides of the previous displayed equation yields

$$(e^x)^{(m+1)} = e^x = (a_0 + a_1 x + \cdots + a_m x^m)^{(m+1)} = 0$$

which is absurd. Consequently, the exponential function is not a polynomial.

(2) Obviously $P_n(x) = \left(1 + \frac{x}{n}\right)^n$, $n \geq 1$, is a polynomial sequence. So, it only remains to show that $d(P_n, e^x) \to 0$ as $n \to \infty$, and hence by uniqueness of the limit, $e^x$ is the only limit of $(P_n)$ (see the remark below for a comment on this).

Let us show now $P_n \to e^x$ in $(X, d)$. To this end, we compute $\sup_{x \in [0,1]} |P_n(x) - e^x|$. We can write

$$\sup_{x \in [0,1]} |P_n(x) - e^x| \leq e^1 \sup_{x \in [0,1]} |P_n(x)e^{-x} - 1|.$$

Hence

$$0 \leq |P_n(x)e^{-x} - 1| \leq 1 - P_n(1)e^{-1} = 1 - \left(1 + \frac{1}{n}\right)^n e^{-1}.$$

Taking the supremum and passing to the limit as $n \to \infty$ yield

$$0 \leq d(P_n, e^x) \leq e\left(1 - \left(1 + \frac{1}{n}\right)^n e^{-1}\right) \longrightarrow 0.$$

REMARK. We would like to give a comment on this answer. We did not consider this reasoning, i.e. the use of the uniqueness of the limit, as a correct one in the "True or False" Section corresponding to the question related to Example 7.1.11.

The reason here differs from the other one. The minimum one suggests for the integral metric to be well-defined is that the functions be Riemann-integrable (in such case the limit is not unique as we do not have a distance) whilst in this case (the supremum metric) what one suggests is that at least the functions be bounded and $d$ in this case remains a metric. Hence the limit is unique.

REMARK. In Exercise 8.3.3, we give a different way of showing that $(P_n)$ converges uniformly to $e^x$.

(3) Take any continuous function $f$ which is not a polynomial ($e^x$, $\sin x$,...etc) and hence $f \notin X$. By the Weierstrass Theorem (Theorem 7.1.14), we know that there is a polynomial sequence $(P_n)_n$ converging uniformly to $f$, i.e. $P_n \to f$ with respect to $d$. Hence $(P_n)_n$ is Cauchy but converging to $f$ which is outside of $X$. Thus $(X, d)$ is not complete.

SOLUTION 7.3.18. The answer is no! Take any continuous function $f$ on $[0, 1]$ which is *not* differentiable, i.e. it does not belong to $X$ (e.g. $x \mapsto |2x - 1|$). The Weierstrass Theorem then implies that there exists a polynomial sequence $(P_n)$ such that

$$\lim_{n \to \infty} d_\infty(P_n, f) = 0.$$

Since $P_n$ are all polynomials, for all $n$: $P_n \in X$. Since $(P_n)$ converges to $f$ in $(X, d_\infty)$, $(P_n)$ is Cauchy in $(X, d_\infty)$. But $f \notin X$ and hence $(X, d_\infty)$ is not complete.

SOLUTION 7.3.19.

(1) This is a simple consequence of Test 5, say.
(2) $A$ is not closed and hence it is not complete.
(3) Let $(x_n)$ be a Cauchy sequence in $(A, d')$, i.e.

$$\forall \varepsilon > 0, \exists N \in \mathbb{N}, \forall n, m \in \mathbb{N} \, (n, m \geq N \Rightarrow d'(x_n, x_m) = |\ln x_n - \ln x_m| < \varepsilon).$$

We immediately see that $(\ln x_n)$ becomes a Cauchy sequence in usual $\mathbb{R}$ which is obviously complete. Thus there exists some $y \in \mathbb{R}$ such that $|\ln x_n - y| \to 0$ as $n \to \infty$. Set $x = e^y$. Then $x > 0$, i.e. $x \in A$. Besides,

$$d'(x_n, x) = |\ln x_n - \ln x| = |\ln x_n - y| \longrightarrow 0$$

as $n \to \infty$. Consequently, $(A, d')$ is complete.
(4) We show that $id : (A, d) \to (A, d')$ is a homeomorphism. The bijectivity of $id$ is evident. It is continuous since the logarithm function is continuous on $(0, \infty)$. The inverse function $id^{-1} =$

$id : (A, d') \to (A, d)$ is also continuous as the exponential function is in $(0, \infty)$.

(5) We deduce from the previous answers that completeness is not a topological property.

## SOLUTION 7.3.20.

(1) Let $A$ be a complete metric subspace. We want to show that $A$ is closed, so let $x \in \overline{A}$. By Theorem 4.1.37, there is a sequence $(x_n)$ in $A$ such that $x_n \to x$. Since $(x_n)$ is convergent, it is Cauchy. But $A$ is complete, hence the (*unique*) limit of $(x_n)$, i.e. $x$ is in $A$, and this establishes the closedness of $A$.

(2) Assume that $X$ is complete and let $A \subset X$ be closed. To show that $A$ is complete, let $(x_n)$ be a Cauchy sequence in $A$. Hence $(x_n)$ is Cauchy in $X$ too. Since $X$ is complete, $(x_n)$ must converge to an $x \in X$. By Theorem 4.1.37 again, $x \in \overline{A}$. But $A$ is closed and so $x \in A$. Accordingly, the Cauchy sequence $(x_n)$ in $A$ converges to $x \in A$, proving the completeness of $A$.

## SOLUTION 7.3.21.

(1) Let $\{X_i\}_{1 \leq i \leq k}$ be complete metric spaces. To show that $\bigcup_{1 \leq i \leq k} X_i$ is complete, let $(x_n)$ be a Cauchy sequence in this union. If the sequence $(x_n)$ is such that $n$ belongs to a finite set, then the result is evident. If the sequence $(x_n)$ is such that $n$ is in $\mathbb{N}$, then there is at least one $X_j$ which contains infinitely many $n$ of the sequence. Therefore, we have a subsequence $(x_{n_j})_j$ in $X_j$ which is also Cauchy. Since $X_j$ is complete, the subsequence $(x_{n_j})_j$ converges to $x \in X_j \subset \bigcup_{1 \leq i \leq k} X_i$. Proposition 7.1.16 does the remaining job, i.e. it allows us to say that $(x_n)$ too converges to $x$, and this proves the completeness of $\bigcup_{1 \leq i \leq k} X_i$.

(2) No! Take $A_n = [\frac{1}{n}, 1]$, $n \in \mathbb{N}$. Then all $A_n$ are complete because they are closed in the complete $\mathbb{R}$. Nonetheless,

$$\bigcup_{n \in \mathbb{N}} \left[\frac{1}{n}, 1\right] = (0, 1],$$

which is not complete for it is not closed.

(3) Let $(X_i)_{i \in I}$ be a class of complete spaces. Then they are all closed. Hence their arbitrary intersection too is closed. Since (for one $j$, in fact for all $j \in I$!)

$$\bigcap_{i \in I} X_i \subset X_j,$$

$\bigcap_{i \in I} X_i$ becomes a closed subspace in a complete space. Accordingly, $\bigcap_{i \in I} X_i$ is complete.

**SOLUTION 7.3.22.**

(1) Suppose that $X$ and $Y$ are complete. To show that $Z$ is complete, let $(x_n, y_n)$ be a Cauchy sequence in $(Z, \delta)$. Hence for all $\varepsilon > 0$, there is $N \in \mathbb{N}$ such that for all $n, m \geq N$ we have:

$$d(x_n, x_m) + d'(y_n, y_m) = \delta((x_n, y_n), (x_m, y_m)) < \varepsilon.$$

This implies that $(x_n)$ is Cauchy in $X$ and that $(y_n)$ is Cauchy in $Y$. Since $X$ and $Y$ are complete, $(x_n)$ and $(y_n)$ converge to $x \in X$ and $y \in Y$ respectively. We can guess that $(x, y) \in Z$ should be the limit in $Z$ of $((x_n, y_n))_n$. Let us corroborate this. We have

$$\delta((x_n, y_n), (x, y)) = d(x_n, x) + d'(y_n, y) \longrightarrow 0 \text{ as } n \longrightarrow \infty$$

and this finishes the proof of one implication.

(2) Assume that $Z = X \times Y$ is complete and let's show that so are its factors. Let $(x_n)$ be a Cauchy sequence in $X$ and let $(y_n)$ be a Cauchy sequence in $Y$. Then for all $\varepsilon > 0$,

$$\exists n_1 \in \mathbb{N}, \ \forall n, m : \ (n, m \geq n_1 \implies d(x_n, x_m) < \varepsilon/2)$$

and

$$\exists n_2 \in \mathbb{N}, \ \forall n, m : \ (n, m \geq n_2 \implies d'(y_n, y_m) < \varepsilon/2).$$

Now, take $N = \max(n_1, n_2)$. Then for all $n, m \geq N$,

$$\delta((x_n, y_n), (x_m, y_m)) = d(x_n, x_m) + d'(y_n, y_m) < \varepsilon.$$

Hence $((x_n, y_n))_n$ is Cauchy in $X \times Y$ and so it converges to some $(x, y) \in X \times Y$ (that is, $x \in X$ and $y \in Y$). But for all $n$, we have

$$0 \leq d(x_n, x) \leq d(x_n, x) + d'(y_n, y) = \delta((x_n, y_n), (x, y))$$

and hence

$$\lim_{n \to \infty} d(x_n, x) = 0,$$

proving the completeness of $(X, d)$. Similarly, we can show that

$$\lim_{n \to \infty} d'(y_n, y) = 0$$

and so $(Y, d')$ is complete as well, completing the proof of the other implication.

**SOLUTION 7.3.23.** Let $f$ be such a function with $|f'(x)| \leq k$ for some $k$ and all $x$. By the Mean Value Theorem applied to $f$ on $[x, y]$ $(x, y \in I)$ we have

$$\exists c \in (x, y): \ f(y) - f(x) = (y - x)f'(c)$$

and by the hypothesis on the derivative of $f$

$$|f(y) - f(x)| \leq k|y - x|,$$

i.e. $f$ is a contraction.

Conversely, if $f$ satisfies the Lipschitz condition, then for some $k$ and all $x, x_0 \in I$ (with $x \neq x_0$) we have

$$\left| \frac{f(x) - f(x_0)}{x - x_0} \right| \leq k.$$

Passing to the limit as $x \to x_0$, we find that $|f'(x_0)| \leq k$, as required.

**SOLUTION 7.3.24.**

(1) <u>Existence:</u> Start with an arbitrary $x_0$ in $X$ and define $(x_n)$ inductively by

$$x_{n+1} = f(x_n), \ n \in \mathbb{N}.$$

If we can show that $(x_n)$ is Cauchy in $X$, then it will converge for $X$ is complete.

Let $n \in \mathbb{N}$. Then

$$d(x_{n+1}, x_n) = d(f(x_n), f(x_{n-1})) \leq kd(x_n, x_{n-1})$$

so that

$$d(x_{n+1}, x_n) \leq kd(x_n, x_{n-1}) \leq k^2 d(x_{n-1}, x_{n-2}) \leq \cdots \leq k^n d(x_1, x_0).$$

Now, we proceed to show that $(x_n)$ is Cauchy. Let $n, m \in \mathbb{N}$ (WLOG take $m > n$). By the generalized triangle inequality, we may write

$$d(x_n, x_m) \leq d(x_n, x_{n+1}) + d(x_{n+1}, x_{n+2}) + \cdots + d(x_{m-1}, x_m).$$

Hence

$$d(x_n, x_m) \leq (k^n + k^{n+1} + \cdots + k^{m-1})d(x_1, x_0)$$
$$= k^n(1 + k \cdots + k^{m-n-1})d(x_1, x_0)$$
$$= k^n \frac{1 - k^{m-n}}{1 - k} d(x_1, x_0)$$
$$< \frac{k^n}{1 - k} d(x_1, x_0).$$

Since $k \in [0, 1)$, $k^n$ goes to zero when $n$ approaches infinity. Therefore, $(x_n)$ is a Cauchy sequence. By the completeness of $X$, $(x_n)$ converges to some $x \in X$.

The limit $x$ is in fact the fixed point of $f$. Indeed, since $f$ is continuous (by Proposition 7.1.34), we have as $n \to \infty$:

$$f(x) \longleftarrow f(x_n) = x_{n+1} \longrightarrow x.$$

It follows by uniqueness of the limit that $f(x) = x$, that is, we have shown that $f$ has a at least a fixed point.

(2) Uniqueness: Assume that $f$ has two fixed points $x$ and $y$, say. That is, $f(x) = x$ and $f(y) = y$. Then

$$d(x, y) = d(f(x), f(y)) \leq kd(x, y).$$

So

$$(1 - k)d(x, y) \leq 0.$$

The latter cannot occur as $k < 1$, unless $d(x, y) = 0$, i.e. unless $x = y$. Thus our assumption that $f$ has two fixed points has led to a contradiction. Therefore, $f$ possess one, and only one, fixed point.

**SOLUTION** 7.3.25. Transform the given equation to

$$\frac{1}{32}x^4 + \frac{x^3}{2} - x + \frac{1}{4} = 0 \text{ or } \frac{1}{32}x^4 + \frac{x^3}{2} + \frac{1}{4} = x.$$

Then set $f(x) = \frac{1}{32}x^4 + \frac{x^3}{2} + \frac{1}{4}$. If $f$ has $[0, 1/2]$ as its domain, we must check that $f$ has as its range a subset of $[0, 1/2]$ as well. For all $x$

$$f'(x) = \frac{1}{8}x^3 + \frac{3}{2}x^2 \geq 0.$$

Therefore, $f$ is increasing on $[0, 1/2]$. Since it is also continuous, we have

$$f([0, 1/2]) = [f(0), f(1/2)] = [1/4, 161/512] \subset [0, 1/2].$$

Now, it can easily be seen that

$$|f'(x)| \leq \frac{25}{64} < 1 \text{ for all } 0 \leq x \leq \frac{1}{2}.$$

Hence $f$ is a contraction (cf. Proposition 7.1.35) and so it admits a unique fixed point $x \in [0, 1/2]$, i.e. $f(x) = x$ for a unique $0 < x < 1/2$. Thus the given equation (which is just $f(x) - x = 0$) has a unique solution between 0 and $\frac{1}{2}$.

**REMARK.** For the curious ones, the solution is $\simeq 0.2588...$ (for example, using the method of successive approximations).

SOLUTION 7.3.26.

(1) Set $g = f^{(n)}$. Since $g$ is a contraction, we know that for some $k \in [0, 1)$ and all $s, t \in X$

$$d(g(s), g(t)) \le kd(s, t).$$

Hence by the Fixed Point Theorem, there *exists a unique* $x \in X$ such that $g(x) = x$. Next, we can write

$$d(x, f(x)) = d(g(x), f(x))$$
$$= d(g(x), (f \circ g)(x))$$
$$= d(g(x), (g \circ f)(x))$$
$$\le kd(x, f(x))$$

because $g$ is a contraction. Since $k < 1$, we immediately deduce that $f(x) = x$, i.e. $f$ has a fixed point.

In other words, we have shown that if $f^{(n)}$ is a contraction (then it admits a unique fixed point), then $f$ has the same fixed point.

To show its uniqueness, assume $y$ is another fixed point of $f$. Then $y$ is necessarily a fixed point for $f^{(n)}$ for

$$f^{(n)}(y) = f^{(n-1)}(f(y)) = f^{(n-1)}(y) = \cdots = f(y) = y.$$

But the fixed point of $f^{(n)}$ is unique and hence $x = y$. This finishes the proof.

(2) (a) Define $\cos x : \mathbb{R} \to \mathbb{R}$. Then this function cannot be a contraction since if it were, we would have for some $k < 1$ and all real $x, y$ (with $x \ne y$ say)

$$|\cos x - \cos y| \le k|x - y| \text{ or } \left| \frac{\cos x - \cos y}{x - y} \right| \le k.$$

But then passing to the limit as $y \to x$ would give us $|\sin x| \le k$ (for all $x$), which is false!

(b) Let $f(x) = \cos(\cos x)$. Then $f'(x) = \sin x \sin(\cos x)$. Hence

$$|f'(x)| \le |\sin(\cos x)| \le \sin 1 < 1$$

(as $|\sin x| \le 1$ and the function sine is increasing on $[-1, 1]$).

Thus the Mean Value Theorem gives for all $x$ and $y$

$$|\cos^{(2)} x - \cos^{(2)} y| \le (\sin 1)|x - y|,$$

i.e. $x \mapsto \cos^{(2)} x$ is a contraction and so by the first question $\cos x$ too has a unique fixed point.

(c) The way of finding the fixed point is well described in the proof of the Fixed Point Theorem. We start with a point $x_0$. Then we calculate successively $x_2 = \cos x_1$, $x_3 = \cos x_2$,...etc. We start with $x_1 = 0.7$ in radians and not degrees! (we could have started by any other value in $(0, 1)$). Then

$$x_2 = \cos(x_1) = 0.7648,$$
$$x_3 = \cos(x_2) = 0.7215,$$
$$x_4 = \cos(x_3) = 0.7508 \cdots$$

and we carry on until this process stabilizes at $x_{18} = 0.7391$, accurate to four decimal places (this is a little slow and other methods give here a better result like Newton's).

## SOLUTION 7.3.27.

(1) First we can show that $f$ takes its values in $X$ (you may check that $\inf f(x) = \sqrt{2}$ for $x \in X$). Next, let $x, y \in X$. Then $x, y \geq 1$ and hence

$$|f(x) - f(y)| = |x - y| \left| \frac{1}{2} - \frac{1}{xy} \right|.$$

But

$$0 < \frac{1}{xy} \leq 1 \text{ and so } \left| \frac{1}{2} - \frac{1}{xy} \right| \leq \frac{1}{2}.$$

Therefore,

$$\forall x, y \in X : |f(x) - f(y)| \leq \frac{1}{2}|x - y|.$$

(2) Assume there is $x \in X$ such that $f(x) = x$. Then

$$\frac{x}{2} + \frac{1}{x} = x \text{ or } x^2 = 2$$

and the previous equation clearly does not hold for any $x \in X$ since $\pm\sqrt{2} \notin \mathbb{Q}$!

(3) It does not contradict the Fixed Point Theorem since $X$ is not complete even though $f$ is a contraction. This tells us that the hypothesis on the space being complete cannot be merely dropped.

**SOLUTION 7.3.28.**

(1) Let $x, y \geq 1$. We could use similar estimates to those in the previous exercise, but it is important here to show that 1 is the best constant. To achieve this aim, we use the Mean Value Theorem which, when applied to $f$ on $[x, y]$, gives

$$\exists c \in (x, y) : \ f(x) - f(y) = f'(c)(x - y).$$

The reader may check that $\sup_{x \geq 1} |f'(x)| = 1$. Hence

$$\forall x, y \geq 1 : \ |f(x) - f(y)| < |x - y|.$$

(2) It is plain that

$$\forall x \geq 1 : \ f(x) \neq x.$$

(3) It does not contradict the Fixed Point Theorem since $f$ is not a contraction despite the fact that $X$ is complete. This shows the importance of the contraction's hypothesis.

**REMARK.** This example does not contradict the result of Exercise 5.3.36 either because $[1, \infty)$ is not compact.

**SOLUTION 7.3.29.** Let $f : (X, d) \to (X', d')$ be onto and such that

$$\forall x, y \in X : \ d'(f(x), f(y)) = d(x, y).$$

(1) Suppose that $(X', d')$ is complete and let us show that $(X, d)$ is complete. Let $(x_n)$ be a Cauchy sequence in $X$. Then $d(x_n, x_m) \to 0$ as $n, m \to \infty$ and so

$$\lim_{n, m \to \infty} d'(f(x_n), f(x_m)) = 0.$$

This means that $(f(x_n))$ is Cauchy in $X'$, assumed complete, and hence it must converge to some $y \in X' = f(X)$. Hence there exists one $x \in X$ such that $y = f(x)$. Thus

$$d'(f(x_n), y) = d'(f(x_n), f(x)) = d(x_n, x) \longrightarrow 0$$

as $n$ goes to infinity. Accordingly, $(X, d)$ is complete.

(2) The proof of the other implication is very similar to the previous one and we leave it to the reader.

**SOLUTION 7.3.30.** Recall that a metric space is compact iff it is complete and totally bounded. We already know that $[a, b]$ is complete for it is closed in $\mathbb{R}$ which is complete. We need only show that it is totally bounded. Let $\varepsilon > 0$. Choose $n$ such that $\frac{2}{n} \leq \varepsilon$. Consider now the cover $\{B_n\}$ defined as

$$B_0 = \left[a, a + \frac{1}{n}\right), \ B_i = \left(x_i - \frac{1}{n}, x_i + \frac{1}{n}\right) \text{ and } B_n = \left(b - \frac{1}{n}, b\right].$$

We may easily check that the diameter of $B_i$ is less than or equal to $2/n$. So $[a, b]$ is totally bounded.

**SOLUTION** 7.3.31. Assume that $(X, d)$ is complete and let $(A_n)$ be satisfying the given properties. We are required to show that $\bigcap_{n=1}^{\infty} A_n \neq \varnothing$. For each $n$, select $x_n$ in $A_n$. Since the family $(A_n)$ is decreasing, we obtain that

$$\{x_k : \ k \geq n\} \subset A_n.$$

But by assumption $d(A_n) \to 0$. An appeal to Proposition 7.1.9 then allows us to infer that

$$\lim_{n \to \infty} d(\{x_k : \ k \geq n\}) = 0.$$

By Proposition 7.1.10, $(x_n)$ is Cauchy which, by the completeness of $X$, converges to some $x \in X$. To finish the proof of this part, observe that $(x_{n+p})_p$ is a subsequence of $(x_n)$ and so it also converges to $x$. But

$$\forall p : \ x_{n+p} \in A_n.$$

Hence $x \in \overline{A_n} = A_n$ for $A_n$ is closed. Since this is valid for all $n$, we finally get

$$x \in \bigcap_{n \in \mathbb{N}} A_n,$$

as expected.

Conversely, assume that the intersection of any decreasing sequence of closed nonvoid subsets of $X$ whose diameter go to zero is nonvoid. To show that $(X, d)$ is complete, let $(x_n)$ be Cauchy. For each $n$, set

$$A_n = \overline{\{x_k : \ k \geq n\}}.$$

Then all $A_n$ are clearly closed and decreasing. Since $(x_n)$ is Cauchy, Proposition 7.1.10 asserts that

$$\lim_{n \to \infty} d(\{x_k : \ k \geq n\}) = 0$$

and so Proposition 7.1.9 yields

$$\lim_{n \to \infty} d(A_n) = 0.$$

Therefore, by assumption we must have

$$\bigcap_{n=1}^{\infty} A_n \neq \varnothing.$$

Let $x$ be a common point to each $A_n$. Since also $x_n$ is in $A_n$ ($x_n \in \{x_k : \ k \geq n\} \subset A_n$), we finally have:

$$\forall n \in \mathbb{N} : \ d(x_n, x) \leq d(A_n)$$

and so
$$\lim_{n\to\infty} d(x_n, x) = 0,$$
i.e. the Cauchy sequence $(x_n)$ in $X$ converges to $x \in X$. Thus $(X, d)$ is complete and this completes the proof.

**SOLUTION** 7.3.32. Both examples are in the setting of $\mathbb{R}$ endowed with the usual topology ($\mathbb{R}$ is complete).

(1) Let $A_n = [n, \infty)$. Then each $A_n$ is closed, non-empty and decreasing (with respect to $\subset$). However, we see that $\bigcap_{n=1}^{\infty} A_n = \varnothing$ which is not a contradiction for $d(A_n) \not\to 0$ as $n \to \infty$.

(2) Consider $A_n = (1, 1+\frac{1}{n}]$. Then $(A_n)$ is decreasing and satisfies
$$d(A_n) \to 0 \text{ as } n \text{ goes to infinity. However, } \bigcap_{n=1}^{\infty} A_n = \varnothing. \text{ This}$$
is also not a contradiction for $A_n$ is not closed.

**SOLUTION** 7.3.33. Let $\{U_n\}_n$ be a collection of open and dense parts of $X$. To show that $\bigcap_{n\in\mathbb{N}} U_n$ is dense, it suffices to show that

$$U \cap \left( \bigcap_{n\in\mathbb{N}} U_n \right) \neq \varnothing$$

for any open (nonvoid) set in $X$. So, let $U$ be any open (nonvoid) set in $X$. Since $U_1$ is dense, it follows that

$$U \cap U_1 \neq \varnothing.$$

This means that there is $x_1 \in U \cap U_1$ which, since $U \cap U_1$ is open, means that

$$\exists r_1 > 0, \ B(x_1, r_1) \subset U \cap U_1.$$

Since $U_2$ is dense, $U_2$ intersects each open set and in particular,

$$U_2 \cap B(x_1, r_1) \neq \varnothing.$$

So this intersection contains a point $x_2$, say. Now, since $U_2$ is open, we know that

$$\exists r_2 > 0, \ B(x_2, r_2) \subset U_2.$$

We may also take $r_2 < \frac{1}{2} r_1$ and also smaller than $r_1 - d(x_1, x_2)$ so that we obtain

$$B_c(x_2, r_2) \subset B(x_1, r_1).$$

Continuing in this way, we get a sequence of open balls $(B(x_n, r_n))$ such that

$$B_c(x_n, r_n) \subset B(x_{n-1}, r_{n-1}), \ B(x_n, r_n) \subset U_n \text{ and } \lim_{n\to\infty} r_n = 0.$$

Now the sequence $(x_n)$ is Cauchy as if $n, m \geq p$, then $x_n, x_m \in B(x_p, r_p)$ and hence

$$d(x_n, x_m) \leq d(x_n, x_p) + d(x_p, x_m) < 2r_p.$$

But $X$ is complete. Hence $x_n$ converges to $x \in X$. Finally, for $n > p$, we have $x_n \in B(x_{p+1}, r_{p+1})$ and hence its limit is in its closure. In other words,

$$x \in \overline{B(x_{p+1}, r_{p+1})} \subset B_c(x_{p+1}, r_{p+1}) \subset B(x_p, r_p) \subset U_p.$$

Since this is true for all $p$, we have that $x$ is in all $U_p$ or in their intersection. Since $(B(x_p, r_p))_p$ is decreasing, it follows that $x \in B(x_1, r_1) \subset U$. Thus

$$U \cap \left( \bigcap_{n \in \mathbb{N}} U_n \right) \neq \varnothing,$$

as required.

**SOLUTION** 7.3.34. Let $W_n = V_n \setminus \overset{\circ}{V}_n = V_n \cap (\overset{\circ}{V}_n)^c$ and take $U_n = W_n^c$. Since $W_n$ is closed, $U_n$ is open. Besides, $W_n$ has empty interior for

$$\overset{\circ}{W}_n = (V_n \cap (\overset{\circ}{V}_n)^c)^\circ = \overset{\circ}{V}_n \cap ((\overset{\circ}{V}_n)^c)^\circ \subset \overset{\circ}{V}_n \cap (\overset{\circ}{V}_n)^c = \varnothing.$$

This implies that $U_n$ is dense in $X$ as:

$$\overline{U_n} = \overline{W_n^c} = (\overset{\circ}{W}_n)^c = X.$$

By assumption, $\bigcup_{n=1}^{\infty} V_n = X$. So, by observing that $V_n = W_n \cup \overset{\circ}{V}_n$, we then easily see that

$$X = \bigcup_{n \in \mathbb{N}} V_n = \left( \bigcup_{n \in \mathbb{N}} W_n \right) \cup \left( \bigcup_{n \in \mathbb{N}} \overset{\circ}{V}_n \right)$$

or that

$$\varnothing = \left( \bigcup_{n \in \mathbb{N}} W_n \right)^c \cap \left( \bigcup_{n \in \mathbb{N}} \overset{\circ}{V}_n \right)^c.$$

Therefore,

$$\bigcap_{n \in \mathbb{N}} U_n = \bigcap_{n \in \mathbb{N}} W_n^c = \left( \bigcup_{n \in \mathbb{N}} W_n \right)^c \subset \left( \bigcup_{n \in \mathbb{N}} \overset{\circ}{V}_n \right)^{cc} = \bigcup_{n \in \mathbb{N}} \overset{\circ}{V}_n.$$

By the Baire Theorem (Theorem 7.1.29), we know that $\bigcap_{n \in \mathbb{N}} U_n$ must be dense. Hence

$$\bigcap_{n \in \mathbb{N}} U_n \subset \bigcup_{n \in \mathbb{N}} \overset{\circ}{V}_n \subset X \implies X = \overline{\bigcap_{n \in \mathbb{N}} U_n} \subset \overline{\bigcup_{n \in \mathbb{N}} \overset{\circ}{V}_n} \subset \overline{X} = X \implies \bigcup_{n \in \mathbb{N}} \overset{\circ}{V}_n = X,$$

as required.

**SOLUTION 7.3.35.** Assume $\mathbb{R}$ is countable. Hence

$$\mathbb{R} = \{x_1, x_2, \cdots, x_n, \cdots\} = \bigcup_{n=1}^{\infty} A_n$$

where $A_n = \{x_n\}$ for all $n \in \mathbb{N}$.

In usual $\mathbb{R}$, each $A_n$ is closed and $\overset{\circ}{A}_n = \varnothing$. Since $\mathbb{R}$ is complete, Corollary 7.1.30 tells us that $\bigcup_{n=1}^{\infty} \overset{\circ}{A}_n$ must be dense but this is clearly not the case as one can clearly see that this union is the empty set. Thus $\mathbb{R}$ is not countable, as desired.

**SOLUTION 7.3.36.** The proof is based on the Baire Theorem (remember that it states that if $X$ is a complete metric space, then the countable intersection of open and dense sets in $X$ is dense in $X$).

Since $C$ is closed in usual $\mathbb{R}$, it is complete. Assume that $C$ does not have any isolated points, that is, it has limit points only. Let $x \in C$. Since $C$ is separated, $\{x\}$ is closed in $C$ and so $C \setminus \{x\}$ is open in $C$. Set $U_x = C \setminus \{x\}$. Since $C$ does not have any isolated point, $\overline{U_x} = C$, that is, $U_x$ is dense in $C$. According to the Baire Theorem, the intersection $\bigcap_{x \in C} U_x$ (which is countable as $C$ is) must be dense in $C$. But

$$\bigcap_{x \in C} U_x = \varnothing$$

which is never dense in $C$. Therefore, this contradiction means that $C$ has at least an isolated point.

**SOLUTION 7.3.37.** We need only show that $\overline{f(A)} \subset f(A)$. Let $y \in \overline{f(A)}$, then there exists a sequence $(y_n)$ in $f(A)$ such that $y_n \to y$. Hence, there exists $(x_n)$ such that $x_n \in A$ with $y_n = f(x_n)$. Since $(f(x_n))_n$ converges, it is Cauchy in $Y$. As a consequence of the hypothesis, we have for all $n, m$:

$$d'(f(x_n), f(x_m)) \geq d(x_n, x_m).$$

Therefore, $(x_n)_n$ becomes in its turn a Cauchy sequence but in $A$. Since $A$ is closed in the complete $X$, it follows that $A$ is complete. This guarantees the convergence of $(x_n)$ in $A$, i.e.

$$d(x_n, x) \longrightarrow 0 \text{ as } n \to \infty,$$

with $x$ in $A$. Now, since $f$ is continuous, one obtains

$$y = \lim_{n \to \infty} f(x_n) = f(x) \in f(A),$$

establishing the result.

**SOLUTION 7.3.38.** First, it is clear that the given problem is equivalent to the integral equation (what is its type?):

$$y(x) = \int_0^x 2t(1 + y(t))dt.$$

Now, pick $y_0(x) = 0$. Then

$$y_1(x) = \int_0^x 2t(1 + y_0(t))dt = \int_0^x 2t(1 + 0)dt = x^2,$$

$$y_2(x) = \int_0^x 2t(1 + y_1(t))dt = \int_0^x 2t(1 + t^2)dt = x^2 + \frac{x^4}{2},$$

$$\vdots$$

$$y_n(x) = \int_0^x 2t(1 + y_{n-1}(t))dt = x^2 + \frac{x^4}{2!} + \cdots + \frac{x^{2n}}{n!}.$$

(which we have guessed and we should check its trueness by a proof by induction). Observe that $y_n(x)$ is the $n$th partial sum of the Taylor series

$$\sum_{n=1}^{\infty} \frac{x^{2n}}{n!}.$$

Accordingly,

$$\lim_{n \to \infty} y_n(x) = e^{x^2} - 1 := y(x).$$

We can finally check that this is the exact solution of the given problem.

**SOLUTION 7.3.39.**

(1) We know from the course of integration that if $u$ and $K$ are continuous, then $x \mapsto \int_a^b K(x,t)u(t)dt$ is continuous. Hence $T(u)$, being a sum of two continuous functions, is continuous, i.e. $T(u) \in X$ whenever $u \in X$.

(2) (a) No! $M$ exists and it is always finite because $K$ is continuous on the compact $[a,b] \times [a,b]$ and so $K$ is bounded (it even attains its bounds so that we can change "sup" by "max").

(b) Let $|\lambda| < \frac{1}{M(b-a)}$. To show that $T$ is a contraction, let $u, v \in X$, $x \in [a,b]$. Then

$$T(u)(x) = g(x) + \lambda \int_a^b K(x,t)u(t)dt$$

and

$$T(v)(x) = g(x) + \lambda \int_a^b K(x,t)v(t)dt.$$

Now we may write

$$d_\infty(T(u), T(v)) = \sup_{a \le x \le b} |T(u)(x) - T(v)(x)|$$

$$= \sup_{a \le x \le b} \left| \lambda \int_a^b K(x,t)u(t)dt - \lambda \int_a^b K(x,t)v(t)dt \right|$$

$$= \sup_{a \le x \le b} \left| \lambda \int_a^b K(x,t)(u(t) - v(t))dt \right|$$

$$\le \sup_{a \le x \le b} |\lambda| \int_a^b |K(x,t)||u(t) - v(t)|dt.$$

Now, it is clear that for all $x$ and all $t$ (both in $[a,b]$), we have

$$|K(x,t)| \le M \text{ and } |u(t) - v(t)| \le d_\infty(u,v)$$

so that (for all $x$ and all $t$)

$$|K(x,t)||u(t) - v(t)| \le M d_\infty(u,v).$$

Hence

$$\int_a^b |K(x,t)||u(t) - v(t)|dt \le \int_a^b M d_\infty(u,v)dt = M d_\infty(u,v)(b-a).$$

Therefore

$$d_\infty(T(u), T(v)) \le |\lambda|M(b-a)d_\infty(u,v).$$

Since by assumption $|\lambda|M(b-a) < 1$, it follows that $T$ is a contraction.

(3) Since $(X, d_\infty)$ is complete (Theorem 7.1.12) and $T : X \to X$ is a contraction, Theorem 7.1.37 guarantees that $T$ has one, and only one, fixed point $u \in X$. In other words, $(E)$ has a unique solution $u \in C([a,b], \mathbb{R})$.

**REMARK.** The reader must bear in mind that the uniqueness of the solution $u$ is understood to be among the continuous functions!

(4) Using the notation of Equation $(E)$ for Equation $(E')$, we see that:

$$a = 0, \ b = 1, \ \lambda = \frac{1}{2}, \ K(x,t) = xt \text{ and } g(x) = \frac{5x}{6}.$$

Hence $|K(x,t)| \leq 1$ for all $x, t \in [0, 1]$. We know that Equation $(E)$ has a unique solution if $M(b-a)|\lambda| < 1$. In our case:

$$M(b-a)|\lambda| = 1 \times 1 \times \frac{1}{2} = \frac{1}{2} < 1$$

so that $(E')$ too has a unique solution. To find it, we use the method of successive approximations. Pick $u_0(x) = 1$ with $x \in [0, 1]$. Then

$$u_1(x) = \frac{5x}{6} + \frac{1}{2} \int_0^1 xtu_0(t)dt = \frac{5x}{6} + \frac{x}{2} \int_0^1 tdt = \frac{13x}{12},$$

$$u_2(x) = \frac{5x}{6} + \frac{1}{2} \int_0^1 xtu_1(t)dt = \frac{73x}{72},$$

$$u_3(x) = \frac{5x}{6} + \frac{1}{2} \int_0^1 xtu_2(t)dt = \frac{433x}{432} \ldots$$

We may conjecture that

$$u_n(x) = \frac{1 + 2 \times 6^n}{2 \times 6^n}x, \ n \in \mathbb{N}.$$

Let's show that the previous statement holds using a proof by induction. It is true for $n = 1$. Assume the formula holds for $n$. We then have

$$u_{n+1}(x) = \frac{5x}{6} + \frac{1}{2} \int_0^1 xtu_n(t)dt = \frac{5x}{6} + \frac{x}{2} \times \frac{1 + 2 \times 6^n}{2 \times 6^n} \int_0^1 t^2 dt$$

and so

$$u_{n+1}(x) = \frac{5x}{6} + \frac{x}{2} \times \frac{1 + 2 \times 6^n}{2 \times 6^n} \times \frac{1}{3} = \frac{5 \times 2 \times 6^n + 1 + 2 \times 6^n}{2 \times 6^n \times 6}x,$$

that is,

$$u_{n+1}(x) = \frac{12 \times 6^n + 1}{2 \times 6^{n+1}}x = \frac{1 + 2 \times 6^{n+1}}{2 \times 6^{n+1}}x,$$

as desired. In the end, we know that the solution of $(E')$ is the (uniform) limit of $u_n$, i.e. the solution of $(E')$ is given by

$$u(x) = x, \ 0 \leq x \leq 1.$$

**SOLUTION 7.3.40.**

(1) Let $x \in [a, b]$ and let $u, v \in X$. Then

$$T(u)(x) - T(v)(x) = \lambda \int_a^x K(x,t)u(t)dt - \lambda \int_a^x K(x,t)v(t)dt$$

and so

$$T(u)(x) - T(v)(x) = \lambda \int_a^x K(x,t)(u(t) - v(t))dt.$$

But for all $x$ and all $t$, we clearly have

$$|K(x,t)||u(t) - v(t)| \leq Md_\infty(u,v)$$

where $M = \sup_{a \leq x, t \leq b} |K(x,t)| < \infty$ (by a similar argument used in Exercise 7.3.39). Hence

$$|T(u)(x) - T(v)(x)| \leq |\lambda| Md_\infty(u,v) \int_a^x 1.dt = |\lambda| M(x-a) d_\infty(u,v).$$

Next, we have

$$T^2(u)(x) = T[T(u)(x)] = f(x) + \lambda \int_a^x K(x,t) Tu(t) dt$$

and hence

$$T^2(u)(x) - T^2(v)(x) = \lambda \int_a^x K(x,t)(Tu(t) - T(v)(t))dt.$$

Whence

$$|T^2(u)(x) - T^2(v)(x)| \leq |\lambda| \int_a^x |K(x,t)||Tu(t) - T(v)(t))|dt$$

$$\leq |\lambda| \int_a^x |K(x,t)||\lambda| M(t-a) d_\infty(u,v) dt$$

$$= |\lambda|^2 Md_\infty(u,v) \int_a^x (t-a)|K(x,t)|dt$$

$$\leq |\lambda|^2 M^2 d_\infty(u,v) \int_a^x (t-a)dt$$

$$= |\lambda|^2 M^2 \frac{(x-a)^2}{2!} d_\infty(u,v).$$

Continuing this process, we obtain for an arbitrary $n$

$$|T^n(u)(x) - T^n(v)(x)| \leq |\lambda|^n M^n \frac{(x-a)^n}{n!} d_\infty(u,v)$$

Therefore, as $x \leq b$ we obtain

$$|T^n(u)(x) - T^n(v)(x)| \leq |\lambda|^n M^n \frac{(b-a)^n}{n!} d_\infty(u,v).$$

In the end,

$$d_\infty(T^n(u), T^n(v)) \leq \alpha d_\infty(u,v)$$

where

$$\alpha = |\lambda|^n M^n \frac{(b-a)^n}{n!}.$$

Accordingly, $T^n$ is a contraction if $\alpha < 1$. However, it is clear that for any $\lambda$, we can always choose a large enough $n$ such that

$$|\lambda|^n M^n \frac{(b-a)^n}{n!} < 1$$

(this is mainly due to the factorial "$n!$" in the dominator which makes this possible).

(2) Since $(X, d_\infty)$ is complete and since $T^n$ is a contraction, we may rely on Proposition 7.1.39 to say that $T$ has a unique fixed point $u$, i.e. $Tu = u$. This amounts to say that $(E)$ has a unique solution $u$ in $X$.

(3) Take as a starting point $u_0(x) = 0$. Then

$$u_1(x) = x - \int_0^x (x-t)u_0(t)dt = x,$$

$$u_2(x) = x - \int_0^x (x-t)u_1(t)dt = x - \int_0^x (x-t)t dt = x - \left[\frac{xt^2}{2} - \frac{t^3}{3}\right]_0^x,$$

i.e.

$$u_2(x) = x - \frac{x^3}{3!},$$

$$u_3(x) = x - \int_0^x (x-t)u_2(t)dt = x - \frac{x^3}{3!} + \frac{x^5}{5!}.$$

It is clear that if carry on, then the $n$th term would be given by:

$$u_n(x) = x - \frac{x^3}{3!} + \frac{x^5}{5!} + \cdots + (-1)^{n-1}\frac{x^{2n-1}}{(2n-1)!}$$

(a guess needing a corroboration by a proof by induction!). The previous does remind us of the infinite power series defining the sine function, i.e.

$$\sin x = \sum_{n=1}^{\infty}(-1)^{n-1}\frac{x^{2n-1}}{(2n-1)!}.$$

The solution of $(E')$ is therefore given by

$$u(x) = \lim_{n \to \infty} u_n(x) = \sin x.$$

**SOLUTION** 7.3.41. We use the method of the fixed point, so we need a complete metric space as well as a contraction on this space. Consider $(\mathbb{R}^2, d_\infty)$ which is complete. Recall that

$$d_\infty((x_1, x_2), (y_1, y_2)) = \max(|x_1 - y_1|, |x_2 - y_2|).$$

Then define $f : (\mathbb{R}^2, d_\infty) \to (\mathbb{R}^2, d_\infty)$ by

$$f(x_1, x_2) = \left( \frac{1}{5}(2\sin x_1 + \cos x_2), \frac{1}{5}(\cos x_1 + 3\sin x_2) \right).$$

The aim now is to show that $f$ is a contraction, i.e. for some $k \in [0, 1)$ and all $(x_1, x_2)$ and $(y_1, y_2)$ in $\mathbb{R}^2$ we have

$$d_\infty(f(x_1, x_2), f(y_1, y_2)) \leq k d_\infty((x_1, x_2), (y_1, y_2)).$$

Let $(x_1, x_2)$ and $(y_1, y_2)$ be in $\mathbb{R}^2$. Then

$$\left| \frac{1}{5}(2\sin x_1 + \cos x_2) - \frac{1}{5}(2\sin y_1 + \cos y_2) \right|$$

$$= \left| \frac{2}{5}(\sin x_1 - \sin y_1) + \frac{1}{5}(\cos x_2 - \cos y_2) \right|$$

$$\leq \frac{2}{5} |\sin x_1 - \sin y_1| + \frac{1}{5} |\cos x_2 - \cos y_2|$$

$$\leq \frac{2}{5} |x_1 - y_1| + \frac{1}{5} |x_2 - y_2| \text{ (we used the Mean Value Theorem twice!)}$$

$$\leq \frac{2}{5} d_\infty((x_1, x_2), (y_1, y_2)) + \frac{1}{5} d_\infty((x_1, x_2), (y_1, y_2))$$

$$= \frac{3}{5} d_\infty((x_1, x_2), (y_1, y_2)).$$

Very similarly, we may show that

$$\left| \frac{1}{5}(\cos x_1 + 3\sin x_2) - \frac{1}{5}(\cos y_1 + 3\sin y_2) \right| \leq \frac{4}{5} d_\infty((x_1, x_2), (y_1, y_2)).$$

Therefore,

$$d_\infty(f(x_1, x_2), f(y_1, y_2)) \leq \frac{4}{5} d_\infty((x_1, x_2), (y_1, y_2)),$$

showing that $f : (\mathbb{R}^2, d_\infty) \to (\mathbb{R}^2, d_\infty)$ is a contraction. Finally, the Fixed Point Theorem guarantees that $f$ has a unique fixed point $(x_1, x_2)$, i.e. $f(x_1, x_2) = (x_1, x_2)$ and so the given system has a unique solution $(x_1, x_2)$.

## 7.4. Hints/Answers to Tests

**SOLUTION** 55. No! (Why?)...

**SOLUTION** 56. Use the sequence defined by $x_n = n$...

**SOLUTION** 57. Apply the Baire Theorem to $[0, 1]$...

**SOLUTION** 58. Use Exercises 4.5.7 & 7.3.36.

**SOLUTION** 59.

(1) It is easy to see that for any $n \neq m$, we have

$$d_2(e_n, e_m) = \sqrt{2}.$$

Hence none of the subsequences can be Cauchy...

(2) We deduce that the unit sphere in $(\ell^2, d_2)$ is not compact even though it is closed and bounded...

# CHAPTER 8

# Function Spaces

## 8.2. True or False: Answers

(1) False! We provide a counterexample. Let

$$f_n(x) = \frac{1}{1 + nx}, \quad 0 < x < 1.$$

It can be shown that $(f_n)$ is decreasing (w.r.t. $n$). Also, it is plain that

$$\lim_{n \to \infty} \frac{1}{1 + nx} = 0 = f(x)$$

and so $f$ is continuous on $(0, 1)$. Nonetheless,

$$\sup_{0 < x < 1} |f_n(x) - f(x)| = \sup_{0 < x < 1} \left| \frac{1}{1 + nx} \right| = 1 \not\longrightarrow 0,$$

i.e. the convergence is not uniform.

This example also shows that compactness is needed in the Dini Theorem.

(2) True! Let $x \leq y$. Since $x \mapsto f_n(x)$ is increasing, $f_n(x) \leq f_n(y)$ for each $n$. Hence

$$\lim_{n \to \infty} f_n(x) \leq \lim_{n \to \infty} f_n(y) \implies f(x) \leq f(y),$$

i.e. $f$ is increasing.

**REMARK.** The result is no longer true if "strictly increasing" replaces "increasing". The reader should try to give a counterexample.

(3) If $(f_n)$ is a convergent sequence of bounded functions, then its pointwise limit $f$ need not be bounded. Let $(f_n)$ be defined as follows

$$f_n(x) = \begin{cases} e^x, & |x| \leq n, \\ 0, & |x| > n. \end{cases}$$

Then every $(f_n)$ is bounded (why?). Its pointwise limit, which is clearly seen to be $f(x) = e^x$ (with $x \in \mathbb{R}$), is not bounded on $\mathbb{R}$.

But, if the convergence is uniform, then the uniform limit must be bounded. we give a proof although it is similar in core to that of Exercise 7.3.16.

Since $(f_n)$ converges uniformly to $f$, for all $\varepsilon > 0$ and in particular for $\varepsilon = 1$,

$$\exists N \in \mathbb{N}, \ \forall n \in \mathbb{N}, \ n \geq N \implies |f_n(x) - f(x)| \leq 1 \ \text{ for all } x \in \mathbb{R}.$$

But all $f_n$ are bounded and in particular $f_N$ is. Whence for some positive $C$ and all $x \in \mathbb{R}$: $|f_N(x)| \leq C$. Therefore, for each $x \in \mathbb{R}$

$$|f(x)| \leq |f(x) - f_N(x)| + |f_N(x)| \leq 1 + C,$$

i.e. $f$ is bounded.

(4) False! $f_n(x) = x^n$ on $[0, 1]$ is again a counterexample. Then $(f_n)$ converges pointwise to the zero function on $[0, 1)$. The convergence is uniform on $[0, a]$ for each $0 < a < 1$ as

$$\lim_{n \to \infty} \sup_{0 \leq x \leq a} |x^n - 0| = \lim_{n \to \infty} a^n = 0$$

and we can easily see that the convergence is not uniform on $[0, 1)$.

(5) False! The following is a counterexample: $f_n(x) = (1-x)^n$ on $(0, 1]$. Details are left to the reader...

(6) Although continuity implies Riemann integrability, the result does not hold in general (that is, we still need uniform continuity).

As a counterexample, let $(f_n)$ be defined on $[0, 1]$ as follows

$$f_n(x) = \begin{cases} 4n^2x, & 0 \leq x \leq \frac{1}{2n}, \\ -4n^2x + 4n, & \frac{1}{2n} < x < \frac{1}{n}, \\ 0, & \frac{1}{n} \leq x \leq 1. \end{cases}$$

It is clear that $(f_n)$ converges pointwise to $f(x) = 0$ for all $x \in [0, 1]$. Drawing a graph tells us that $\int_0^1 f_n(x)dx$ is actually the area of a triangle of apexes $(0, 0)$, $(\frac{1}{n}, 0)$ and $(\frac{1}{2n}, 2n)$. So

$$\int_0^1 f_n(x)dx = \frac{1}{n} \times (2n) \times \frac{1}{2} = 1.$$

However,

$$\lim_{n \to \infty} \int_0^1 f_n(x)dx = 1 \neq \int_0^1 \lim_{n \to \infty} f_n(x)dx = \int_0^1 f(x)dx = 0.$$

(7) False! Consider $f_n(x) = x^n/n$ defined on $[0,1]$. Then it is evident that $f_n$ converges pointwise to $f(x) = 0$ for every $x \in [0,1]$, though

$$\lim_{n\to\infty} f_n'(x) = \lim_{n\to\infty} x^{n-1} = \begin{cases} 0, & 0 \leq x < 1, \\ 1, & x = 1, \end{cases} \neq f'(x) = 0.$$

## 8.3. Solutions to Exercises

**SOLUTION 8.3.1.**

(1) It is clear that $(f_n)$ converges pointwise to the zero function (defined on $[0,1]$). Call this limit $f$.
(2) We can also easily show that $(f_n)$ converges uniformly to the $f$, i.e. that

$$\lim_{n\to\infty} \sup_{x\in[0,1]} |f_n(x) - f(x)| = 0.$$

**SOLUTION 8.3.2.**

(1) We have

$$\lim_{n\to\infty} f_n(x) = \lim_{n\to\infty} \frac{e^{nx}}{\sqrt{n}} = 0$$

for every $x \in [-1,0]$. Hence $(f_n)$ converges to the zero function. Designate this limit by $f$.
(2) Let us verify that the convergence is uniform. We have

$$\lim_{n\to\infty} \sup_{x\in[-1,0]} |f_n(x) - f(x)| = \lim_{n\to\infty} \sup_{x\in[-1,0]} \frac{e^{nx}}{\sqrt{n}} = \lim_{n\to\infty} \frac{1}{\sqrt{n}} = 0.$$

(3) We obviously have

$$f_n'(x) = \sqrt{n}e^{nx}, \ \forall x \text{ s.t. } -1 \leq x \leq 0$$

which does not converge pointwise on $[-1,0]$ for $f_n'(0) = \sqrt{n}$. A fortiori, it does not converge uniformly.

**SOLUTION 8.3.3.** Let

$$f_n(x) = \left(1 + \frac{x}{n}\right)^n, \ x \in [0,1], n \in \mathbb{N}.$$

It is clear that $(f_n)$ converges pointwise to $e^x$ in $\mathbb{R}$ (and in particular in $[0,1]$). Let's show that this is actually a uniform convergence. As indicated in the exercise, we use the Dini Theorem. So, we need to check that the conditions of this theorem are fulfilled. First, $[0,1]$ is compact. Also, it is plain that all $f_n$ are continuous. Now, let $x \in [0,1]$.

Recall from Exercise 1.3.13 that if $a_1, a_2, \cdots, a_n$ are positive real numbers and if

$$A_n = \frac{a_1 + a_2 + \cdots + a_n}{n} \quad \text{and} \quad G_n = \sqrt[n]{a_1 a_2 \cdots a_n},$$

then (for all $n$)

$$A_n \geq G_n \text{ and so } A_{n+1} \geq G_{n+1}.$$

Setting $a_1 = 1$ and $a_2 = a_3 = \cdots = a_{n+1} = 1 + (x/n)$ in the previous inequality gives

$$\frac{1 + (1 + \frac{x}{n}) + \cdots + (1 + \frac{x}{n})}{n+1} = \frac{1 + n(1 + \frac{x}{n})}{n+1} \geq \sqrt[n+1]{\left(1 + \frac{x}{n}\right)^n}$$

or simply

$$1 + \frac{x}{n+1} \geq \left(1 + \frac{x}{n}\right)^{\frac{n}{n+1}}.$$

Therefore,

$$\left(1 + \frac{x}{n+1}\right)^{n+1} \geq \left(1 + \frac{x}{n}\right)^n.$$

This shows that the sequence $(f_n)$ is increasing and so all the conditions of the Dini Theorem are satisfied. Accordingly, $(f_n)$ converges uniformly to $x \mapsto e^x$.

**SOLUTION** 8.3.4. We apply the Dini Theorem. We are hence required to show that $(P_n)$ is monotonic and that it converges pointwise to $|x|$. We first claim $P_n(x) \leq |x|$ for all $n$. To prove it, we use a proof by induction. It is clear $P_0(x) = 0 \leq |x|$ (for all $x$). Assume $P_n(x) \leq |x|$. Then

$$|x| - P_{n+1}(x) = |x| - P_n(x) - \frac{1}{2}(|x| - P_n(x))(|x| + P_n(x))$$

$$= \frac{1}{2}(|x| - P_n(x))(2 - |x| - P_n(x)) \geq 0$$

by the induction hypothesis and $x \in [-1, 1]$ hypotheses.

We can easily check that $P_n(x) \geq 0$ for all $n$ and all $x$. To apply the Dini Theorem, it only remains to check that $(P_n)$ is monotonic. It is in fact increasing as for all $n$ (and all $x \in [-1, 1]$)

$$P_{n+1}(x) - P_n(x) \geq 0.$$

Therefore, $(P_n(x))$ is an increasing and bounded above sequence and so it converges (pointwise) to some $f(x)$ (with $0 \leq f(x) \leq |x|$). Finally, it is clear that $f(x)$ verifies $x^2 = f^2(x)$ or $f(x) = |x|$. Thus, all the hypotheses of the Dini Theorem are fulfilled, implying that $(P_n)$ converges uniformly to $|x|$.

**SOLUTION** 8.3.5. The sequence $(f_n)$ converges pointwise, compactly and uniformly to the zero function.

As for $(g_n)$, it converges to $x \mapsto g(x) = x$; pointwise and compactly, but not uniformly.

**SOLUTION** 8.3.6. To show that $V$ is closed, we take an arbitrary convergent sequence $(x_n)$ in $V$, i.e. $(f_n(x_n))$ is Cauchy in $Y$, and then we must show that its limit $x$ remains in $V$, i.e. $(f_n(x))$ is Cauchy in $Y$.

If $p, q, n$ are positive integers, then we clearly have

$$d'(f_p(x), f_q(x)) \leq d'(f_p(x), f_p(x_n)) + d'(f_p(x_n), f_q(x_n)) + d'(f_q(x_n), f_q(x)).$$

Let $\varepsilon > 0$. Since $(f_n)$ is equicontinuous at $x$, for some $\alpha > 0$ and all $n \in \mathbb{N}$ and all $y \in X$

$$d(x, y) < \alpha \implies d'(f_n(x), f_n(y)) < \frac{\varepsilon}{3}.$$

But $d(x_n, x) \to 0$, and hence there exists $n' \in \mathbb{N}$ for which $d(x_{n'}, x) < \alpha$. Therefore,

$$d'(f_q(x_{n'}), f_q(x)) < \frac{\varepsilon}{3} \text{ and } d'(f_p(x_{n'}), f_p(x)) < \frac{\varepsilon}{3}$$

for all $p, q \geq n'$. We also observe that $(f_n(x_{n'}))_n$ (for fixed $n'$) is Cauchy, so

$$\exists n'' \geq n', \ \forall p, q \in \mathbb{N} : \ (p, q \geq n'' \implies d'(f_p(x_{n'}), f_q(x_{n'})) < \frac{\varepsilon}{3}).$$

Combining the last three "$\varepsilon/3$-inequalities" gives us

$$\forall \varepsilon > 0, \exists n'' \in \mathbb{N} : \ \forall p, q \in \mathbb{N} : \ ((p, q \geq n'' \implies d'(f_p(x), f_q(x)) < \varepsilon),$$

$(f_n(x))$ is Cauchy, i.e. $x \in V$ so that $V$ is closed.

**SOLUTION** 8.3.7.

(1) Since $k$ is continuous on the compact $[0,1]^2$, it is uniformly continuous. This *implies* that (in case of equality of the second coordinates)

$$\forall \xi > 0, \exists \alpha > 0, \forall x, y, t \in [a, b] : \ (|x-t| < \alpha \implies |k(x, y) - k(t, y)| < \xi).$$

But $(f_n)$ is bounded, hence

$$\exists M \geq 0 : \ \forall n \in \mathbb{N} : \ \sup_{x \in [0,1]} |f_n(x)| \leq M.$$

Let $x$ be fixed and let $\varepsilon > 0$. Set $\xi = \frac{\varepsilon}{M}$. Since $k$ is uniformly continuous, we may find $\alpha > 0$ such that $|x - t| < \alpha$ with

$$|k(x, y) - k(t, y)| < \xi = \frac{\varepsilon}{M}.$$

Hence as long as $|x - t| < \alpha$, we have

$$|Kf_n(x) - Kf_n(t)| \leq \int_0^1 |k(x,y) - k(t,y)||f_n(y)|dy$$

$$\leq M \int_0^1 |k(x,y) - k(t,y)|dy$$

$$\leq M \int_0^1 \frac{\varepsilon}{M}dy$$

$$\leq \varepsilon,$$

i.e. $(Kf_n)$ is equicontinuous at each $x \in [0,1]$, hence it is so on $[0,1]$.

(2) We claim that $(Kf_n)$ is bounded. Indeed, for all $n$

$$\left| \int_0^1 k(x,y)f_n(y)dy \right| \leq M \int_0^1 |k(x,y)|dy.$$

Since $(Kf_n)$ is equicontinuous, by the Ascoli Theorem $(Kf_n)$ is relatively compact.

**SOLUTION** 8.3.8. Since all the moments vanish, so does their sum and so

$$\int_a^b f(x)p(x)dx = 0,$$

where $p(x) = a_n x^n + \cdots + a_1 x + a_0$ is any polynomial with real coefficients. Now, the function $f$, being continuous on the compact set $[a,b]$, is bounded and hence for some positive $C$ and all $x \in [a,b]$: $|f(x)| \leq C$. By the Weierstrass Theorem, for every $\varepsilon > 0$, there exists a polynomial $q$ such that

$$|f(x) - q(x)| < \varepsilon$$

for all $x \in [a,b]$. By the above observation, $\int_a^b f(x)q(x)dx = 0$ too and hence we may write

$$0 \leq \int_a^b f^2(x)dx = \int_a^b (f^2(x) - f(x)q(x))dx$$

$$= \int_a^b f(x)(f(x) - q(x))dx$$

$$= \left| \int_a^b f(x)(f(x) - q(x))dx \right| \quad \text{(why?)}$$

$$\leq \int_a^b |f(x)||f(x) - q(x)|dx$$

$$< C\varepsilon.$$

Since this is true for all $\varepsilon$, we immediately deduce that $\int_a^b f^2(x)dx = 0$ which, combined with the fact that $f^2$ is continuous and positive, yield $f^2(x) = 0$ for all $x \in [a,b]$ and thus $f = 0$.

**SOLUTION 8.3.9.** Set

$$d_\infty(f,g) = \sup_{x \in [0,1]} |f(x) - g(x)| \text{ for all } f,g \in C([0,1],\mathbb{R}).$$

First remember that we have to find a dense and countable set in $C([0,1],\mathbb{R})$. Let $f \in C([0,1],\mathbb{R})$. If we come to show that for every $\varepsilon > 0$, there exists a polynomial $q$ with rational coefficients such that $d_\infty(f,g) < \varepsilon$, then Exercise 1.2.23 (which tells us that the set of polynomials with rational coefficients is countable) will allow us to conclude that $C([0,1],\mathbb{R})$ is separable.

Since $f$ is continuous on $[0,1]$, there exists -by means of the Weierstrass Theorem- a polynomial $p$ such that

$$d_\infty(f,p) < \varepsilon.$$

If all the coefficients of $p$ are rational, then we have our polynomial. Otherwise, $p$ will have irrational coefficients (not necessarily all). So if we assume $p$ has degree $n$, then

$$p(x) = \alpha_n x^n + \cdots + \alpha_1 x + \alpha_0 = \sum_{i=0}^n \alpha_i x^i$$

where *not all* $\alpha_i$ $(i = 0, 1, \ldots ; n)$ are rational. Now, it is clear that every interval centered at $\alpha_i$ with length $2\varepsilon$ contains a rational $\beta_i$ (and hence $|\alpha_i - \beta_i| < \varepsilon$ for each $i$). Thus we have recovered a polynomial $q$ with rational coefficients $\beta_i$ written as

$$q(x) = \beta_n x^n + \cdots + \beta_1 x + \beta_0 = \sum_{i=0}^n \beta_i x^i.$$

Therefore, it only remains to approximate $p(x)$ by $q(x)$ (in the supremum distance). We have for all $x$ in $[0,1]$

$$|p(x) - q(x)| \leq \sum_{i=0}^n |\beta_i - \alpha_i| x^i < \varepsilon(n+1)$$

which implies that $d_\infty(p,q) < \varepsilon(n+1)$. Thus

$$d_\infty(f,q) \leq d_\infty(f,p) + d_\infty(p,q) < \varepsilon(n+1) + \varepsilon = \varepsilon(n+2)$$

and the proof is complete.

## 8.4. Hints/Answers to Tests

**SOLUTION** 60. The sequence $(f_n)$ converges pointwise to $\frac{1}{1-x}$ on $(-1, 1)$...The convergence is not uniform on $(-1, 1)$ but it is so on $[-a, a]$ where $a < 1$.

**SOLUTION** 61. The given sequence converges uniformly to the function defined by $f(x) = x$...

**SOLUTION** 62. $f$ is continuous...

**SOLUTION** 63. Yes (prove it!)...

# Bibliography

1. C. D. Aliprantis, O. Burkinshaw, *Problems in Real Analysis*, Academic Press, 1999.
2. F. Beukers, A note on the irrationality of $\zeta(2)$ and $\zeta(3)$, *Bull. London Math. Soc.*, **11/3** (1979), 268-272.
3. G. Birkhoff, G-C. Rota, *Ordinary differential equations*. Fourth edition. John Wiley & Sons, Inc., New York, 1989.
4. V. Bryant, *Metric Spaces: iteration and application*, Cambridge University Press, 1985.
5. W. Cheney, *Analysis for applied mathematics*. Graduate Texts in Mathematics, **208**. Springer-Verlag, New York, 2001.
6. G. Cohen, *A Course in Modern Analysis and Its Applications*, AMSLS **17**, Cambridge University Press, 2003.
7. J. B. Conway, A Course in Functional Analysis, *Springer*, 1990 (2nd edition).
8. J. Doboš, Sums of closed graph functions. Real functions (Liptovský Jàn, 1996). *Tatra Mt. Math. Publ.*, **14** (1998), 9-11.
9. B. J. Gardner, M. Jackson, The Kuratowski closure-complement theorem, *New Zealand J. Math.*, **38** (2008), 9-44.
10. A. Granas, J. Dugundji, *Fixed point theory*, Springer Monographs in Mathematics. Springer-Verlag, New York, 2003.
11. A. J. Jerri, *Introduction to integral equations with applications*. Second edition. Wiley-Interscience, New York, 1999.
12. I. Kaplansky, *Set Theory and Metric Spaces*, AMS Chelsea Publishing, 1977 (second edition), reprinted by the AMS in 2001.
13. A. N. Kolmogorov, S. V. Fomin, *Introductory real analysis*. Translated from the second Russian edition and edited by Richard A. Silverman. Corrected reprinting. Dover Publications, Inc., New York, 1975.
14. E. Kreyszig, *Introductory functional analysis with applications*, Wiley Classics Library. John Wiley & Sons, Inc., New York, 1989.
15. C. Kuratowski, Sur l'Opération A de l'Analysis Situs (French), Fund. Math., **3** (1922), 182-199.
16. S. Lipschutz, *Schaum's Outline of Theory and Problems of General Topology*, Schaum's Outline Series: McGraw-Hill Book Company, 1965.
17. P. E. Long, Classroom Notes: Functions with Closed Graphs, *Amer. Math. Monthly*, **76/8** (1969), 930-932.
18. J. M. Møller, *General Topology*, http://www.math.ku.dk/~moller.
19. J. R. Munkres, *Topology*, Prentice Hall, 2000 (2nd edition).
20. W. Rudin, *Principles of Mathematical Analysis*, McGraw-Hill International Editions, Mathematics Series, Third Edition 1976.

21. W. Rudin, *Functional Analysis*, McGraw-Hill Book Co. Second edition, International Series in Pure and Applied Mathematics, McGraw-Hill, Inc., New York, 1991.

22. D. Sherman, Variations on Kuratowski's 14-set theorem, *Amer. Math. Monthly*, **117/2** (2010), 113-123.

23. W. Sierpinski, Un théorème sur les continus, *Tôhoku Math. J.*, **13** (1918), 300-303.

24. G. Skandalis, *Topologie et Analyse* (French), Dunod, 2001.

25. Y. Sonntag, *Topologie et Analyse Fonctionnelle* (French), Ellipses, 1998.

26. L. A. Steen, J. A. Seebach, Jr, *Counterexamples in Topology*, Dover, 1995.

27. E. M. Stein, R. Shakarchi, Real analysis. Measure theory, integration, and Hilbert spaces. Princeton Lectures in Analysis, III. Princeton University Press, Princeton, NJ, 2005.

28. M. Stoll, *Introduction to Real Analysis*, Addison-Wesley Higher Mathematics, 2001 (2nd edition).

29. W. A. Sutherland, *Introduction to Metric and Topological Spaces*, Oxford University Press, 1975.

30. K. Yosida, *Lectures on differential and integral equations*. Translated from the Japanese. Reprint of the 1960 translation. Dover Publications, Inc., New York, 1991.

31. S. M. Zemyan, *The classical theory of integral equations*, A concise treatment. Birkhäuser/Springer, New York, 2012.

# Index

Printed in the United States
By Bookmasters